POLYMER RESEARCH DEVELOPMENTS

POLYMER RESEARCH DEVELOPMENTS

ROBERT K. BREGG
EDITOR

Nova Science Publishers, Inc.
New York

For permission to use material from this book please contact us:
Telephone 631-231-7269; Fax 631-231-8175
Web Site: http://www.novapublishers.com

NOTICE TO THE READER

LIBRARY OF CONGRESS CATALOGING-IN-PUBLICATION DATA

Polymer research developments / Robert K. Bregg (editor).
p. cm.
Includes index.
ISBN 1-59454-742-4
1. Polymers--Research. I. Bregg, Robert K.
QD281.P6P6338 2005
547'.7'072--dc22 2005024548

Published by Nova Science Publishers, Inc. ✢New York

CONTENTS

PREFACE

Polymers are substances containing a large number of structural units joined by the same type of linkage. These substances often form into a chain-like structure. Starch, cellulose, and rubber all possess polymeric properties. Today, the polymer industry has grown to be larger than the aluminum, copper and steel industries combined. Polymers already have a range of applications that far exceeds that of any other class of material available to man. Current applications extend from adhesives, coatings, foams, and packaging materials to textile and industrial fibers, elastomers, and structural plastics. Polymers are also used for most composites, electronic devices, biomedical devices, optical devices, and precursors for many newly developed high-tech ceramics. This new book presents leading-edge research in this rapidly-changing and evolving field.

A classification of the main types of organometallic polymers is proposed and those with organometallic groups anchored to the polyphosphazenes are revised. This chapter describes the preparation, characterization, spectroscopic, thermal and electrical conductivity properties of a series of polyphosphazenes containing diverse organometallic fragments anchored to the main chain. Chapter 1 attempts to summarize what is known about this type of polymer with particular emphasis on their particular properties and their utility in material science.

The time-resolved, fast transient fluorescence technique (FTRF) which uses Strobe Master System (SMS), is employed for studying swelling of disc-shaped polymeric gels in chapter 2. In this technique the photo multiplier tube (PMT) is gated or strobed by a voltage pulse that is synchronized with the pulsed light source. The intensity of fluorescence emission of an organic dye is measured in a very narrow time window. Three series of polymeric gels based on methlymethacrylate (MMA), styrene (S) and the combination of methlymethacrylate and styrene monomers with ethylene glycol dimethacrylate (EGDM) as a crosslinker agent in the presence of 2,2′-azobisisobutyronitrile (AIBN) were synthesized by using free-radical crosslinking copolymerization (FCC). In the first series of these polymeric gels, five different gels were prepared with various EGDM contents in the presence of MMA monomers. After drying these gels, swelling experiments were performed in chloroform and controlled swelling experiments were performed under chloroform vapor at room temperature. Then, one of the dry gel in the first series was used to study gel swelling in various solvent mixtures. In the second series, heterogels synthesized from MMA and S monomers with constant EGDM content were used to monitor the swelling behaviour of these gels under chloroform vapor. In order to measure the swelling activation energy under chloroform vapor, polystyrene gels were used at various temperature in the last series. In all these swelling experiments, lifetimes of extrinsic fluorescence probe P, which was introduced during polymerization, in and out of polymeric gels were measured from fluorescence decay traces during in-situ swelling processes. The effect of the solvent and organic vapor penetration have been interpreted by applying quenching mechanisms to the produced data where the Stern-Volmer type of quenching mechanism may be assumed for the fluorescence decay of P in the gel sample during swelling process. It is observed that as the swelling time increases, excited pyrenes decay faster and faster by indicating that quenching of excited pyrenes increase in the swelling gel. The Li-Tanaka equation was used to determine the swelling time constant, τ_c and the cooperative diffusion coefficients, D_c which were found to be decreased as the crosslinker content was increased in chloroform, under chloroform vapor and in the solvent mixtures by increasing chloroform content. On the other hand, heterogels with high S content under chloroform vapor swell much more slowly than heterogels with low S content. The activation energy for the swelling of polystyrene gel was measured by using Arrhenius law and found to be 80.0 kJ mol^{-1}.

Research concerning polymeric materials containing tetraazamacrocycles is of great interest for many applications because their complexes show exceptional kinetic and thermodynamic properties, and has attracted increasing attention during the last four decades. The elaboration of these polymeric materials implies first the synthesis of bifunctional tetraazamacrocycles. The methods for the formation of these structures have been extensively reviewed in chapter 3. However, their development is limited by the difficulty to obtain these macrocycles with a high yield. Classical strategy for this kind of reactions involves the use of high dilution conditions. Under these conditions the probability of intramolecular reactions is significantly greater than the probability of intermolecular reactions. The authors were particularly interested in synthetic procedures as short as possible to prepare *N, N'*-difunctionalized tetraazacyclododecanes in order to immobilize the tetraazamacrocyclic structure in a polymer chain by polycondensation with difunctional comonomers such as diols or diamines. Two different procedures in one step using high dilution conditions or using a catalyst system have been investigated for the synthesis of bisamide tetraazamacrocycles. In order to explain the low reaction yield of ring closure, the side-products have been characterized by MALDI-TOF MS to identify their cyclic or linear structure and their size. The influence of experimental parameters on the nature of the structures (cyclic or linear oligomers) has also been discussed. The results have confirmed the competition between cyclization and chain-growth polymerization occurring during synthesis of tetraazamacrocycles. This explains the presence of cyclic and linear oligomers which prevents a better reaction yield for ring closure. Nevertheless, the authors report their investigation on the best conditions to favour obtaining of these two macrocycles.

The oxidative polycondenzation reaction conditions and optimum parameters of 2-[(pyridine-2-yl-methylene) amino] pyridine-3-ol were determined by oxidants such as air oxygen, H_2O_2, and NaOCl in an aqueous alkaline solution between 40 and 90 °C. Synthesized 2-[(pyridine-2-yl-methylene) amino] pyridine-3-ol (2-PMAP) and oligo-2-[(pyridine-2-yl-methylene) amino] pyridine-3-ol (O-2-PMAP) were characterized by ^{1}H-NMR, ^{13}C-NMR, FT-IR, UV-Vis, size exclusion chromatography (SEC) and elemental analysis techniques as described in chapter 4. At the optimum reaction conditions, the yield of oligo-2-[(pyridine-2-yl-methylene) amino] pyridine-3-ol was found to be 47.00% (for air O_2 oxidant), 43.00% (for H_2O_2 oxidant) and 73.50% (for NaOCl oxidant). The number-average molecular weight (M_n), weight-average molecular weight (M_w) and polydispersity index (PDI) values of (O-2-PMAP) were found to be 9708 g $mol^{-1}$and 12235 g mol^{-1} and 1.260, using NaOCl, 3476 g mol^{-1} and 4126 g mol^{-1} and 1.187, using air O_2, and 18466 g $mol^{-1}$and 22318 g mol^{-1} and 1.209, using H_2O_2, respectively. DTA and TG analyses were shown to be stable of oligo-2-[(pyridine-2-yl-methylene) amino] pyridine-3-ol through to temperature and thermo-oxidative decomposition.

Molecular imprinting is a rapidly developing technique for designing and preparing host compounds for molecular recognition. Although molecular imprinting technique has demonstrated great potential for specific recognition of small molecules, the reports on the imprinting of biological macromolecules such as proteins, whole cells and viruses have been scarce. However, there is still a strong incentive to prepare bio-imprinted polymers for the use of diagnostics, sensors, bioseparation, etc. In chapter 5 the authors analyzed the principle and characteristic of protein molecular imprinting technique and reviewed the methods of preparing the protein-imprinted polymer that have been reported. The difficulties of protein molecular imprinting technique were also presented. Moreover, the research team, has done some work on how to prepare protein-imprinted polymer, especially how to prepare protein-imprinted soft-wet gel beads which is fit to adsorb and recognize template protein. So they introduced the preparation method, the recognition mechanism of protein imprinted soft-wet gel beads. In the end, the prospect of protein-imprinted technique was also given.

A brief up-to-date review on recent developments regarding structural explanations of conduction mechanism in polymers, kinetic and optical properties, methods for preparation thin conductive polymer films and some of their applications referred to one of the main direction in this field, chemical sensing are presented in chapter 6.

Asymptotic methods based on the slenderness ratio are used to obtain the leading--order equations which govern the fluid dynamics of the draw--down zone of axisymmetric, nonisothermal, Newtonian, compound liquid jets such as those employed in the manufacture of textile fibers, composite fibers and optical fibers, at low Reynolds numbers. These equations are first used to determine the steady state, linear stability, and time--dependent dynamics of isothermal compound fibers for draw ratios greater than the critical value provided by a linear stability analysis of the viscous flow regime. An analytical solution of the temperature distribution in the absence of axial conduction is then presented before undertaking time--dependent studies of the nonlinear dynamics of nonisothermal, compound fibers subject to axial velocity perturbations at either the die's exit or the take--up point. The effects of molecular orientation and crystallization on polymer optical fibers are then accounted for by adding two more equations to the leading--order equations derived asymptotically in chapter 7, and some numerical results are presented. Finally, models for the air drag, latent heat effects and viscous dissipation are described.

In chapter 8, the mathematical models for exogenous and endogenous depolymerization processes are described. Inverse problems of initial value problems to determine degradation rates are formulated, and numerical techniques to solve those problems are illustrated with examples. In particular, the Simpson's rule is applied, whenever it is appropriate, to solve equations whose solutions are the degradation rates. The numerical results to simulate the transition of the weight distributions are also presented.

In: Polymer Research Developments
Editor: Robert K. Bregg, pp. 1-22

ISBN 1-59454-742-4

Chapter 1

Coordination of Organometallic Fragments to Polyphosphazenes Containing Side Groups with Donor Atoms

C. Díaz* and M. L. Valenzuela
Departamento de Química, Facultad de Ciencias, Universidad de Chile, Santiago, Chile

Abstract

A classification of the main types of organometallic polymers is proposed and those with organometallic groups anchored to the polyphosphazenes are revised. This chapter describes the preparation, characterization, spectroscopic, thermal and electrical conductivity properties of a series of polyphosphazenes containing diverse organometallic fragments anchored to the main chain. This review attempts to summarize what is known about this type of polymer with particular emphasis on their particular properties and their utility in material science.

Introduction

Organometallic compounds as well as polymers are of much interest in material science [1]. The link between these two types of compounds gives rise to organometallic polymers [2]. These materials exhibit interesting properties specifically because they combine the electronic properties associated with both transition metal compounds and the processing advantages of polymers. Characteristics routinely associated with transition metals such as multiple oxidation states and the magnetisms [2d] are rarely achieved only in organic or inorganic polymers.

* E-mail: cdiaz@uchile.cl

The incorporation of transition metal into a polymer main chain [2d] therefore clearly offers considerable potential for the preparation of processable materials with properties that differ significantly from those of conventional organic or inorganic polymers. However because of the limited synthesis method, the most common metallic derivatives are those with the organic polymer backbone. This type of polymers have, however, some disadvantages respect to their inorganic. They usually have a relatively low monoxidative as well as low thermal stability. Such difficulties are in part overcomed by some polymer with inorganic backbone such as polyphosphazene. Therefore organometallic polymers with inorganic bakbone chain may be expected to function as convenient and processable to metal-containing ceramic, films, fibers and coatings with high stability and desirable useful physical properties.

The organometallic function in organometallic polymers can arise either from the MLn fragment (which may contain one or more M-C bonds) or from the backbone linkage -[spacer]-C-M.

In general organometallic polymers can be conveniently classified in the following types as is shown in scheme 1:

I. Organometallic polymers with transition metals in the main chain.
II. Polymers with organometallic moieties in the side group structure.
III. Polymers with the metal-metal bond in the main chain.

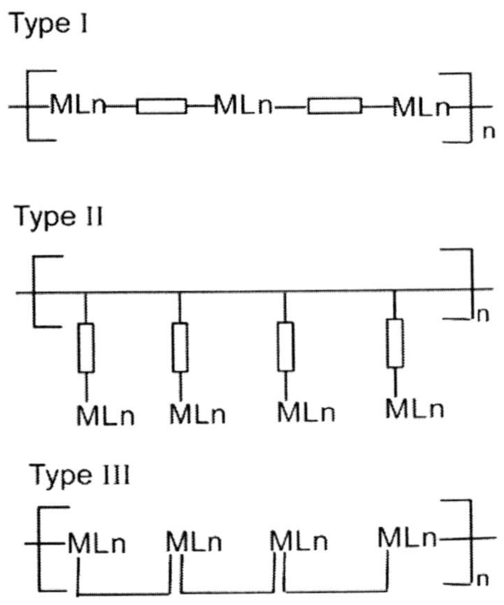

Scheme 1

Type I of organometallic polymers containing covalent M-spacer bonds have been recently revised [2a], but those with polyphosphazenes as the polymeric chain are scarce. On the other hand there are only some examples of type III of organometallic polymers having metal-metal bonds, but they are with organic polymeric chain [1d, 2a]. Another more general recent review [2c] on organometallic polymers treats mainly with type I polymers and coordination polymers but organometallic poyphosphazenes are mentioned only briefly. Also some organometallic polyphosphazenes have been briefly discussed in some selected books [1d, 1e].

Type II of polyphosphazene polymers are scarce mainly due to the few available synthesis methods.To our best knowledge the first polymers of this type were prepared by Allcock [3a] in 1983. Starting from the previously known polymer $\{[NP(OC_6H_5)(OC_6H_4PPh_2)]_x[NP(OC_6H_5)_2]_y\}_n$ [3b,3c] several organometallic derivatives were prepared being some of them insolubles suggesting a cross-linking by the organometallic fragment.

Oligomer Models

One of the approaches to study macromolecular systems has been the previous study with small molecules that may mimic the structure and reactivity of the most complex compounds [4]. Macromolecular substitution reactions are more complicated than the corresponding reactions at the small-molecule level. Thus for instance in a typical macromolecular reaction of poly(dichlorophosphazene) the replacement of 30.000 chlorine atoms per polymer molecule must be induced. On the other hand the small-molecule systems allow characterization by high resolution NMR, mass spectrometry and X-ray techniques that are difficult or impossible for high molecular weight polymers.

In the research of polyphosphazenes a prudent protocol has usually been to examine prospectively polymer reactions and structures first at the small-molecule model level. Cyclic and linear short chain species have been used as small models for polyphosphazenes [5]. While reactivity features are well mimic by cyclic models, structural aspects are only partially reproduced [6]. Specifically, polymers bearing pendant organometallic groups may be modeled by some organometallic derivatives of the cyclotriphosphazene (see scheme 2).

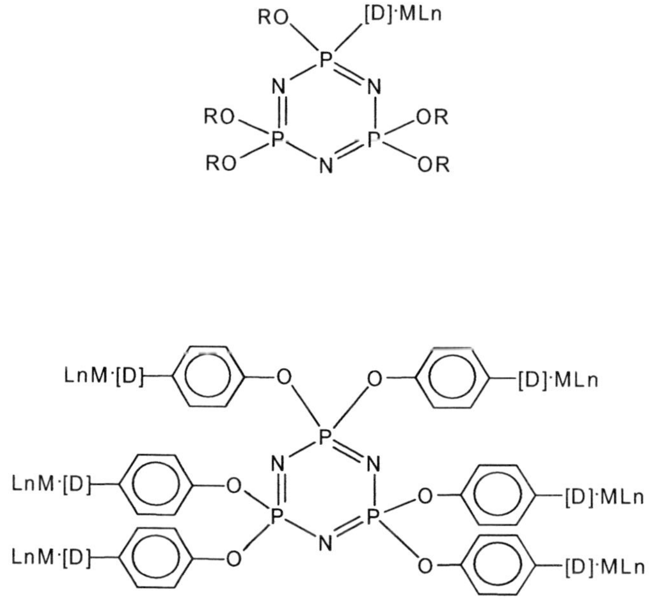

Scheme 2. MLn = Organometallic Fragment, D = Atom or Dador Group

However only few examples have been reported until now. For instance the models: $[N_3P_3(OC_6H_5)_{6-n}(OC_6H_4CN)_n(Mn(CO)_3L_2)_n](X_n)$ n=1,6 mimic well the main features of the respective polymers [7]: $[-N=P(OC_6H_5)_2N=P(OC_6H_4CN)(Mn(CO)_3L_2)(OC_6H_5)]_n$.

The small molecules $N_3P_3(OC_6H_5)_5(OC_6H_4(ML_n))$ and $N_3P_3(OC_6H_4ML_n)_6$ ML_n= $CpFe(CO)_2$ are adequate models for the respective polymers:

$[NP(OC_6H_4(ML_n))_x\text{-}(OC_6H_4X)_{2-x}]_n$, X= H or Br, ML_n= $CpFe(CO)_2$, although the polymer product as expected are insoluble [8].

On the other hand the spectroscopic properties of the small models:

$[N_3P_3(OC_6H_5)_5OC_6H_4PPh_3ML_n]$ and $[N_3P_3(OC_6H_5)_5OC_6H_4PPh_3]_2ML_n$, ML_n=AuCl, $H_2Os_3(CO)_{10}$, $CpMn(CO)_3$, $Fe(CO)_3(Ph=CHC(O)CH_3)$, and $[RhCl(CO)_2]_2$ mimic well the respective polymers [3]:

$\{[NP(OC_6H_5)(OC_6H_4PPh_3ML_n)]_x[NP(OC_6H_5)_2]_y\}_n$

Other phosphazenes with metallo units directly bonded to the phosphorous atoms of the skeleton have been reported [9].

Substitution of the chlorine atoms in the oligomer $N_3P_3Cl_{6-n}(OC_6H_5)_n$ (n= 0 and 5) by $HOC_6H_4CH_2CN$ in presence of NH_4PF_6 or $TlPF_6$ to give the ligands $N_3P_3(OC_6H_5)_{6-n}(OC_6H_4CH_2CN)_n$ (n=6 or 1) followed by their reaction with CpFe(dppe)I give the respective complexes $\{N_3P_3(OC_6H_5)_{6-n}(OC_6H_4CH_2CN{\cdot}CpFe(dppe))_n\}\{PF_6\}_n$. These were useful as reactivity models for designing the methods for obtaining the corresponding polymers [10]. Most recently several models of the type showed in Scheme 2 have been reported but not all their analogues polyphosphazenes are known [10, 11].

Synthesis of Polyphosphazenes Having Organometallic Fragments

As mentioned by Manners [2a, b], the development of the area of inorganic polymer science has been held up by considerable difficulties associated with the synthesis of these materials. In particular the scarcely and hardness of methods of synthesis of organometallic polyphosphazenes. However new methods to prepare these polymers reported recently have solved in part this problem [7b].

Three main synthetic methods have been used to prepare polymers with pendant organometallic fragments (see scheme 3):

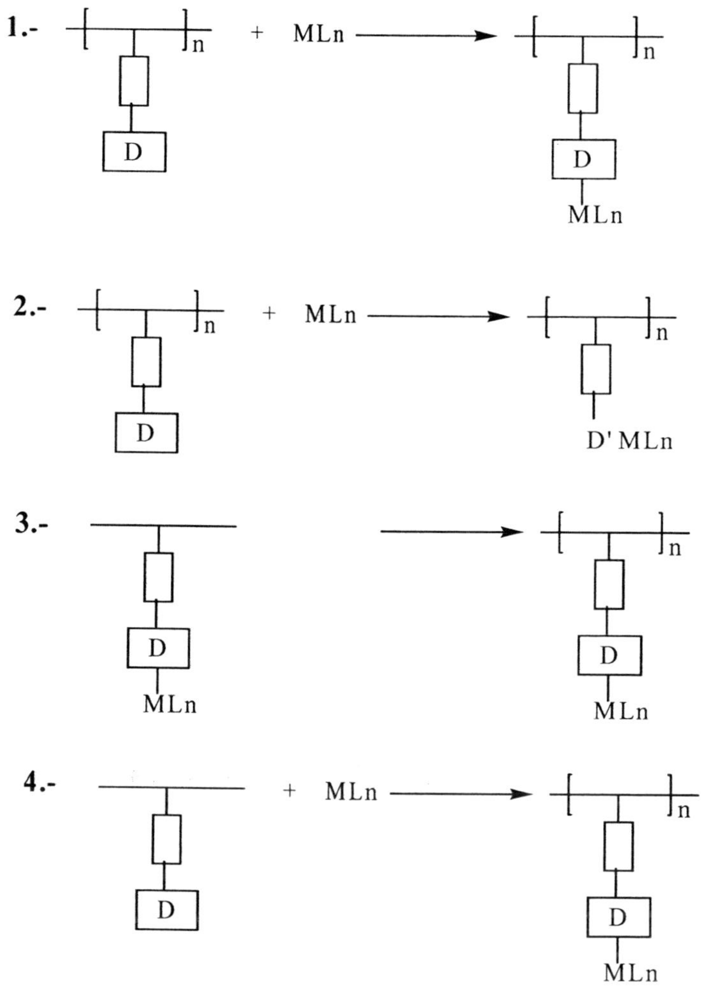

Scheme 3

1. By coordination of the organometallic fragment MLn to polymer ligands functionalized with a donor group, [D].
2. By chemical reaction of a pendant group [D] of the polymers with the organometallic MLn fragment to give a D′ group coordinated to the metal leading a new bond [spacer]–D'.
3. By ring-opening polymerization of the organometallic monomers.

A fourth but less common method (4) involves the simultaneous in situ-polymerization /coordination.

The most common synthetic route (route 1) involves the preparation of a polymer bearing the donor group D, followed by a susbstitution reaction with the appropiate organometallic compounds having a labile ligand. However, apart from the synthetic difficulties in the preparation of the polymeric ligand itself, this approach implies two main complications, namely (a) incomplete coordination of the organometallic fragment, frecuently by steric reasons; and (b) crosslinking reactions by a double substitution on the precursor organometallic by two ligands coming from different polymeric chains.

The method 2.- is a very convenient preparation when the method 1.- is unsuccessful. The method consists in the reaction of a reactive group [D] (for instance $\{NPCl_2\}_n$) with the organometallic precursor MLn coordinated to the appropiate spacer for instance HOC_6H_4X, X= CN, CH_2CN, PPh_2 or HOC_5H_4N (namely $LnM{\cdot}XC_6H_4OH$) in the presence of K_2CO_3 in acetone as solvent. Sometimes the inconvenient of this route is the dificulty in the preparation of the precursor $LnM{\cdot}XC_6H_4OH$.

The most specific method 3.- is restricted to cyclotriphosphazenes having strained organometallic fragments coordinated to the ring as the $N_3P_3(OCH_2CF_2)_4(N\text{-}C_5H_4)_2Fe$ and $N_3P_3R(OCH_2CF_2)_3(N\text{-}C_5H_4)_2Fe$, R = OC_6H_5, Me, Ph series. Although several attemps to obtain organometallic polyphosphazenes by this ring-opening method -to best of our knowledge - these are the only examples reported.

Polyphosphazenes having Phosphines Groups Coordinated to Organometallic Derivatives

The coordinative properties of the phosphine-bearing poly((aryloxy)phosphazenes) $[NP(OC_6H_5)_x(OPPh)_{1-x}]_n$ (X= 0.3 or 0.6) see chart 1, were investigated for the first time in 1983 by Allcock et al. [3a]. The polymer without metal was first prepared by Allcock [3b,3c] in 1980.

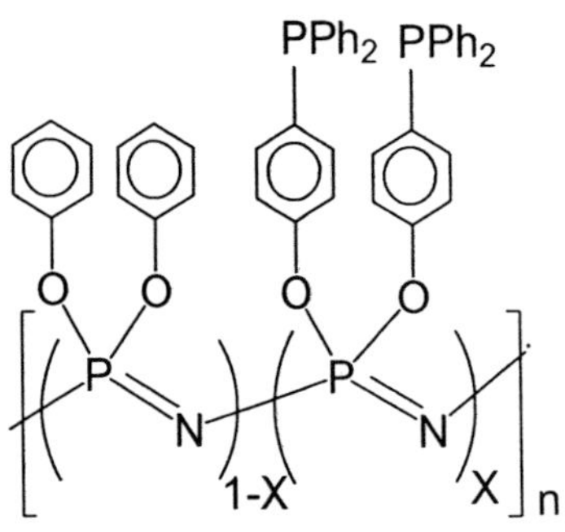

Chart 1. (1a) X= 0.3 (1b) X= 0.6

For organometallics such as $H_2Os_3(CO)_{10}$, $Mn(CO)_3(\eta\text{-}C_5H_5)$, $Fe(CO)_3(PhCH{=}CHC(O)CH_3$ and $[RhCl(CO)_2]_2$, coordination of the pendent phosphine units to all five metals occurred. However some of the products were insoluble due to a matrix crosss-linking provoked by the organometallic fragments. For several of these polymers an anomalous behavior was observed after being isolated in the solid state. When first prepared, the polymer complexes were soluble in tetrahydrofuran. However after precipitation into hexane, this solubility was lost. It was speculated that the close proximity of functional groups in the solid state favors reorganization and further decarbonylation to yield a cross-linked polymer. Thus the interaction of polymers 1a, 1b with $Co_2(CO)_8$ affords a polymer similar to one above described which catalyzes hydroformylation reactions [12].

Further organometallic derivatives of phosphines containing polyphosphazenes were reported by Carriedo et al in 1999 [13]. The polyphosphazene containing the $W(CO)_5$ anchored to the polimeric chain and shown in scheme 4, was prepared by the method II and exhibited little polydispersity with a M_w of 1.700.000. Similar soluble polyphosphazene containing the anchored organometallic $(\eta^5\text{-}C_5H_4Me)Mn(CO)_2$ 2b was prepared using the same route and starting from the precursor $(\eta^5\text{-}C_5H_4Me)Mn(CO)_2[PPh_2C_6H_4OH]$ [13b]. The product had a M_w of 1,000,000 and a M_w/M_n of 4.

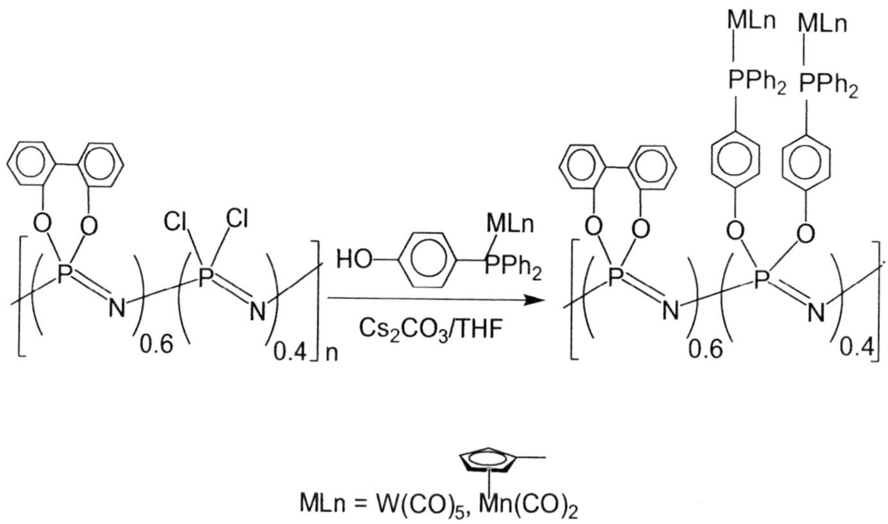

Scheme 4. (2a) : MLn = $W(CO)_5$, (2b) : MLn= $CH_3\text{-}C_5H_4Mn(CO)_2$

Polyphosphazenes having CH_2CN Groups Coordinated to Organometallic Derivatives

Several organometallic derivatives of the polyphosphazene containing the benzylcyanide spacers have been reported [14a-e]. Formulas and approximate composition of the polymers are displayed in Table 1.

The polymers with the organometallic fragments MLn= $Cr(CO)_5$ [14a], $(CH_3\text{-}C_5H_4)Mn(CO)_2$ [14b], Cp*Fe(dppe)[14d], CpFe(dppe) and $CpRu(PPh_3)_2$ [14e], were prepared using the method I. They are insoluble except 7a having Cp*Fe(dppe) pendent groups for which a M_w = 1,260,000 was found. For the polymers 3a and 3b, as found for another organometallic polyphosphazenes, when prepared these polymer were soluble in tetrahydrofuran but after precipitation in hexane its solubility was lost. In this case the insolubility was attributed to the deprotonation of the diphenylphosphinoethane ligand of CpFe(dppe) which provoke a cross-linking of the polymer chains [14e]. Solubility of the organometallic polymer 7a was achieved by the presence of the Cp*Fe(dppe) groups anchored to the phosphazene chain [14d].

Table 1. Formulas and approximate composition for the organometallic polymer containing the Benzylcyanide spacer

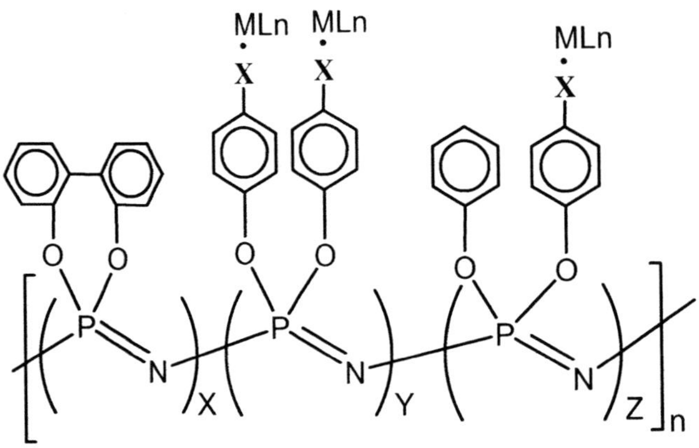

	X	MLn	X	Y
3a	$-CH_2CN$	$C_pFe(dppe)$	0.80	0.18 [a]
3b	$-CH_2CN$	$C_pFe(dppe)$	0.55	0.20 [b]
4a	$-CH_2CN$	$C_pRu(PPh_3)_2$	0.80	0.18 [a]
4b	$-CH_2CN$	$C_pRu(PPh_3)_2$	0.55	0.20 [b]
5a	$-CH_2CN$	$Cr(CO)_5$	0.80	0.18 [a]
6b	$-CH_2CN$	$(\pi\text{- }CH_3\text{-}C_5H_4)Mn(CO)_2$	0.55	0.20 [b]
7a	$-CH_2CN$	$C_5(CH_3)_5Fe(dppe)$	0.80	0.18 [a]

[a] Z = 0.2 ; [b] Z = 0.25

Polyphosphazenes having Pyridine Groups Coordinated to Organometallic Derivatives

Although there are several examples of organometallic derivatives of phosphine and benzylcyanide spacers containing polyphosphazene are known, those with pendent pyridine groups are scarce [15]. To date the only reported example is the derivative containing the organometallic $W(CO)_5$ [15a] see scheme 5.

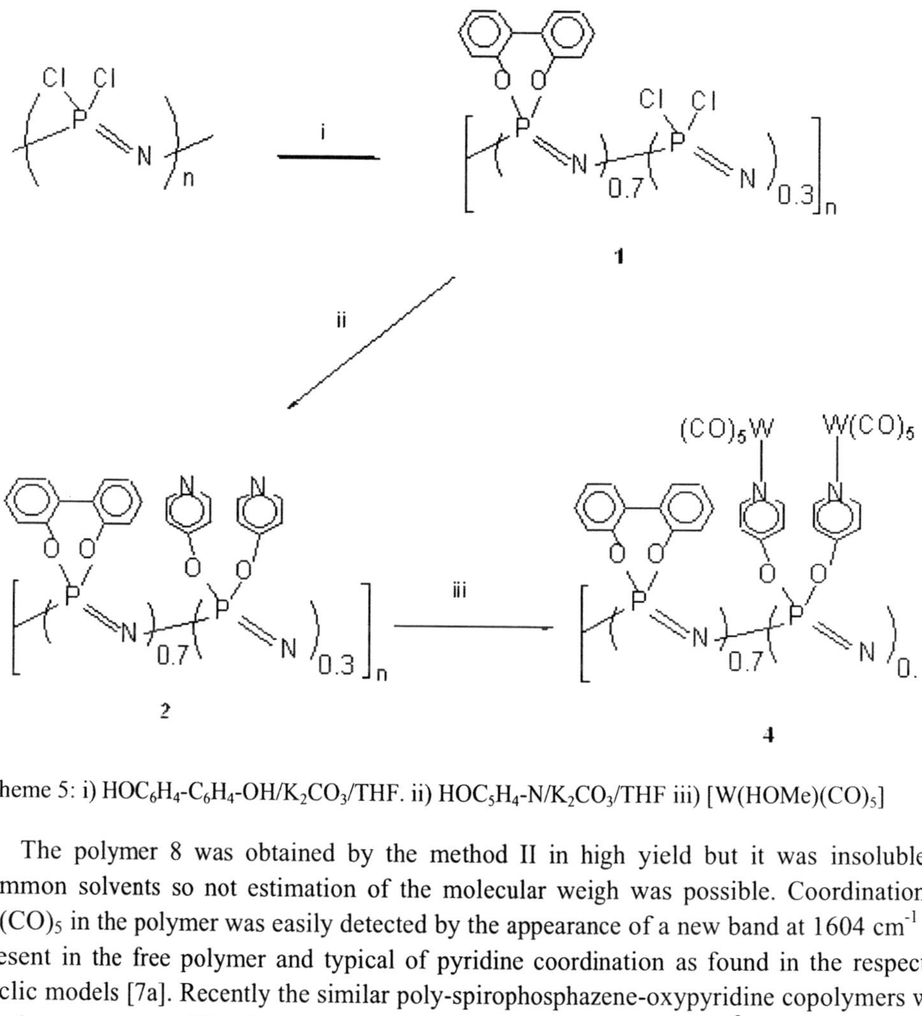

Scheme 5: i) HOC_6H_4-C_6H_4-$OH/K_2CO_3/THF$. ii) HOC_5H_4-$N/K_2CO_3/THF$ iii) $[W(HOMe)(CO)_5]$

The polymer 8 was obtained by the method II in high yield but it was insoluble in common solvents so not estimation of the molecular weigh was possible. Coordination of $W(CO)_5$ in the polymer was easily detected by the appearance of a new band at 1604 cm^{-1} not present in the free polymer and typical of pyridine coordination as found in the respective cyclic models [7a]. Recently the similar poly-spirophosphazene-oxypyridine copolymers with pendent organometallics fragments CpFe(dppe), $CpRu(PPh_3)_2$, and $(n^5$-$CH_3C_5H_4)Mn(CO)_2$ have been prepared [16]. The formulas and approximate composition of these polymers are displayed in Table 2. These polymer were obtained as insoluble powders and all exhibit the expected IR pyridine coordination band around 1600 cm^{-1} typical for coordinated oxypyridine This absorption band was clearly observed for all the polymers in spite of their relatively low metal content. Thus the expected ν(CO) bands for the organometallic moiety were observed normally. Additionally the cationic nature of polymer 10 was observed by the presence of $ν(PF_6)$.

Table 2. Genaral Structure of the Polimers (9) – (11)

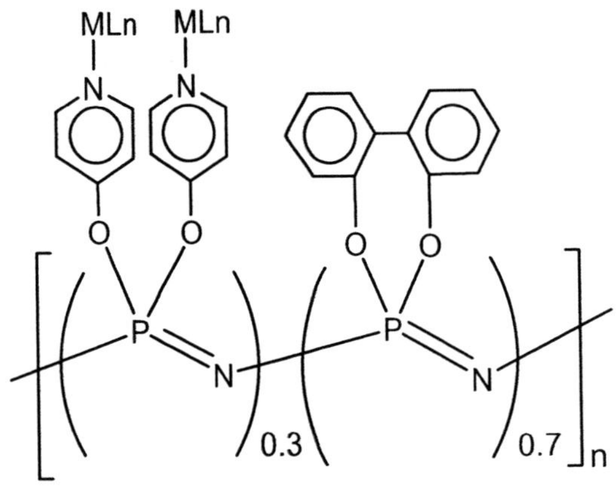

MLn	Polymer N°
CH_3-$C_5H_4Mn(CO)_2$	(9)
CpFe(dppe)	(10)
$CpRu(PPh_3)_2$	(11)

Polyphosphazenes having Nitrile Groups Coordinated to Organometallic Derivatives

To our best knowledge the only reported [14c] organometallic polymer of this type is 12, see chart 2:

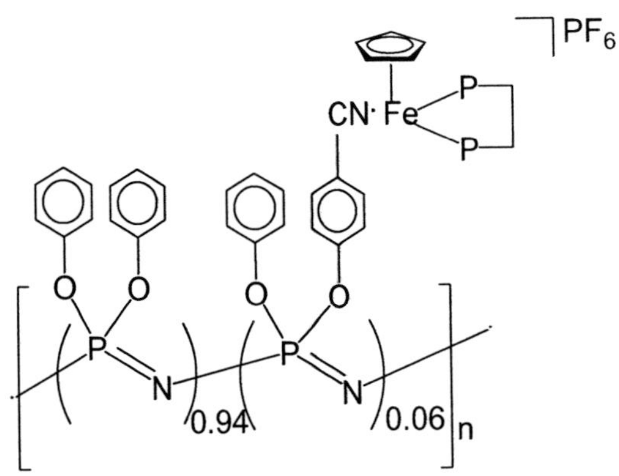

(12)

Chart 2

This polymer was synthesized from reaction of the respective copolymer without metal with the precursor $[CpFe(dppe)CH_3CN][PF_6]$ by complete elimination of the acetonitrile from the equilibrium mixture that they form in dichloromethane solution shown in scheme 6.

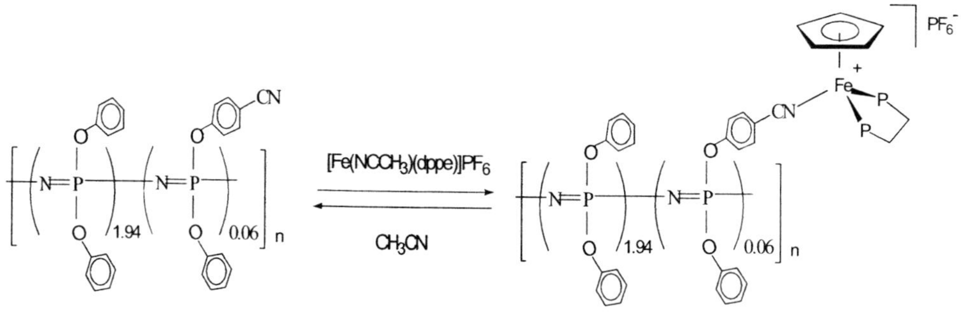

Scheme 6

Due to the solubility of the polymer it was possible to make an electrochemical study of the polymer. As shown in figure 1c the cyclic voltammogram exhibited a one-electron oxidation wave centered at 0.84 V, with an ΔEp that varies from 118 to 124 mV when the scan rate changes from 20 to 300 mV s^{-1}. This corresponds to an electrochemically reversible oxidation. The $E_{1/2}$ of 0.84 V is almost identical to that measured in he CV of the cyclic phosphazene complex models $\{N_3P_3(OC_6H_5\text{-}R)_5\ [OC_6H_4\ CNFe(dppe)Cp]\}[PF_6]$R= H, Bu^t, (fig.1b) and in that of the simple benzonitrile complex $[Fe(C_6H_5CN)(dppe)Cp][PF_6]$ (fig1a). The oxidation of 12 is, however, chemically irreversible, because the ratio I_a/i_c was of the order of 0.91 at 20 mVs^{-1}. In fact, the voltammogram clearly showed a reduction wave centered near 0 V (see Figure 1c), whose intensity increases with the number of cycles, corresponding to a decomposition product formed during the oxidation process. As shown in figures 1a and 1b, a similar decomposition product is also formed after the oxidation of the cyclic model, although this occurs with a less extent in the oxidation of the complex $[Fe(C_6H_5CN)(dppe)Cp][PF_6]$ (Figure 1a).

Thus, ΔEp measured in the CV of polymer 12 at 20 mV s^{-1} (118) is only slightly higher that those corresponding to the cyclic phosphazene models (74 and 88 mV respectively). This shows that there is no kinetic difficulties for the electrode oxidation of the $[Fe(dppe)Cp]^+$ fragments attached to the polymer. This behaviour is in sharp contrast with that observed for the analogous complex polymers with $[Mn(CO)_3(dppe)]^+$ fragments [7].

SPECTROSCOPIC PROPERTIES

NMR Data

As can be seen from the section 3, few organometallic polymers are soluble. In these cases they have been fully characterized by NMR spectroscopy, and their data are displayed in Table 3.

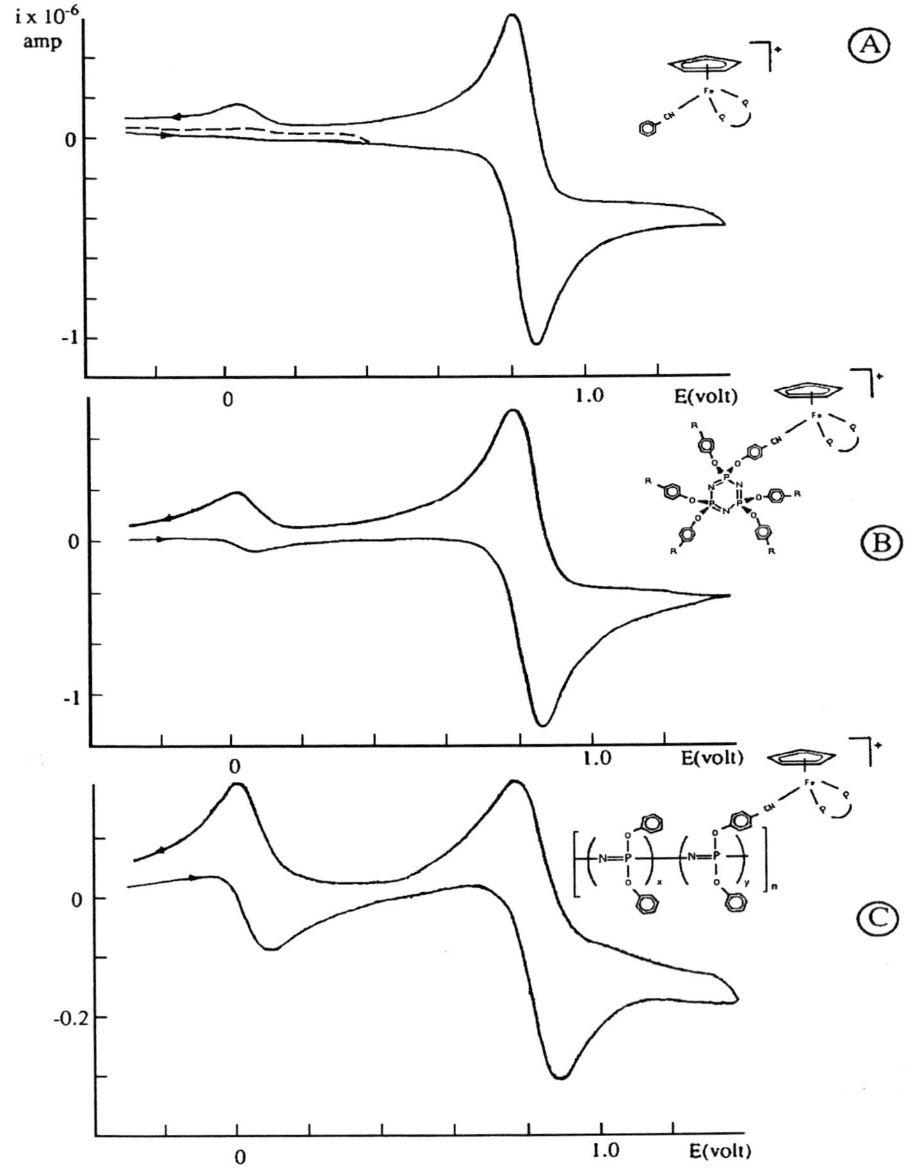

Figure 1 Cyclic voltammogram of polymer 12 (C) and of their models $[N_3P_3(OC_6H_4R)_5(OC_6H_4CN \bullet Fe(dppe)(Cp)][PF_6]$(B)and $HOC_6H_4CN \bullet Fe(dppe)(Cp)[PF_6]$ (A).

From Table 3 it can be viewed that the chemical shift values of the phosphorus atoms linked to the bis-spiro units (P_1) that lies in the range of -3.8 ppm to -6.4 ppm are much higher than the -20 ppm found for the phosphorus atoms connected to the aryloxy units (with exception of the polymer 7a). A similar tendency was found for the series polymers without metal [7b] and the variation was attributed to the deshielding effect arising from the formation of a ring containing a phosphorus atom, thus this effect being dependent on the ring size.

Table 3. NMR data for organometallic Polymers

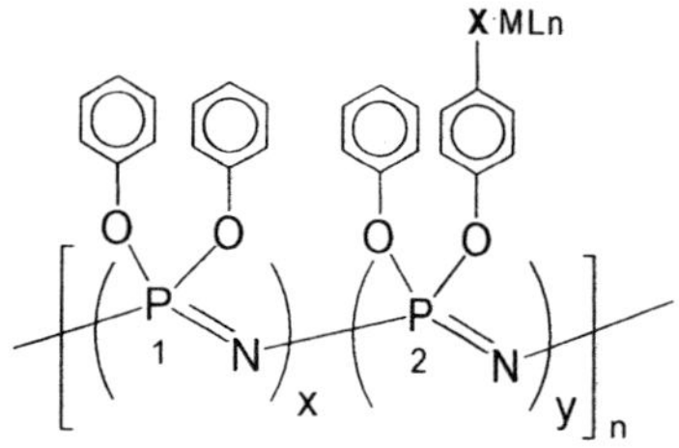

MLn	X	x	y	^{31}P-NMR P$_1$	P$_2$	X-M	MLn
AuCl	PPh$_2$	0.3	0.7	-	-19	33	-
		0.6	0.4	-	-19	33	-
$H_2Os_8(CO)_{10}$	PPh$_2$	0.3	0.7	-	-20	-3	-
		0.6	0.4	-	-20	-3	-
CpMn(CO)$_2$	PPh$_2$	0.3	0.7	-	-19	92	-
		0.6	0.4	-	-20	-	-
RhCl(CO)	PPh$_2$	0.3	0.7	-	-20	28	-
		0.6	0.4	-	-19	29	-
CpFedppe	CN	0.94	0.06	-19.6	-21.1	-	97.4

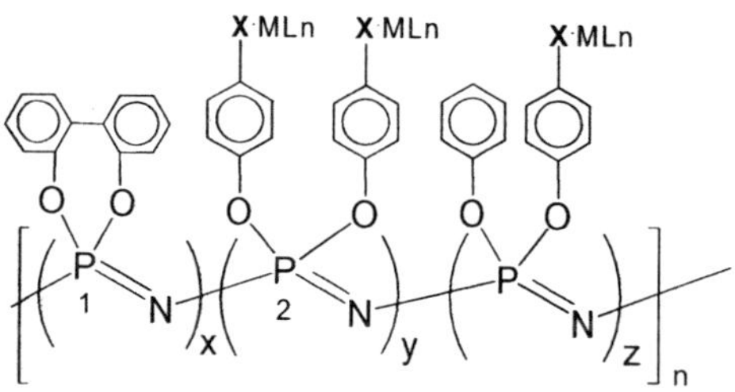

MLn	X	x	y	^{31}P-NMR P$_1$	P$_2$	X-M	MLn
W(CO)$_5$	PPh$_2$	0.65	0.35	-5.0	-23.5	20	-
(Me-C$_5$H$_4$)CpMn(CO)$_2$	PPh$_2$	0.6	0.4	-6.4	-23.6	92.1	-
W(CO)$_5$	OC$_5$H$_4$N	0.3	0.7	-3.8	-25	-	-
Cp*Fedppe	CH$_2$CN	0.8	0.18	-19.6	-21.1	-	34.53

On the other hand the phosphorus skeletal linked to the metal ligand (phosphorous P$_2$) appears in the range -25--9 ppm. For the insoluble polymers, NMR data can be obtained by solid state NMR using the CPMAS (cross-polarization magic angle spinning) technique. However to our best knowledgements the only solid state study on an organometallic polyphosphazene was reported for $\{[NP(OC_6H_4Br)(OC_6H_4FeCp(CO)_2]_{0.5}[NP(OC_6H_5)(OC_6H_4FeCp(CO)_2)]_{0.5}\}_n$ [8]. However only ^{12}C-CPMAS data were reported. For instance the cyclopentadienyl carbon atoms appears at 86 ppm, similarly to solution spectrum data for related CpFe(CO)$_2$ compounds .

IR Data

Several IR data have been informed for organometallic containing metals. Assignments of these polymers are aided by vibrational studies on the polyphosphazenes without metals [17]. As can be viewed from table 4.

Table 4. IR data for Polyphosphazene containg organometallic grups[a]

Polymer	ν(C-OP)	ν(P=N)	ν(P-OC)	ν(CN)	(OC_5H_4N)	ν(M-Ln)
2a		1246, 1197		2217		2071,1981, 1936[b]
2b	1263	1246, 1198	1170			1927, 1860[c]
3a	1268	1245	1120	2216		1097[d]
3b	1246	1189, 1174	1122	2216		1095[d]
4a	1271	1246, 1196	1160	2250		1070[d]
4b	1267	1246, 1198	1118	2249		1095[d]
5a	1270	1245, 1203		2251		2056,1967,1938[e]
6b	1260	1246, 1196	1171	2253		1929, 1857[c]
7a	1247	1192	1181	2252		
8		1246, 1193	1096		1604[f]	2071,1977, 192[b]
9	1289	1243, 1190	1118		1652[f]	1951, 1839[c]
10	1281	1247, 1197	1096		1584[f]	
11	1231	1275, 1186	1116		1589[f]	

[a] KBr solid, unless otherwise.
[b] ν(CO) of $W(CO)_5$ group.
[c] ν(CO) of n^5-$CH_3C_5H_4$-$Mn(CO)_2$ group.
[d] δ(C-H) bending in-plane vibrations of the C_5H_5 group.
[e] ν(CO) of $Cr(CO)_5$ group.
[f] ν(C-C) of C_5H_4N ring.

The identity of the polyphosphazene chain can be confirmed from the two normally observed ν(PN) bands: one around 1240 cm^{-1} and the another at 1200 cm^{-1}. Additionally the bands ν(PO-C) and ν(P-OC) can be observed in the range 1230-1280 cm^{-1} and 1090-1160 cm^{-1} respectively. In some cases the characteristic bands of the polymer spacers X were observed: ν(CN) in the range 2216 cm^{-1}-2250 cm^{-1} for the polymers 3-7 as well as the characteristic ν(CC) pyridine ring bands in the zone 1584 cm^{-1}-1650 cm^{-1} for the polymers 8-11. Thus this band is typical of pyridine coordination because its appears only when a metal coordination of the nitrogen pyridine occurs. In some cases the IR spectra of the organometallic fragments exhibit some characteristic absorption bands as the $W(CO)_5$, $Cr(CO)_5$ and (n^5-$CH_3C_5H_4$) $Mn(CO)_2$ fragments in the organometallic polymers see Table 3.

Ir spectra gave valuable information of the percentage of metal content in solid insoluble polymers where the organometallic fragment has a characteristic band. For instance this technique has been used in the chromium [14a] and manganese [14b] polymers to estimate the charge degree of the metal. This was made by comparing the intensity areas ratio ν(CO)/ν(PN) of some simple model cyclic phosphazene of known content of metal carbonyl, with the ratio ν(CO)/ν(PN) of the polymer. This is illustrated for polymer 5a in figure 2.

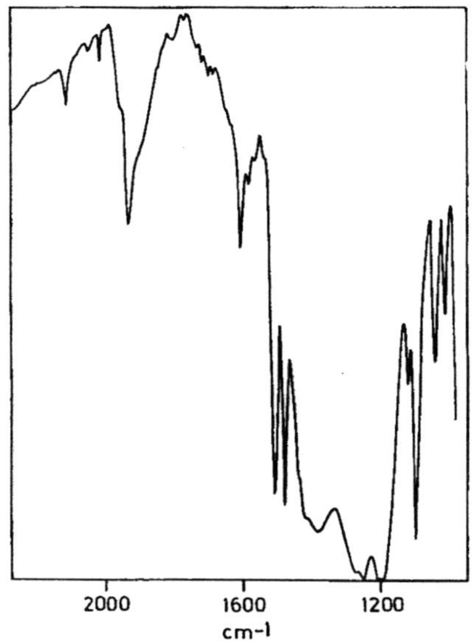

Figure 2. IR spectrum (KBr disk) from 2500-1000 cm^{-1} of copolymer 5a that contains the $Cr(CO)_5$ moiety.

UV-Visible Data

Polyphosphazenes without metals do not absorb in the visible region [14]. Then the organometallic fragment is often an adequate chromosphere in this spectral zone. Notwithstanding this, few Uv-visible data for organometallic polyphosphazene polymers have been reported [14a, 14b, 14d, 14e]. The series of polymers 3-7 present in solid state characteristic absorptions, , in the Uv region. The chromophore CpFe(dppe) exhibit a characteristic absorption bands in the range 400-460 nm depending of the π-acceptor properties of the ligand [14d]. Thus in the polymers 3a and 3b, the metal chromophores absorb around 438 nm. On the other hand the chromophore $Cr(CO)_5$ present absorptions around 600nm typical of the chromium fragment. The cymantrene group exhibits a broad and not typical absorption band in the region 400nm-700nm [14b].

Thermal Properties

Studies of the Pyrolysis of the Organometallic Polymers under Dinitrogen: Thermogravimetric and DSC Analysis

Thermogavimetric analysis (TGA) studies of the organometallic derivatives of polyphosphazenes are in general scarce [8,14]. Thus the major data arise from the organometallic derivatives from the polyphosphazene series containing the benzylcyanide spacer [14]. Data obtained under N_2 using heating rates of 10°C/min are summarized in Table 5.

Table 5. TGA and DSC[e] data for polyphosphazene and their organometallic derivatives

Polymer	Temp. Desc. (°C)	% Residue at (°C)	Tg (°C)
P—CN[a]	396	35(960)	123
P—CN[b]	400	46(980)	85.4
P—PPh_2[c]	500	38(800)	115
P—OC_5H_4N[d]	450	42(800)	174
2a	-	56(800)	127
2b	350	41(800)	128
3a	298	20(980)	-[f]
3b	280	18(980)	-[f]
4a	280	40(980)	112.6
4b	260	35(980)	106.2
5a	375	32(980)	140
6b	-	64(980)	-[f]
7a	335	11(980)	149
8	-	71(800)	-[f]
9	300	50(800)	122
10	300	30(800)	-[f]
11	220	10(800)	-[f]

[a] Polymer 3a without metal. [b] Polymer 3b without metal. [c] Polymer 12 without metal
[d] Polymer 8 without metal. [e] Measurements Conditions (Netzch STA 409) Heated rate, 10 °C /min, from ambient temperature to 1000 °C under constant flow of nitrogen.
[d] T_g not observed

The thermogramms obtained, show different decomposition patterns depending on the organometallic fragment as well as on the spacer containing the donor X. For instance, the organometallic polymers 2a, 3a, 3b, 6b and 7a, undergo an initial and gradual loss of weight attributed to the loss of some organometallic residues while polymers 2a, 4a, 4b, 5a, 8, 9,10 and 11 exhibit an initial weight loss around 260–400°C depending on the organometallic fragment. High pyrolytic residues were observed for the polymers 2a, 2b, 4a, 4b, 6b, 8, and 9 as is observed from table 5.

Its appears to be that the highest pyrolitic residues are obtained from polymers having anchored organometallic fragments containing labile carbonyl groups or other labile ligands, which by loss of these groups crosslink a high degree of polymer chains. Consistently with this, organometallic polymers having, fragments with no labile ligands such as $CpFe(dppe)^+$ or $Cp^*Fe(dppe)^+$ exhibit the lowest pyrolytic residues. The organometallic fragments $CpRu(PPh_2)_2^+$ exhibit an intermediate percentage of pyrolytic residue due to the moderate lability of the phenylphosphine ligands.

Thus regarding the nature of the metal, the polymers with most high pyrolytic residues content are the W, Mn and Cr independent of the polymer spacer.

Glass transition temperature (Tg) data of the organometallic poly(organophosphazenes) are scarse. Some compiled values are shown in table 5. Usually in this type of polymer the glass transition is not detected in DSC curves. In general an enhancing of the T_g in going from the free polymer to the organometallic derivative has been observed.[8,14] which have been attributed to most restricted flexibility of the chains due to the presence of the cross-linking capacity of the organometallic moiety. In fact for the polymers 2a, 2b, 4a, 4b and 7a an increase of the Tg of the organometallic derivative in comparison to the one without metal polymer was observed. On the other hand a decrease of Tg was observed in the case of the polymers 4a and 9 with respect to the free polymer which can be due, in the case of 4a, to low cross-linking ability of the Ru organometallic fragment and in 9 to the enhancing of the flexibility of the P=N-P=N chains due to the decrease of the steric effect of the aromatic ring in the polyspirophosphazenes by the proximity of the organometallic groups [7b].

Electrical Conductivity Properties

General Considerations

With the exception of type Hagihara polymers [2a, 19], ie $[M(L)_2\text{-C}\equiv\text{C-X-C}\equiv\text{C-}]_n$ M=Pt, Pd and Ni L=n-P(Bu)$_3$, X=C_6H_5, the electrical conductivity of organometallic polymers has not been fully investigated. Relatively few studies have been reported [2c, 20, 21]. However, some data suggest that the conductivity depends on three factors:

a) The conductive properties of the polymer backbone [20]
b) The π -conjugated behavior of the spacer that is connected to the polymer backbone[2c] and c) The electronic 'richness' of the organometallic fragment.

Poyphosphazenes and their Organometallic Derivatives

To explore the possibility of electron transfer behavior of the series of copolyphosphazenes after incorporation of an organometallic fragment, electrical conductivity measurements were carried out by impedance spectroscopy. Values in the range of $\sigma = 10^{-9}$ Scm^{-1} to 10^{-11} Scm^{-1} were obtained for both, the copolymer and the copolymer that contains the organometallic fragments, see Table 6.

Table 6. Impedance spectroscopy conductivity's for the copolyphosphazenes and their organometallics derivatives

Polymers	σ(Scm^{-1})	
(1)	1.2	10^{-11}
(2)	1.6	10^{-10}
(3a)	4.6	10^{-10}
(3b)	4.5	10^{-10}
(4b)	1.16	10^{-9}
(5a)	1.2	10^{-11}
(6b)	8.1	10^{-11}

These values are typical of insulators. Similar electrical conductivies have been reported for other polyphosphazenes (e.g., $[PN(OC_6H_5)_2]_n$) [22]. Although the N=P backbone in copolymers 1a and 1b is insulating, electrical conductivity may be expected due to charge hopping between the metal centers; namely, the external sphere mechanism that was proposed for polyphosphazenes that contain pendant ferrocene groups [23]. In this example, the fragment, Cp_2Fe, is electrochemically active as well as electron rich. In contrast, the insulator behavior of the iron and ruthenium organometallic polymers may well be due to a coil conformation of the polymer that does not permit external sphere electron transfer between the metal centers. In fact, solution properties studied by size exclusion chromatography using simultaneously multiangle light scattering and differential refractive index detectors, indicated that polyspirophosphazenes, including random copolymers similar to 3, behave as a random coil [24a, 24 b]. Further studies to correlate electrical conductivity with polymer structure and conformation are needed and are currently underway. Alternatively, another explanation of the insulator behavior of the organometallic polymer may be the low metal content of the polymer and the non-conductive polyphosphazene matrix.

Applications

Catalysis

A difficulty normally associated with homogeneous catalytic systems is the separation of the products from the reactants and the catalyst and the concentration of the actives sites in some insoluble compounds in heterogeneous catalysis. On the other hand homogeneous catalysts offer the advantage of high selectivity and mild reactions conditions, while heterogeneous catalysts although generally more active and less selective are instead more durable and may be separated more easily from the reaction products.

By anchoring a homogeneous catalyst to a polymer, the advantages of both types of catalysts can be combined. Few examples of catalysis by polyphosphazenes having anchored organometallic fragments have been reported.

The phosphazene polymer 1 with X = 0.15 anchored with the organometallic $Co_2(CO)_8$ catalyze reactions of hydroformylation and isomerization to give aldehydes [12]. No other examples of catalysis by polyphosphazenes containing anchored organometallic fragments have been reported. Most recently, chiral synthesis using polyphosphazenes having organometallic fragments in a "chiral pocket" have been reported [25].

Nanostructured Materials from Polyphosphazenes Containing Anchored Organometallic Fragments

Materials with submicron dimension as nanoparticles represent an exciting new class of materials [26]. As a consequence of their tiny size, nanomaterials often display unique physical and chemical properties that are atypical of the bulk materials; for instance optical, magnetic and electrical properties are sensitive to size effects. Furthermore nanosized particles are equally very efficient in the field of catalysis [27] due to high ratio of surface to volume. Consequently numerous processes of nanomaterials synthesis have been investigated aiming to control their size, morphology, structure and chemical composition. A large numbers of studies concerning the production of nanoparticles have been published [26,28]. Chemical techniques are widely used to produce nanostructured materials due to their straightforward nature and the potential for producing large quantities of the final product. The chemical method to obtain metal nanoparticles generally involves the reduction in solution of a metal salt in presence of some stabilizing of the nanoparticle. As stabilize the most frequently used have been surfactants, polymers and mesoporous silica matrix.

Polyphosphazenes could be suitable host polymer to control the growth of nanometer size particle in solution. Nanoparticles of CdS matrix encapsulated in the polymers $[NP(OCH_2CH_2OCH_2CH_2OCH_3)_2]_n$ have been reported [29a]. On the other hand gold nanoparticles stabilized by poly(methylphenylphosphazene) were obtained by reduction of the gold salt $AuCl_4^-$ with $NaBH_4$ in presence of the polyphosphazene [29b].

Nanoparticles in the range 5-8nm were obtained. Stabilization of the nanoparticles was achieved by the interaction of the basic nitrogen atom of the polyphosphazene moiety with the metal.

Another approximation to nanostructures in solid state has been recently proposed [30]. Pyrolysis in air of the polyphosphazenes 2-7 affords nanostructured (see scheme 7) materials as revealed by their SEM and TEM images.

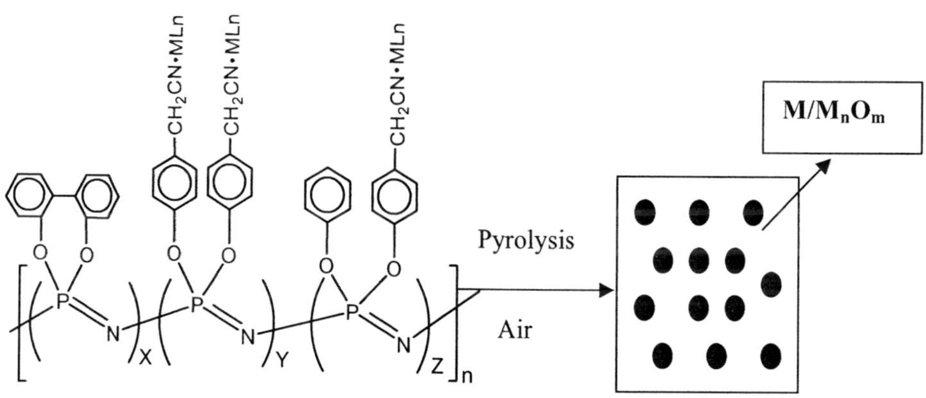

Scheme 7

For instance, as is shown in Figure 3 the pyrolytic residue from the pyrolysis of the organometallic polyphosphazene 5a exhibits nanostructures of chromium oxide of about 74 nm. The EDAX shown in the inset of the figure 3 clearly indicates the presence of chromium, phosphorus and oxygen atoms. The phosphorus and oxygen atoms probably arise from the solid state phosphorus oxides which act as matrix, stabilizating the metallic nanostructures.

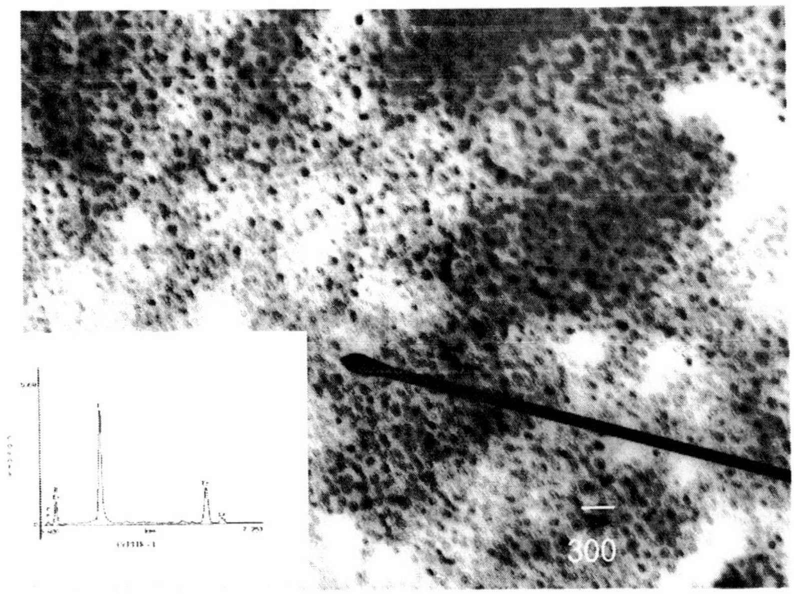

Figure 3. TEM image of the product from the pyrolysis of the polymer 5a on a holes gold grid.The darker circular region are chromium nanoparticles. The inset is an EDAX analysis of a portion of the material.

Future Prospects

Organometallic polymers that combine the processing advantages and exceptional good physical properties of an inorganic backbone (as polyphosphazenes or polysiloxanes) with the interesting properties of an organometallic fragment (magnetism, redox, electrical conduction, luminescence, etc.), can give rise to materials with unprecedented and exciting properties. However, in order to enhance the scarce amounts of this type of compounds the improvement and the developments of novel appropriate synthesis methods are required. As pointed by Manners [2a], this could be not facile. This could afford in a near future, organometallic polymers with properties as yet now not obtained. For instance paramagnetic organometallic polyphosphazenes have not still been prepared. Another less explored field is the use of polyphosphazenes having anchored organometallics active catalysts in catalysis. The design of a polyphosphazenes having an organometallic inserted in a chiral pocket may be useful in asymmetric synthesis. On the other hand the use of polyphosphazene having anchored organometallic fragments as precursor of nanostructures both in solution as well as in solid state is an incipient and promissory field of research. Finally, the use of this type of organometallic polyphosphazene in micro and nanolithography is an interesting but unexplored field. Thus all these new materials, will probably the play a key role in the future development of materials science and technology.

ACKNOWLEDGEMENTS

We thank to Miguel Barbosa for helpful discussions and Professor G.A. Carriedo for its valuable collaboration. We are also grateful to Conicyt (Project 1030515) for financial support.

REFERENCES

[1] (a) D. O'Hare and D.W. Bruce *Inorganic Materials*,Wiley&Sons, Chichester UK ,1992. (b) N.J. Coville and L. Cheng, *J. Organomet. Chem.* 571,(1998),149 (c) P. J. Pickup *J. of Materials Chemistry*, 9,(1999), 1641. (d) F.Ciardelli, E.Tsuchida and D.Wohrle, *Macromolecule-Metal Complexes*. Springer, Berlin, 1996. (e) C.U.Pittman Jr, C.E.Carraher, M.Zeldin, J.E.Sheats and B.M. Culbertson, *Metal-Containing Polymerc Material,* Plenum, NewYork,1996. (f) P.Wisian-Neilson, H.R.Allcock and K.J.Wynne Inorganic and Organometallic Polymers II; Advanced Material and Intermediates, *ACS Sym. Ser.* 572, 1994.

[2] (a) P. Nguyen, P. Gomez-Elipe and I.Manners *Chem. Rev.*99, (1999), 1515. (b) I.Manners *J.Chem. Commun.* (1999), 857. (c) R.P.Kingsborough and T.M.Swager, *Progress in Inorg.Chem.* 48, (1999), 123. (d) I.Manners, *Science* ,294, 2001, 1664.

[3] (a) H.R.Allcock, K.D.Lavin, N.M.Tollefson and T.L.Evans. *Organometallics.* 2, (1983), 267. (b) H.R. Allcock, T.L.Evans and T.J Fuller *Inorg. Chem.* 19, 1980, 1026. (c) H.R. Allcock, T.L.Evans and T.J Fuller *Macromolecules* 13, 1980, 1325.

[4] (a) H.R.Allcock, S.Al-Shali, D.C.Ngo, K.B.Visscher and M.Parvez, *I. Chem.Soc. Dalton trans.* (1996), 3549. (b) H.R.Allcock, S.Al-Shali, D.C.Ngo, K.B.Visscher and M.Parvez, *I. Chem.Soc. Dalton trans.* (1996), 3521. (c) H.R.Allcock, *Acc. Chem. Res.* 112, (1979), 351.

[5] H.R.Allcock, N.M.Tollefson, R.A.Arcus and R.R.Whittle. *J.Am. Chem.Soc.* 107, (1985), 9166.

[6] H.R.Allcock, D.C.Ngo, M.Parvez and K.B.Visscher. *Inorg.Chem.* 33, (1994), 2090

[7] (a) G .A. Carriedo, L. Fernández-Catuxo, F .I. García Alonso and Gómez- Elipe *J. Organomet. Chem.* 503, (1995), 59. (b) G..A. Carriedo .L. Fernandez-Catuxo, F .J. Garcia Alonso, P .Gómez-Elipe P .A. González, *Macromolcculcs* 29,(1996),5320.

[8] H. R. Allcock, E. N. Silverberg, C. J. Nelson, W. D. Coggio. *Chem.Mater.* 5, (1993), 1307.

[9] (a) H. R. Allcock and P. P. Geiger *J. Am. Chem. Soc.* 101,, (1979), 2492. (b) H. R. Allcock , J. P. Wagner, P.P. Geiger and P.P. Bernheim, *Inorg. Chem.* 20, (1981), 716.

[10] (a) C. Díaz and I. Izquierdo, *Polyhedron*, 18, (1999), 1479. (b) C.Díaz, I.Izquierdo, F.Mendizábal, N.Yutronic, *Inorganica Chimica Acta* 294, (1999), 20.

[11] C.Díaz, M.Barbosa, Z.Godoy, *Polyhedron* 23, (2004), 1027.

[12] R.A.Dubois, P.E.Garrau, K.D.Lavin and H.R.Allcock, *Organometallics* 5, (1986), 460.

[13] (a) G. A. Carriedo, F. J. García Alonso, P. González and P. Gómez-Elipe Polyhedron 18, (1999), 2853 (b) G. A. Carriedo, F. J. García Alonso, P. A. González, C. Diaz, N. Yuronic *Polyhedron* 21, (2002), 2579.

[14] (a) C. Diaz and P.Castillo, *J.Inorg. Organometallic Polymers* 11, (2002), 183 (b) C. Díaz, P. Castillo, G. A. Carriedo, P. Gómez-Elipe, F. J. García Alonso, *Macromolec. Phys. and Chem.* 203,(2002), 1918. (c) G. A. Carriedo, F. J. García Alonso, P. Gómez-Elipe, C. Díaz and N.Yutronic, *Journal of the Chilean Chemical Society* 48, (2003), 25. (d) C. Diaz and P.Castillo, *Polymer Bulletin* 50, (2003), 123. (e) C. Díaz, M. L. Valenzuela, M. Barbosa, *Materials Research Bulletin* 39, (2004), 9.

[15] (a) G. A. Carriedo, F. J. García Alonso, J. L. García Álvarez, C. Díaz and N.Yutronic, Polyhedron 21, (2000), 2587. U. Diefenbach "Phosphazenes with Pyridyl Groups" in "*Phosphazenes: A Worldwide Insight*" M. Gleria and R. DeJaeger Eds. Nova Science Publisher New York, 2001.

[16] C.Díaz et all, unpublished results

[17] G.A.Carriedo, F.J.García Alonso, P.A.Gonzalez and J.R.Menéndez, *J. of Ramman Spectrosc.* 29, (1998), 327.

[18] C.Díaz, N.Yutronic, *Polyhedron* 6, 1987, 503

[19] (a) K. Sonogashira, S. Takahashi and N. Hagihara, *Macromolecules* 10, 1977,879. See also refs: 2a and 2c.

[20] (a) S.Achar, J.D.Scott and R.J.Puddephatt, *Organometallic* 11, (1992), 2325. (b) B.Panunzi, A.Roviello, F.Ruffo and A.Vivo, *Inorg.Chim.Acta* 281, (1998), 141

[21] (a) H.Nishihara, H.Funaki, T.Shinura and K.Aramak, *Synth.Met.* 55, (1993), 942. (b) G.Zotti, S.Zechin, G,Schiavon, A.Berlin, G.Pagani and A.Canavesi, *Chem.Mater.* 7, (1995), 2309. (c) J.Matsuda, K.Aramaki, H.Nishihara, *J.Chem.Soc.Faraday Trans.* 91, (1995), 1477. (d) M. Owolf and M.S.Wrighton, *Chem.Mater.* 6, (1994), 1526.

[22] H.R.Allcock, J.A.Dodge, J.Manners and G.H.Riding, *J.Am.Chem.Soc.* 113, (1991), 9596.

[23] A.L.Crumbless, D.Cooke, J.Castello and P.Wisian-Neilson, *Inorg.Chem.* 32, (1993), 6088

[24] (a) G .A. Carriedo, F. J. García Alonso, P. Gómez-Elipe, J. L. García Álvarez, M. P. Tarazona, M .T. Rodríguez, E. Saiz and J. T. Padrón, *Macromoléculas* 33, (2000), 3671. (b) J. Burdalo, M. P. Tarazona, G. A. Carriedo, F. J. García Alonso, P .A. Gonzales, *Polymer* 40, (1999), 4251.

[25] G. A. Carriedo, F.J.García Alonso and A. Presa *Eur. J. Inorg. Chem.* (2003) 4341.

[26] (a) *Nanomaterials: Síntesis Properties and Applications*, A.S.Edelstein and R.C.Cammarata, Eds, J.W.Arrowsmith Ltda, Bristol, 2000. (b) K.J.Klabunde "*Nanoscale Materilas in Chemistry*" Wiley, Interscience, New York, 2001.

[27] (a) J.Grunes, Ji Gabor and A.Somorjai, *Chem.Comm.* 2257, 2003. (b) A.T. Bell, *Science*, 299, (2003), 1688.

[28] D.L.Leslie-Pelecky and R.D.Rieke, *Chem. Mater.* 8, (1970), 8.

[29] (a) M.A.Olsharsky and H.R.Allcock, *Chem. Mater.* 9, (1997), 1367. (b) C.H.Walter, J.R.St. Jphn and P.Wissian-Nielson, *J.Am.Chem. Soc.* 123, (2001), 3846.

[30] C.Díaz, M.L.Valenzuela, unpublished results.

In: Polymer Research Developments
Editor: Robert K. Bregg, pp. 23-56

ISBN 1-59454-742-4

Chapter 2

FAST TRANSIENT FLUORESCENCE TECHNIQUE (FTRF) FOR STUDYING GEL SWELLING

Matem Erdogan[a] and Önder Pekcan[b]
[a]Department of Physics, Balikesir University, Balikesir, 10100 Turkey
[b]Department of Physics, Işık University, Maslak, Istanbul, 34389 Turkey

ABSTRACT

The time-resolved, fast transient fluorescence technique (FTRF) which uses Strobe Master System (SMS), is employed for studying swelling of disc-shaped polymeric gels. In this technique the photo multiplier tube (PMT) is gated or strobed by a voltage pulse that is synchronized with the pulsed light source. The intensity of fluorescence emission of an organic dye is measured in a very narrow time window. Three series of polymeric gels based on methlymethacrylate (MMA), styrene (S) and the combination of methlymethacrylate and styrene monomers with ethylene glycol dimethacrylate (EGDM) as a crosslinker agent in the presence of 2,2′-azobisisobutyronitrile (AIBN) were synthesized by using free-radical crosslinking copolymerization (FCC). In the first series of these polymeric gels, five different gels were prepared with various EGDM contents in the presence of MMA monomers. After drying these gels, swelling experiments were performed in chloroform and controlled swelling experiments were performed under chloroform vapor at room temperature. Then, one of the dry gel in the first series was used to study gel swelling in various solvent mixtures. In the second series, heterogels synthesized from MMA and S monomers with constant EGDM content were used to monitor the swelling behaviour of these gels under chloroform vapor. In order to measure the swelling activation energy under chloroform vapor, polystyrene gels were used at various temperature in the last series. In all these swelling experiments, lifetimes of extrinsic fluorescence probe P, which was introduced during polymerization, in and out of polymeric gels were measured from fluorescence decay traces during in-situ swelling processes. The effect of the solvent and organic vapor penetration have been interpreted by applying quenching mechanisms to the produced data where the Stern-Volmer type of quenching mechanism may be assumed for the fluorescence decay of P in the gel sample during swelling process. It is observed that as the swelling time increases, excited pyrenes decay faster and faster by indicating that quenching of excited pyrenes increase in the swelling gel. The Li-Tanaka equation was used to determine the swelling time constant, τ_c and the cooperative diffusion coefficients, D_c which were found to be decreased as the crosslinker content was increased in chloroform, under chloroform vapor and in the solvent mixtures by increasing chloroform content. On the other hand, heterogels with high S content under chloroform vapor swell much more slowly than heterogels with low S content. The activation energy for the swelling of polystyrene gel was measured by using Arrhenius law and found to be 80.0 kJ mol^{-1}.

INTRODUCTION

Polymeric gels, consisting of a crosslinked polymer network immersed in liquid undergo a volume-phase transition; when external conditions such as temperature, solvent composition, or osmotic pressure are changed, a gel reversibly swells or shrinks [1]. Gels are known to exist generally in two forms, swollen or shrunken. Volume phase transitions occur between these forms either continuously or sudden jumps between them [2,3]. The swelling kinetics of physical and chemical gels are very important in many technological applications ranging from controlled drug release to gel actuators. Especially in pharmaceutical industries in designing slow-release devices for oral drugs. In agricultural industry for producing storable foods and in medical applications in developing artificial organs the knowledge of the volume transitions of gels are quite important. All these features are directly related to the swelling kinetics of polymeric gels [4]. The kinetics of the gel swelling process has attacted considerable interest in the past decades. The equilibrium swelling of gels in solvent has been extensively studied [5-7]. The degree of swelling of chemically crosslinked polymer is determined by several factors, such as polymer-solvent interaction, crosslinking density, conditions during network formation etc.. The final degree of swelling can also be affected by existing physical crosslinks, and is controlled by temperature, solvents nature and composition.

The continuous swelling is directly related to the viscoelastic properties of a gel. The gel elasticity and the friction between the network and solvent play an important role on the kinetics of the gel swelling [8-10]. It has been known that the relaxation time of swelling is proportional to the square of a linear size of the gel [8] which has been confirmed experimentally [10]. One of the most important features of the gel swelling process is that it is isotropic, i.e. when the radius increases 10%, the axial length increases 10% in a long cylindrical gel. The elastic and swelling properties of permanent networks can be understood by considering two opposing effects, the osmotic pressure and the restraining force. Usually the total free energy of a chemically crosslinked network can be separated into two terms; the bulk and the shear energies. In a swollen network the characteristic quantity of the bulk free energy is the osmotic bulk modulus, K. The other important energy, the shear energy, keeps the gel in shape by minimizing the non-isotropic deformation. The characteristic coefficient of these forces is the shear modulus, μ, which can be most directly evaluated by stress-strain measurements [11,12]. Li and Tanaka [13] have developed a model where the shear modulus plays an important role that keeps the gel in shape due to coupling of any change in different directions. This model predicts that the geometry of the gel is an important factor, and swelling is not a pure diffusion process.

Fluorescence methods, such as steady state spectroscopy, fluorescence anisotropy and fluorescence decay measurements have shown to be quite effective in the investigation for the microscopic environment around a chromophore. Fluorescence dyes can be used to study local environments, basically with two types of experiments. When the dye simply added to the system as a small molecule, the dye is referred as a probe, which is available commercially. As a consequence such experiments are easy to carry out, but often difficult to interpret because one has to know where the dye is located in the system. If one can prepare an experiment which allows the dye to be attached covalently to a specific component of a system, such as polymer chain segment, such dyes are referred to as labels. The question can be raised that, the presence of the dye perturb the system or perturb its own local environments in the system. Perturbation are most common where high dye concentration leads to aggregation, and in crystalline systems where the order in the system can be affected by the dye. Perturbations are much less likely when the fluorescent dye is incorporated into an amorphous fluid or glassy phase. The first fluorescence study on polymer gel was carried out by Horie *et al* to investigate the hydrophobicity and dynamic characteristics of crosslinked polystyrene with dansylamide probe [14]. For about two decades the transient fluorescence (TRF) technique for measuring fluorescence decay has been routinely applied to study many polymeric systems using dyes both as a probe and/or as labels [15-18].

The volume phase transition of acrylamide gel induced by the change in solvent composition and/or pH was investigated using fluorescence depolarization technique. TRF spectroscopy with a non-radiative direct energy transfer (DET) and quenching methods have been used to characterize internal morphologies of composite polymeric materials [15-19]. It has been reported that the local, fractal like structures of interpenetrating network morphology in blend-like particles can be studied by TRF spectroscopy. Film formation from donor and acceptor labeled latex particles was studied using DET in conjunction with TRF [20,21]. A single-photon counting (SPC) technique, which produces decay curves and measures lifetimes in conjunction with DET, was used to study the diffusion of small dye molecules within the interphase domain of anthracene and/or phenantrene labelled poly(methyl methacrylate) (PMMA) particles [22,23].

Many different experimental techniques have been used to study the kinetics of swelling and shrinking of chemical and physical gels among which are quasielastic light-scattering [24], neutron scattering [25] and in situ interferometric [26] measurements. Pyrene derivate was used as a fluorescence molecule to monitor the polymerization, aging and drying of aluminosilicate gels [27]. In situ observation of the sol-gel phase transition in free-radical crosslinking copolymerization (FCC) was studied using the steady-state fluorescence (SSF) technique [28,29]. The same technique was also performed for studying swelling and drying kinetics in disc shape gels [30]. Recently, fast transient fluorescence technique was employed to model the swelling of PMMA gel in chloroform where fluorescence decay profiles were fitted to double exponential function [31]. In these experiments, lifetimes of pyrene (τ_1 and τ_2) were measured which are related to the inside and outside of the disc shaped gel. It was observed that τ_2 increased as the swelling time was increased however τ_1 stayed unchanged during the swelling process. τ_2 values were used to determine the swelling parameters such as the cooperative diffusion coefficient, D_c and the swelling time constant, τ_c.

The purpose of this chapter is to study gel swelling at the molecular level, where excited P molecules are quenched by the penetrant organic molecules/vapor in the range of few angstrom. The penetration of organic molecules/vapor into a disc shaped gels formed by FCC of MMA with various amount of EGDM, S with constant EGDM content and various mixture of MMA and S with constant EGDM content were studied using Fast Transient Fluorescence (FTRF) technique which measures lifetimes. Fluorescence decay profiles of P were measured when the gel was illuminated directly by the exciting light and decay profiles were fitted to the exponential law to obtain lifetimes of P. Strobe Master System (SMS) was used for lifetime measurements of P. Lifetime measurements with SMS takes much shorter than it is done with single photon counting system and Phase instruments. This advantage of SMS allows one to make at least hundreds of measurements during the swelling process of gels. That is the reason we named this technique as Fast Transient Fluorescence (FTRF), which provides many advantages compare to other lifetime measuring techniques. Measuring lifetimes directly provides swelling parameters. It was observed that as gel swells, lifetimes of P inside the gel decrease which can be modeled using low quenching Stern-Volmer equation. Cooperative diffusion coefficients, D_c and swelling time constant, τ_c were determined for all swollen gels by employing Li-Tanaka and Stern-Volmer equations. It is seen that D_c values decreased from 4 to $1.8x10^{-5}$ cm^2s^{-1} under chloroform vapor and from 2.5 to $1.68x10^{-5}$ cm^2s^{-1} in chloroform by increasing EGDM content from 0.015 to 0.035 vol. % for the PMMA gels while D_c values were increased by increasing chloroform content in the solvent mixture for the PMMA gels. Heterogels swollen under chloroform vapor showed that D_c values were decreased by indicating that solvent penetration slows down as styrene content increased in heterogels. Activation energy was measured by using Arrhenius law for the PS gel swollen at various temperature and found to be 80.0 kJ mol^{-1}.

Theoretical Considerations

Mechanism of Swelling

Swelling experiments of disc shaped gels have shown that the relative changes of diameter and thickness are the same, indicating that the gel-swelling processes are not pure diffusional processes. In fact the equality of the relative changes of diameter and thickness comes from the non zero shear modulus, μ which results; the change of total shear energy in response to any small change in shape that maintains constant volume element within the gel should be zero. The high friction coefficient, f, between the network and the solvent overdamps the motion of the network, resulting in a diffusion-like relaxation. The equation of the motion of a network element during the swelling can be given by [13]

$$\frac{\partial \vec{u}}{\partial t} = D_c \vec{\nabla}^2 \vec{u} \tag{1}$$

where $\vec{u}$ is the displacement vector measured from the final equilibrium location after the gel is fully swollen (u=0 at t=∞). $D_c = (K + 4\mu/3)/f$ is the collective diffusion coefficient.

Here t denotes the time and K is the bulk modulus. Eq. (1) has been used with some success to study the swelling of gels [8]. However, the studies did not properly treat the shear deformation that occurs within a gel during swelling, and, hence, can not explain, for example, the isotropic swelling of a cylindrical gel. This shortcoming was due to the shear modulus of the network keeping the system in shape by minimizing the non-isotropic deformation. For a disc shaped gel, any change in diameter is coupled to a change in thickness. The total energy of a gel can be separated into a bulk energy and a shear energy. The bulk energy is related to the volume change, which is controlled by diffusion. The shear energy, F_{sh} on the other hand, can be minimized instantly by readjusting the shape of the gel [13].

$$\delta F_{sh} = 0 \tag{2}$$

Each small diffusion process determined by Eq. (1) must couple to a small shear process given by Eq. (2) producing the following relation for a disc shaped gel

$$\frac{u_r(r,t)}{r} = \frac{u_z(a,t)}{a} \tag{3}$$

where r is the radius and a is the half thickness of the disc gel. Eq. (3) indicates that the relative change in shape of the gel is isotropic, i.e. the swelling rates of a disc in the axial (z) and radial (r) directions are the same.

Simultaneous solution of Eq. (1) and (2) produces the following equations for the swelling of a gel disc in axial and radial directions [13].

$$u_z(z,t) = u_z(z,0)\sum_n B_n \exp(-t/\tau_n) \tag{4a}$$

$$u_r(r,t) = u_r(r,0)\sum_n B_n \exp(-t/\tau_n) \tag{4b}$$

where the axial and the radial displacements are expressed as series of components, each of them decaying exponentially with a time constant, τ_n. The first terms of the expressions are dominant at large t, that is at the last stage of swelling. Eq. (4) can also be written in terms of vapor and solvent uptakes W and W_∞ at time t and at equilibrium respectively as follows

$$\frac{W_\infty - W}{W_\infty} = \sum_{n=1}^{\infty} B_n \exp(-t/\tau_n) \tag{5}$$

In the limit of large t, or if τ_c is much larger than the rest of τ_n, all higher terms ($n \geq 2$) in Eq. (5) can be omitted and the swelling kinetics can be represented by first order kinetics [13]

$$\left[1-\frac{W}{W_\infty}\right]=B_1\exp(-t_s/\tau_c) \tag{6}$$

It should be noted from Eq. (5) that $\sum B_n = 1$, therefore B_1 should be less than 1. B_1 is related to the ratio of the shear modulus, μ, and longitudinal osmotic modulus, $M=(K+4\mu/3)$. Hence, once the value of B_1 is obtained, one can determine the value of $R=\mu/M$. Here we have to note that Eq. (6) can also be obtained by using the theoretical results[13], in the case of $R\to 3/4$ ($\mu/K\to\infty$), time constant $\tau_c\approx(3/4-R)^{-1}$ goes to infinity and all B_n's go to zero except B_1, which goes to unity. The dependence of B_1 on R for a disc can be found in the literature [13]. τ_c is related to the collective diffusion coefficient D_c at the surface of a gel disc by

$$D_c=\frac{3a^2}{\tau_c\alpha_1^{\;2}} \tag{7}$$

where α_1 is a function of R only and is given in the literature [13], and a stands for the half thickness of the gel in the final equilibrium state. Hence, D_c can be calculated. It is important to note that Eq.6 satisfies the following equation

$$\frac{dW(t)}{dt}=\frac{1}{\tau_c}(W_\infty-W) \tag{8}$$

which proves the mentioned first order kinetics. The higher order terms (n≥2) can be considered as fast decaying perturbation to the first order kinetics of swelling in the limit of large t.

Fluorescence Quenching

It is known that the fluorescence and phosphorescence intensities of aromatic molecules are affected by both radiative and non-radiative processes [32]. If the possibility of perturbation due to oxygen is excluded, the radiative probabilities are found to be relatively independent of environment and even of molecular species. Environmental effects on non-radiative transitions, which are primarily intramolecular in nature, are believed to arise from a breakdown of the Born-Oppenheimer approximation [33]. The role of the solvent in such a picture is to add the quasi-continuum of states needed to satisfy energy resonance conditions. The solvent acts as an energy sink for rapid vibrational relaxation that occurs after the rate limiting transition from the initial state. Years ago, Birks *et.al* studied the influence of solvent viscosity on fluorescence characteristics of pyrene solutions in various solvents and observed that the rate of monomer internal quenching is affected by solvent quality [34]. Weber *et.al* reported the solvent dependence of energy trapping in phenanthrene block polymers and explained the decrease in fluorescence yield with the static quenching, caused by the solvent induced trapping states [35]. A matrix that changes little with temperature will enable one to study molecular properties themselves without changing environmental influences. Poly (methyl methacrylate) (PMMA) was used as such a matrix in many studies [36].

Emission of the fluorescence is the radiative transition of an electronically excited molecule from its singlet excited state to its ground state [37,38]. Fluorescence quenching normally refers to any bimolecular process between the excited singlet state of a fluorescence dye and the second species that enhances the decay rate of the excited state. One can schematically represent the process as

$$F^* \xrightarrow{k_f, k_{nr}} F$$

$$F^* \xrightarrow{k_q[Q]} F$$

where F and F^* represent the fluorescent molecule, and its excited form, Q is the quencher. k_f, k_{nr} and k_q represent the fluorescence, non-radiative and quenching rate constants respectively. Many types of processes lead to quenching as pointed out in the previous paragraph. Kinetically, the quenching process can be divided into two main categories; dynamic and static. In dynamic quenching diffusion, to form an encounter pair during the excited state lifetime of the dye, leads to quenching. In static quenching, diffusion does not occur (which is out of our interest). Dynamic quenching is most likely to occur in fluid solution, where the dye or quencher are free to move. If the quenching rate can be characterized in terms of a single rate coefficient (k_q) and the unquenched decay rate of F, in terms of a unique lifetime, τ_0, then the quenching kinetics will follow the Stern-Volmer equation

$$\tau^{-1} = \tau_0^{-1} + k_q[Q] \qquad (9)$$

where $[Q]$ represents the quencher concentration.

EXPERIMENTS

EGDM has been commonly used as crosslinker in the synthesis of polymeric network. The free-radical crosslinking copolymerization of MMA and EGDM was performed in bulk at 60^0C in the presence of 2,2'-azobisisobutyrronitrile (AIBN) as an initiator. P was added as an extrinsic fluorescence probe during the gelation process. AIBN (0.26 wt%) was dissolved in MMA and this stock solution was divided and transferred into round glass tubes of 6.5mm internal diameter. All samples were deoxygenated by bubbling nitrogen for 10 minutes, and then radical copolymerization of MMA with five different amount of EGDM was performed to produce five different gels. Here, P concentration was taken as $4x10^{-4}$M. Here, for our use, the monomers MMA (Merck) and EGDM (Merck) were freed from the inhibitor by shaking with a 10% aqueous KOH solution, washing with water, and drying over sodium sulfate.

They were then distilled under reduced pressure over copper chloride. In order to synthesize polystyrene gels and heterogels, same procedure was followed with S instead of MMA and various amount of MMA and S at 70^0C with 0.015 vol% constant EGDM content, respectively.

Fluorescence decay experiments were performed using Photon Technology International (PTI) Strobe Master System (SMS) shown in Figure 1. In the strobe, or pulse sampling technique [38,39] the sample is excited with a pulsed light source. The name comes about because the Photo Multiplier Tube (PMT) is gated or strobed by a voltage pulse that is synchronized with the pulsed light source. The intensity of fluorescence emission is measured in a very narrow time window on each pulse and saved in a computer. The time window is moved after each pulse. The strobe has the effect of turning of the PMT and measuring the emission intensity over a very short time window. When the data has been sampled over the appropriate range of time, a decay curve of fluorescence intensity versus time can be constructed. Since the strobe technique is intensity-dependent, strobe instrument is much faster than SPC, and even faster than phase instrument. Strobe instrument is much simpler to use than SPC and the data is easier to interpret than phase system. Because of these advantages SMS is used to monitor swelling of gels which takes around several hours.

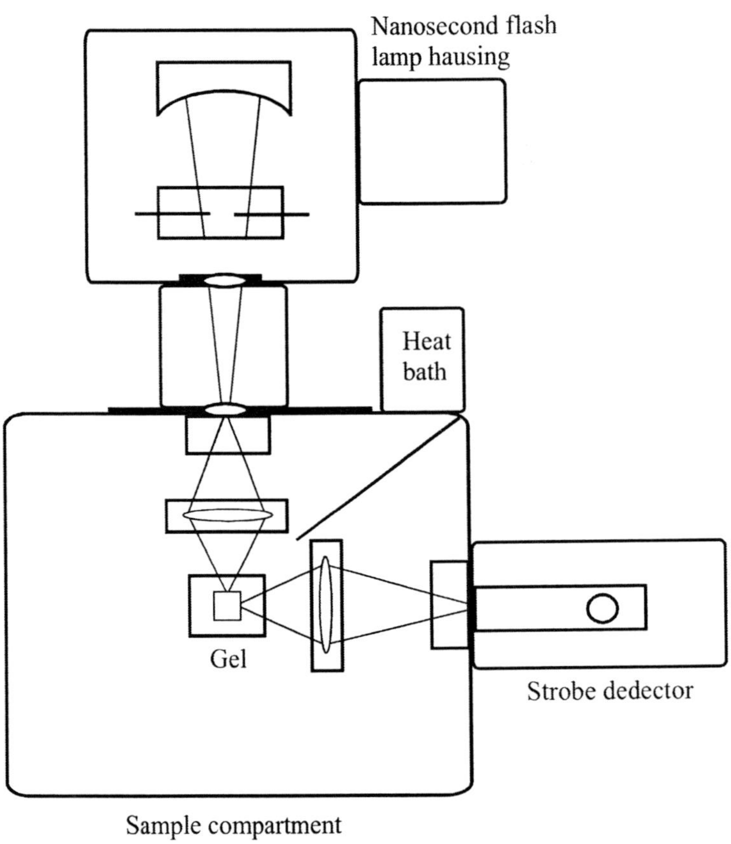

Figure 1: Strobe Master System of Photon Technology International (PTI) for Fluorescence Lifetime measurements.

a-)

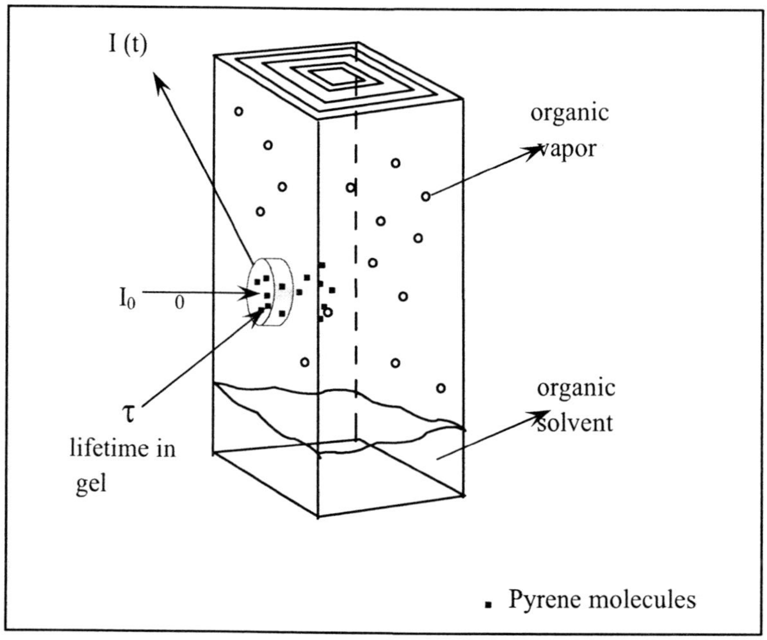

b-)

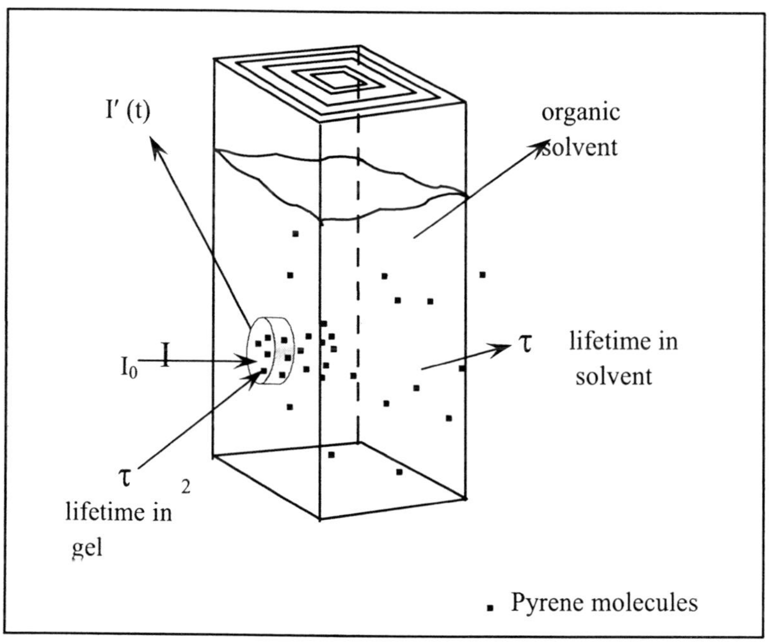

Figure 2: Fluorescence cell in PTI Strobe Master System during a-) vapor, b-) solvent induced swelling . I_0 and I(t) and I′(t) are the excitation and the emission intensities at 345 and 395 nm respectively. τ is the P lifetime for vapor induced gel swelling and τ_2 and τ_1 are lifetimes of P in and out of the gel sample in solvent.

In-situ swelling experiments were carried out in the SMS of PTI, employing a pulsed lamp source (0.5 atm of N_2). Pyrenes were excited at 345 nm and fluorescence decay curves were obtained at 395 nm during in-situ swelling experiments which were performed for various gels in different EGDM content at room temperature in chloroform and under chloroform vapor. No shift was observed at 395nm during all swelling experiments. Five identical disc-shaped PMMA gels were used which were dried and cut from a cylindrical gel obtained from FCC with various EGDM content for swelling experiments in/under chloroform/chloroform vapor at room temperature. Disc shaped gel sample was placed in a 1cm x 1cm quartz cell, where it was attached to one side of a cell by pressing a disc with thick steal wire. For the swelling experiments, at first the bottom of quartz cell was filled with chloroform for vapor induced gel swelling (see Figure 2a) and second, the quartz cell was filled with chloroform itself as shown in Figure2b. Vapor-gel system was also deoxygenated by bubbling nitrogen for 10min. The cell was placed in the SMS system where fluorescence decay measurements were performed at 90^0 angle. Nine disc-shaped PMMA gels containing 0.035 vol. % EGDM content were prepared for the swelling in various percents of chloroform in chloroform-heptane mixtures at room temperature. Nine disc-shaped heterogels was also used for the vapor induced gel swelling under chloroform vapor. Finally, five PS gels were swelled at various temperature to measure swelling activation energy. The fluorescence decay data were collected over 3 decades of decay and fitted by nonlinear least squares using deconvolution method with a dry gel as a scatterer standard in all swelling experiments. The uniqueness of the fit of the data to the model is determined by χ^2 ($\chi^2 \leq 1.10$), the distribution of the weighted residuals and the autocorrelation of the residuals.

RESULTS AND DISCUSSIONS

Swelling of PMMA Gels in/under Chloroform/chloroform Vapor

Decay curves of P obtained from SMS in various swelling times, t_s for the gel sample prepared with 0.015 EGDM vol. % are presented in Figure 3a and b for vapor and solvent induced swellings respectively. In order to probe the swelling process during vapor uptake, the fluorescence decay curves are measured when the gel is in the position of Figure2a and they were fitted to the following exponential law:

$$I(t) = A\, e^{-t/\tau} \tag{10}$$

where τ is the pyrene lifetime and A is the corresponding amplitude of the decay curve. As seen in Figure3a pyrene decay faster as the swelling time, t_s is increased by predicting that pyrenes are quenched by the chloroform molecules. Similarly, to probe the swelling process during solvent uptake, the fluorescence decay curves were measured when the gel was in the position of Figure2b, and they were fitted to the sum of two exponentials:

$$I'(t) = A_1 e^{-t/\tau_1} + A_2 e^{-t/\tau_2} \tag{11}$$

where τ_1 and τ_2 are the long and short components of pyrene lifetimes and A_1 and A_2 are the corresponding amplitudes of the decay curves. In otherwords, τ_2 and τ_1 are the lifetimes of P when the pyrenes are in and out of the gel sample respectively. Figure 3b presents that as the swelling time, t_s is increased excited pyrenes decay faster and faster which indicates that as solvent uptake is increased quenching of excited pyrenes increase. τ values for vapor induced gel swelling and τ_1 and τ_2 values for solvent induced gel swelling are plotted versus swelling time, t_s in Figure 4a and b respectively for the gels prepared with 0.015 EGDM vol.% content. It is seen in Figure 4 that, τ_1 values do not change much however τ and τ_2 values decrease as t_s is increased. Here the role of the vapor and solvent is to add the quasi-continuum of states needed to satisfy energy resonance conditions, i.e. the vapor and solvent act as an energy sink for rapid vibrational relaxation, which occurs after the rate limiting transition from the initial state. Birks *et al.* studied the influence of solvent viscosity on the fluorescence characteristics of pyrene solutions in various solvents and observed that the rate of monomer internal quenching is affected by solvent quality [40].

In order to quantify the results in Figure 4 where exponential decrease in τ and τ_2 are observed as the swelling time, t_s is increased, the Stern-Volmer type of quenching mechanism may be proposed for the fluorescence decay of P in the gel sample during swelling process, where the following relations are given [37]: excitation of ground state pyrene, P to excited state of pyrene, P^*

$$P + h\nu \rightarrow P^*$$

collisional quenching of P^* to P by Q

$$P^* + Q \underset{k_{-1}}{\overset{k_1}{\rightleftarrows}} (P^*Q) \rightarrow P + Q$$

the decay of P^*,

$$P^* \rightarrow P + h\nu$$

where Q presents the quencher molecules, k_1 and k_{-1} are the diffusional rate constants, and k_2 is the rate for internal quenching via electron transfer or spin exchange [40]. Here, quenching efficiency is given by

$$\gamma = \frac{k_2}{k_{-1} + k_2 + \tau_0^{-1} + k_1[Q]} \tag{12}$$

where τ_0 is the lifetime of P in dry gel in which no quenching has taken place. If $k_2 >> (k_{-1} + \tau_0^{-1} + k_1[Q])$ then Eq.(12) result in $\gamma \approx 1$ and k_1 becomes the quenching rate constant, k_q, given by Smouchowski equation as [37]

$$k_q = \frac{4\pi N_A D_M R}{1000} \tag{13}$$

where $D_m = D_p + D_s$ is the sum of the mutual diffusion coefficients of P and solvent molecules, $R = R_p + R_s$ is the sum of their interaction radii, and N_A is the Avagadro's number.

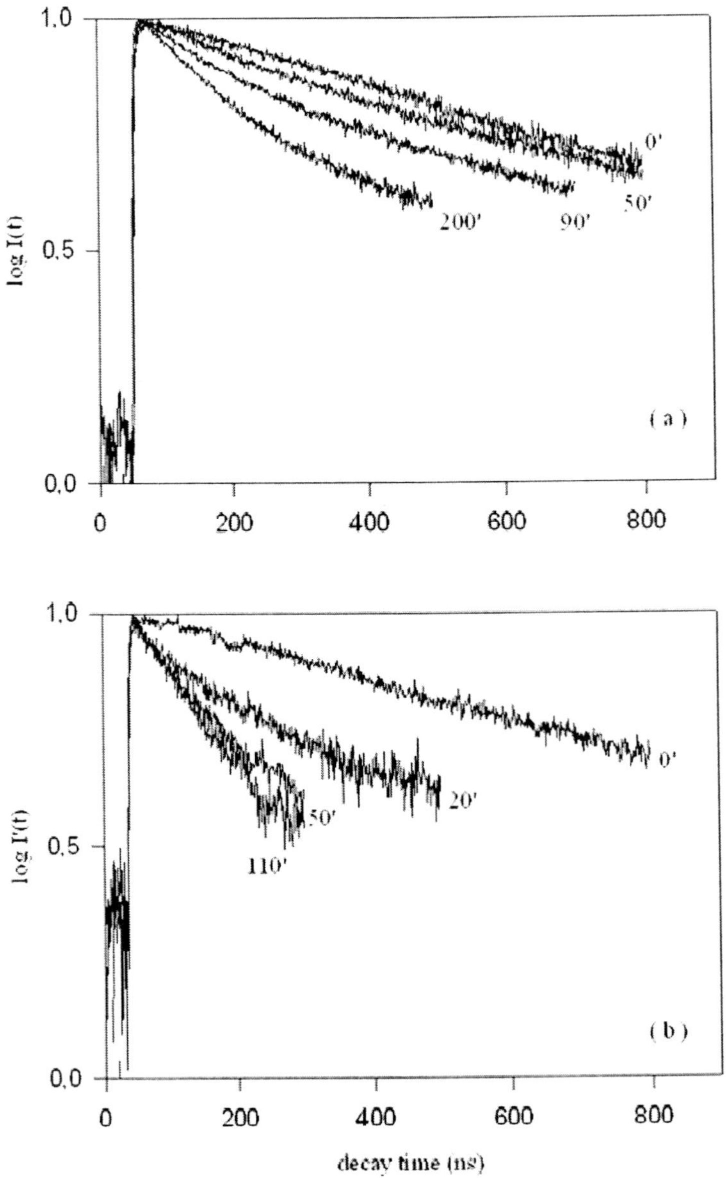

Figure 3: Log of the fluorescence decay profiles, I(t) and I′(t) of P at various swelling times for the gel sample prepared with 0.015 vol. % EGDM content for a-) vapor, b-) solvent, induced gel swelling. The number on each decay curve presents the swelling times, t_s in minute.

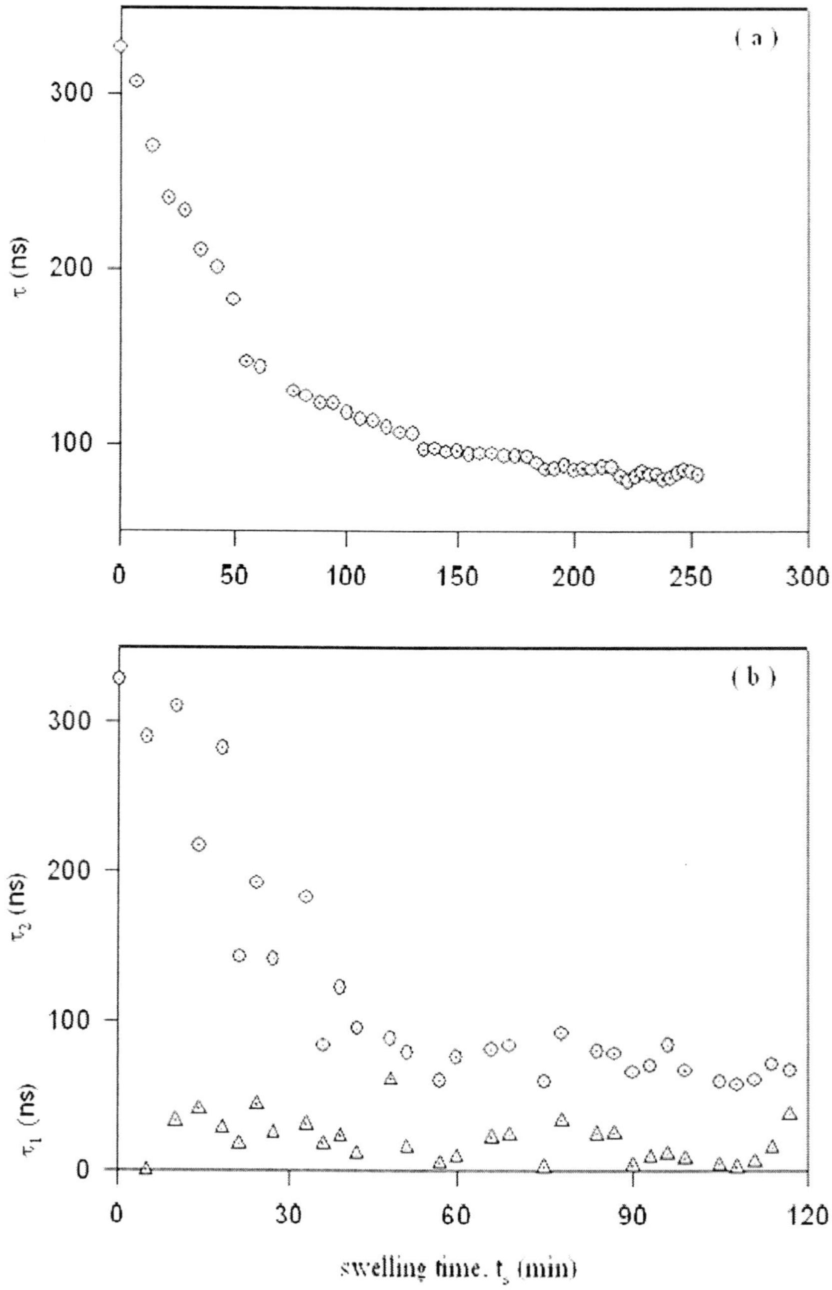

Figure 4: The plots of the measured lifetime values versus swelling time, ts for 0.015 vol.% EGDM content gel, a-) τ for vapor, b-) τ_1 and τ_2 for solvent induced gel swelling.

The rate equation for P^* with delta function excitation can be written as

$$\frac{[P^*]}{dt} = -(\tau_0^{-1} + k_q[Q]) + \delta(t - t_0)[P] \tag{14}$$

Here [ ] represent the concentration of P^* and Q molecules and $\delta(t\text{-}t_0)$is the light pulse of SMS. Solution of Eq.(14) produce Eq.(10), where τ is given by Eq.(9). For low quenching efficiency, where $\tau_0 k_q[Q] \langle\langle\; 1$, Eq (9) becomes

$$\tau \approx \tau_0(1 - \tau_0 k_q[Q]) \tag{15}$$

If one integrates Eq.(15) over the differantial volume (dv) in the gel from the initial a_0 to final a_∞ thickness, the following relation is obtained.

$$W = (1 - \frac{\tau}{\tau_0})\frac{v}{k_q \tau_0} \tag{16}$$

Here, vapor / solvent uptake, W is calculated over differential volume by replacing Q with W as

$$W = \int_{a_0}^{a_\infty} [W] dv \tag{17}$$

where ν is the swollen volume of the gel, which can be measured experimentally and k_q was obtained from separate measurements. Here it is assumed that k_q values do not vary during swelling processes i.e. the quenching process solely originates from the chloroform molecules/chloroform vapor.

The plots of the vapor and solvent uptake, W for the gels prepared with 0.015 vol. % are shown in Figure 5a and b respectively, which are typical vapor and solvent uptake curves, obey Li-Tanaka Equation (Eq. (6)). Logarithmic form of the data in Figure 5 are fitted to the following relation produce from Eq. (6).

$$\ln\left[1 - \frac{W}{W_\infty}\right] = \ln B_1 - \frac{t_s}{\tau_c} \tag{18}$$

The fits are presented in Figure 6a and b for vapor and solvent induced gel swelling respectively, from where B_1 and τ_c values are produced and are listed in Table I and Table II respectively. The plot of τ_c versus EGDM content is shown in Figure 7 for vapor and solvent induced swelling, where it is seen that as the EGDM content is increased penetration of vapor and solvent molecules slow down, as expected. In other words chloroform molecules penetrate into the densely formed gels much slower than loosely formed gels in both cases. Penetration of chloroform in vapor phase is much slower than in solvent phase most probably due to lower osmotic pressure in vapor phase. By knowing B_1 values one can obtain α_1 values and then from Eq. (7) D_c values are obtained and are plotted in Figure 8 versus EGDM content for vapor and solvent induced swelling. The behaviour of D_c in Figure 8 predicts that as EGDM content increase D_c values decrease, i.e. gel segment move much faster in loosely formed gels than in densely formed gels.

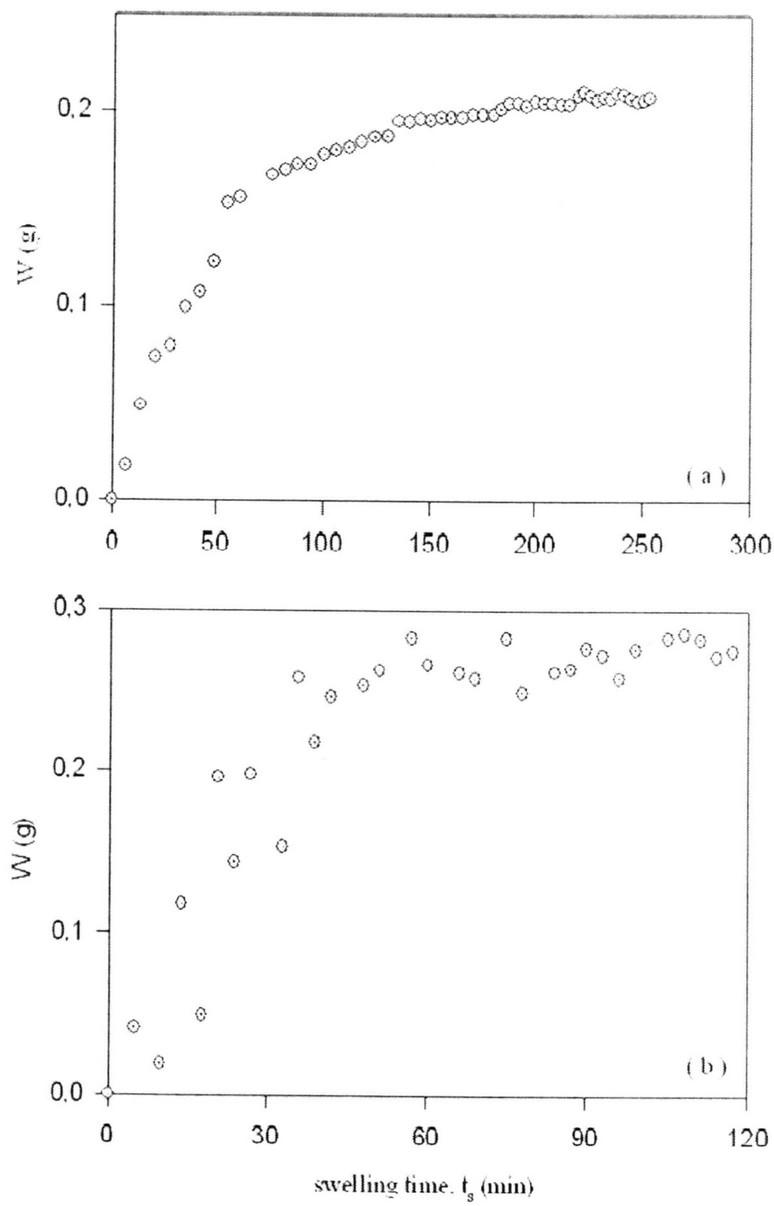

Figure 5: The plots of a-) vapor, b-) solvent, uptake W versus swelling time, t_s for 0.015 vol. % EGDM content gel sample.

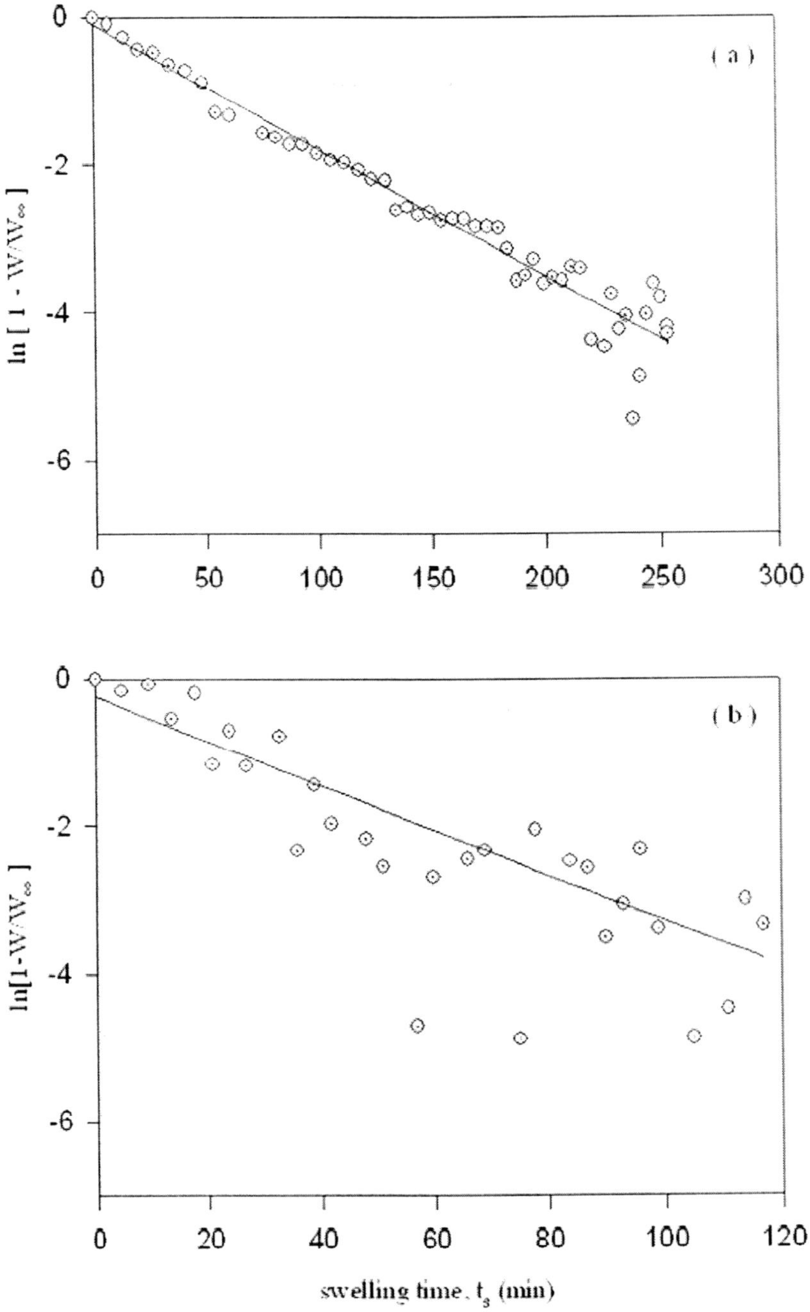

Figure 6: Fit of the data in Figure 5 to Eq. (18),where the slope of the curves produced □c values which are listed in Table I and II for vapor and solvent induced gel swelling, respectively.

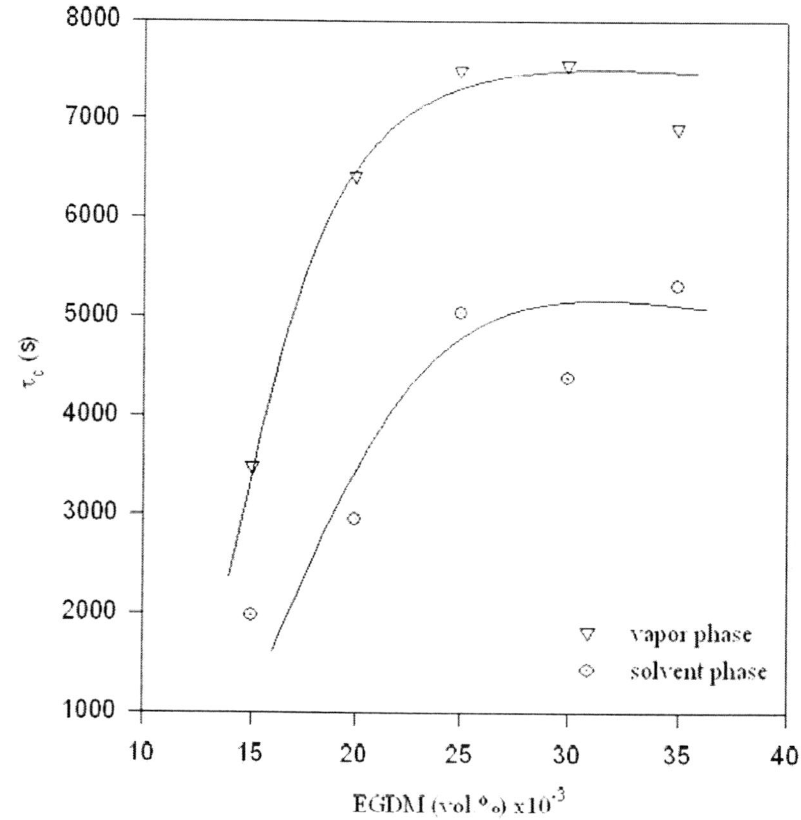

Figure 7: The plots of τ_c values versus vol. % EGDM content for vapor and solvent induced gel swelling. Symbols, ∇ and o stand for vapor and solvent induced gel swelling, respectively.

Table I. Experimentally obtained parameters for the gel swelling under chloroform vapor

EGDM(vol.%)	Δm (g)	Δd (cm)	τ_c (s)	B_1	α_1	D_c ($x10^{-5}$ cm^2s^{-1})
0.015	0.20	0.10	3509	0.92	1,1	2.16
0.020	0.20	0.09	6451	0.84	1.0	1.62
0.025	0.24	0.10	7499	0.93	1.0	1.33
0.030	0.15	0.08	7560	0.93	1.0	1.18
0.035	0.19	0.11	6900	0.87	1.3	1.03

Δm; amount of vapor uptake

Δd; variation in disc thicknesses

τ_c; time constant

D_c; cooperative diffusion coefficient

B_1 and α_1; experimentally determined coefficients

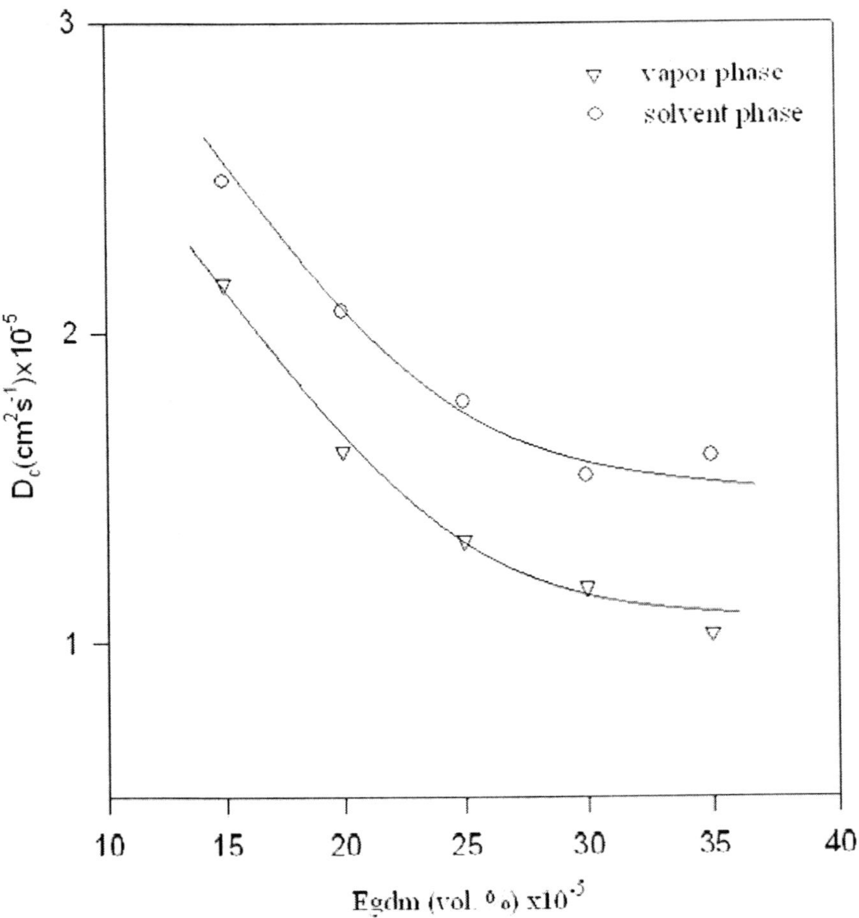

Figure 8: The plots of D_c values versus vol. % EGDM content for vapor and solvent induced gel swelling. Symbols, ∇ and o stand for vapor and solvent induced gel swelling, respectively.

Table II. Experimentally obtained parameters for the gel swelling in chloroform

Egdm (Vol%)	Δm(g)	Δd (cm)	τ_c (s)	B_1	α_1	$D_c(cm^2 s^{-1})$ $x10^{-5}$
0.015	0.27	0.13	1980	0.80	1.6	2.49
0.020	0.40	0.12	2955	0.85	1.4	2.07
0.025	0.23	0.12	5040	0.91	1.1	1.78
0.030	0.28	0.11	4740	0.92	1.1	1.42
0.035	0.16	0.10	5310	0.92	1.1	1.60

Δm; amount of solvent uptake

Δd; variation in disc thicknesses

τ_c; time constant

D_c; cooperative diffusion coefficient

B_1 and α_1; experimentally determined coefficients

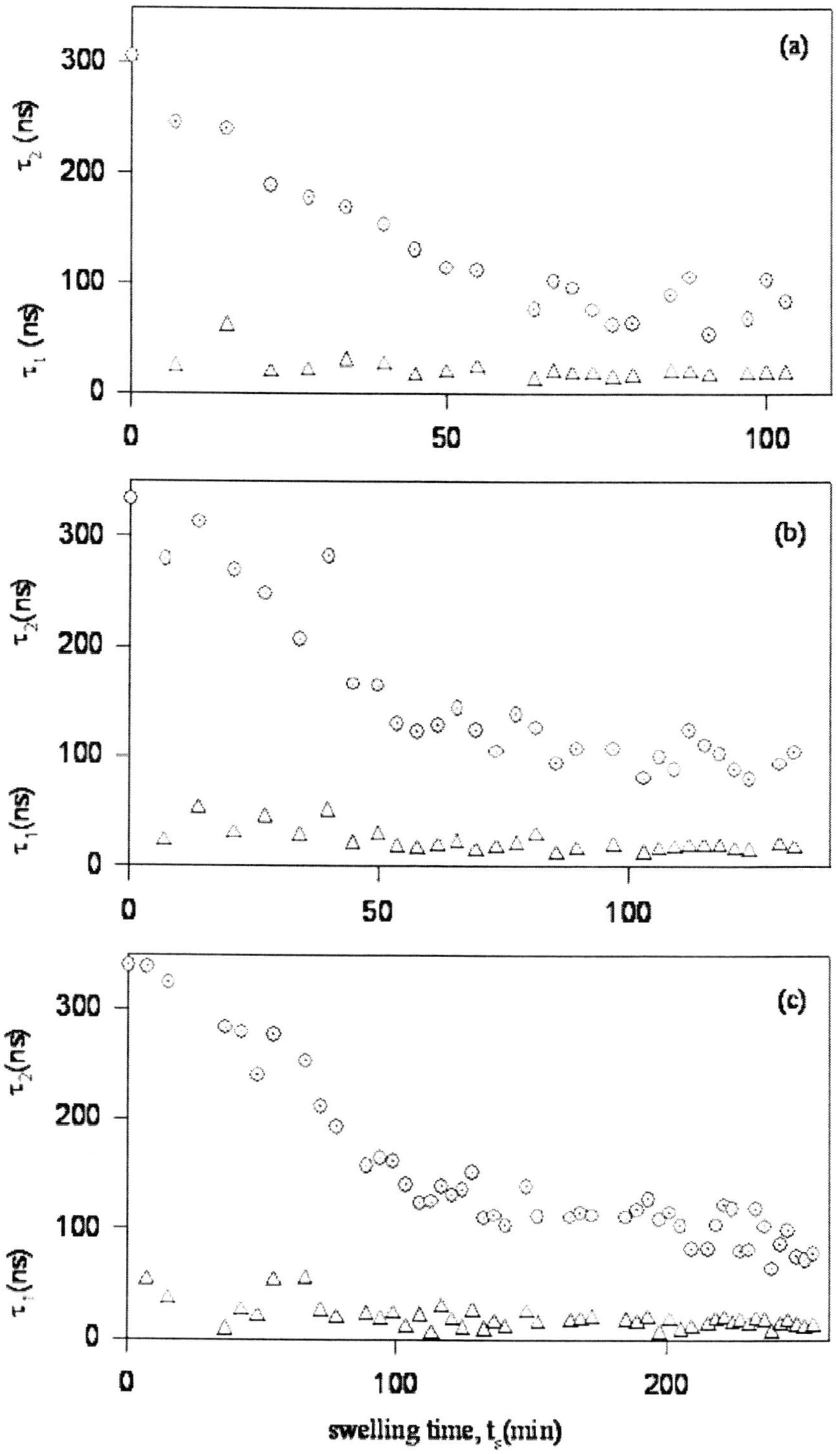

Figure 9. The plot of the measured τ_1 and τ_2 values versus swelling time, t_s for a-100, b-90, c-80 % of chloroform contents in chloroform-heptane mixture.

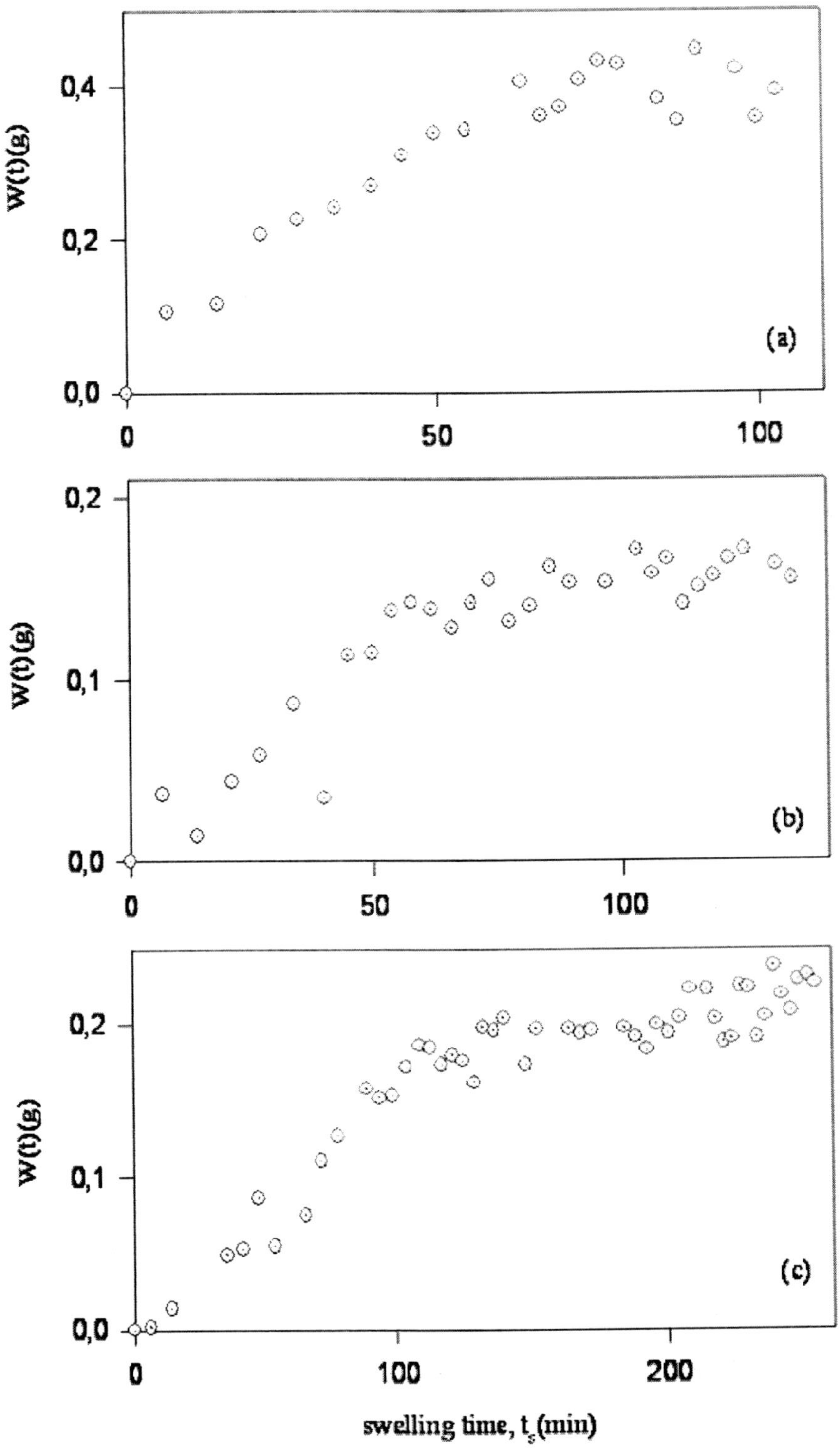

Figure 10. Plot of the solvent uptake, W versus swelling time, t_s for a-100, b-90, c-80 % of chloroform contents in chloroform-heptane mixture.

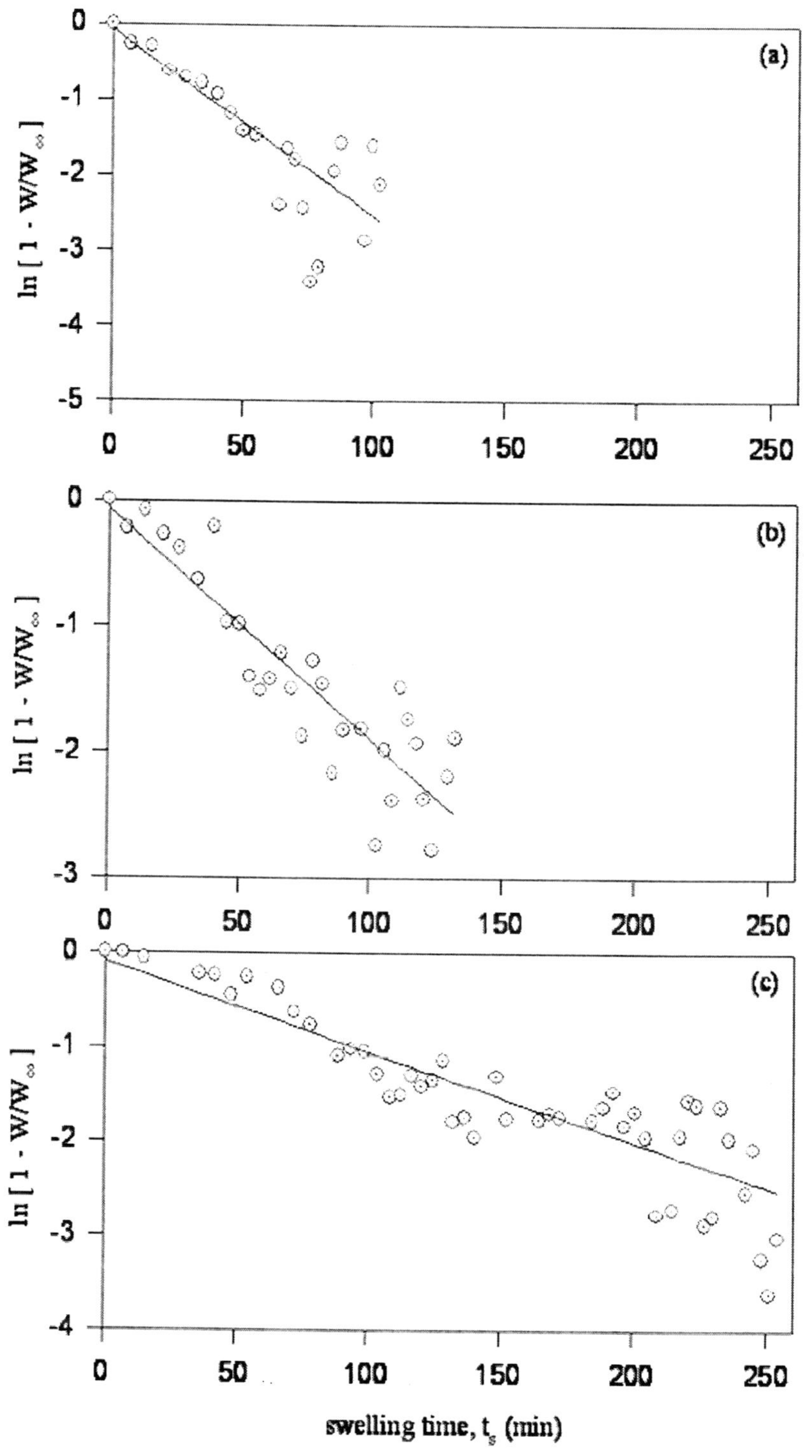

Figure 11. Fit of the data in Figure 10 to Eq.(18). The slope of the curve produced τ_c values which are listed in Table III.

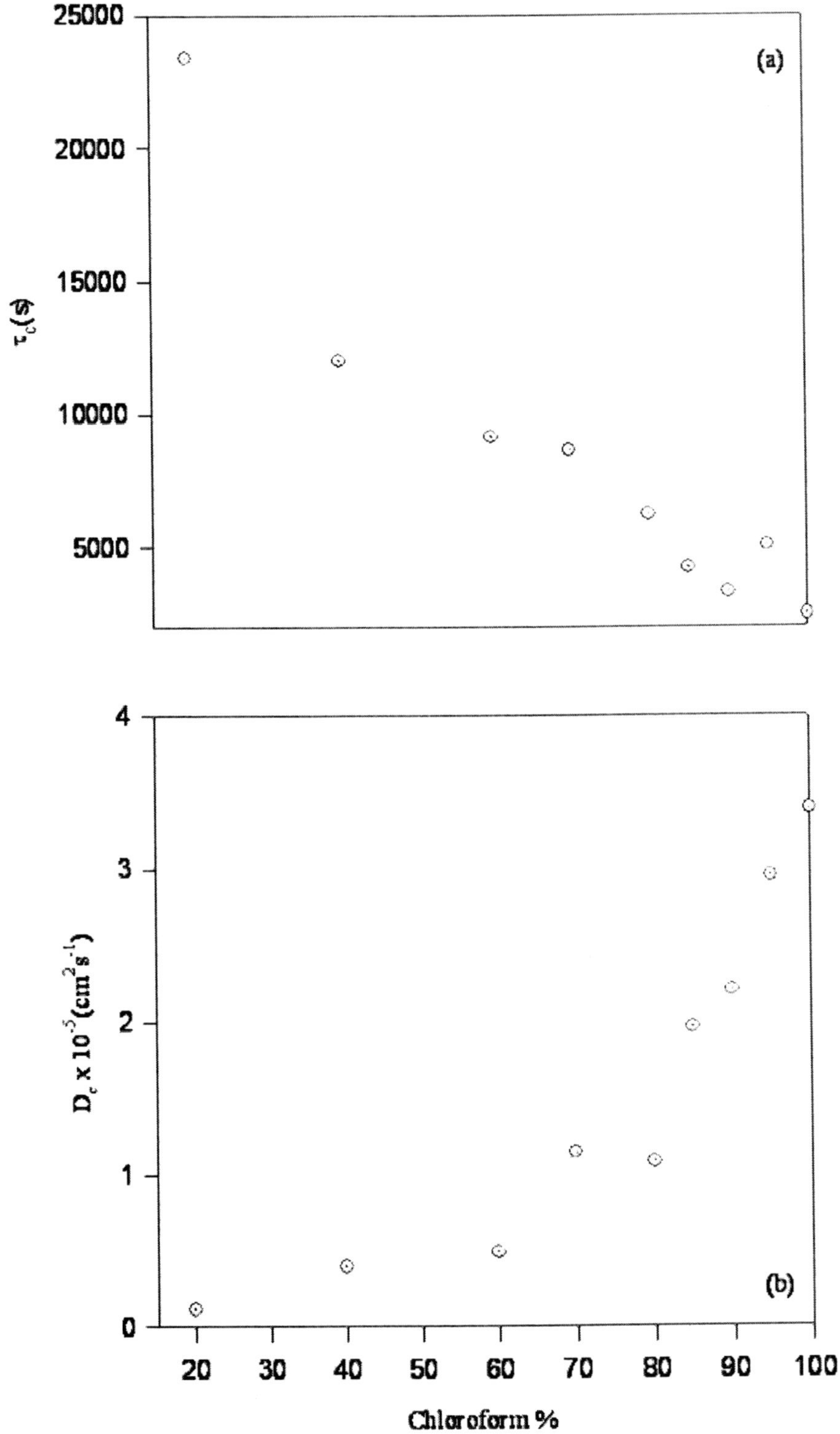

Figure 12. The plot of a- τ_c and b-D_c values versus chloroform % in chloroform-heptane mixture.

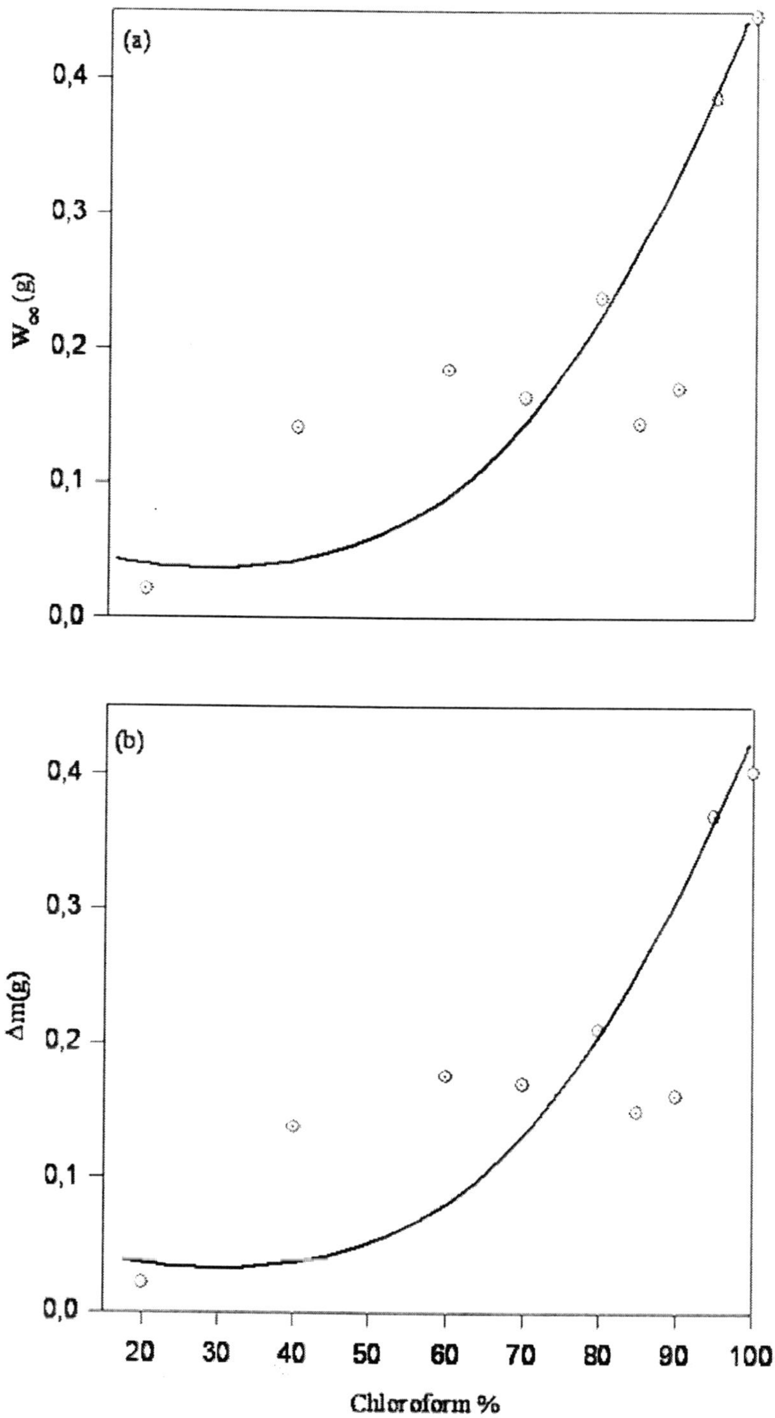

Figure 13. The plot of maximum solvent uptakes measured a) from lifetime measurements, b) from weighting measurements using micro-balance.

Swelling of Pmma Gels in Solvent Mixture

Various percents of chloroform in chloroform-heptane mixtures were used as a swelling agent. In order to probe the swelling process during solvent uptake, the fluorescence decay curves were measured and fitted to the sum of two exponentials given Eq.(11). The measured τ_1 and τ_2 values for the gel samples swollen in various chloroform-heptane mixtures are plotted versus swelling time, t_s in Figure 9. The plots of solvent uptake, W, for the gels swollen in 100, 90 and 80 % chloroform contents are shown in Figure 10a, b and c respectively. The logarithmic data in Figure 10 were fitted to Eq.(18). The fits are presented in Figure 11, from which B_1 and τ_c values were obtained and were listed in Table III. The plot of swelling time constant, τ_c versus the chloroform percentage is shown in Figure 12a. As expected, the penetration of solvent molecules slows down as the heptane content is increased. A decrease in the solvent quality, prevents the gel from swelling. The behavior of D_c in Figure 12b give evidence for gel segments which move faster in a thermodynamically high quality solvent, i.e. in chloroform containing solvents gel segments diffuse considerably faster.

Table III.

% Chloroform	Δm (g)	a_∞ (cm)	τ_c (s)	B_1	α_1	D_c ($x10^{-5}$ cm^2s^{-1})
100	0.40	0.29	2460	0.94	0.87	3.38
95	0.36	0.28	5040	0.97	0.64	2.95
90	0.16	0.27	3276	0.94	0.87	2.20
85	0.14	0.31	4200	0.93	0.95	1.96
80	0.21	0.3	6240	0.91	1.00	1.08
70	0.16	0.34	8640	0.93	0.95	1.14
60	0.17	0.31	9120	0.85	1.27	0.49
40	0.13	0.29	12000	0.89	1.15	0.39
20	0.02	0.23	23370	0.90	1.14	0.12

Experimentally measured; maximum solvent uptake, Δm; final disc thickness , a_∞; time constant, τ_c and cooperative diffusion coefficient, D_c.

W_∞ values could be obtained from the equilibrium regions in Figure 10 and are plotted in Figure 13a. Figure 13b presents the observed Δm values which were obtained by subtracting the final weight of the gel from its initial weight. In fact, the Δm and W_∞ values are the parameters similar to each other but measured by different techniques. As shown in Figure 13a and b they behave quite similarly with regard of the chloroform content. Comparing Figure 12b with Figure 13, it is seen that in poor solvent quality (high heptane content) the gel does not swell much. The segments move rather slowly compared to those in gels swollen in the good solvent quality (high chloroform content).

Swelling of Heterogels under Organic Vapor

The fluorescence decay curves are measured and were fitted to mono-exponential function given Eq.(10) to probe swelling process during vapor induced heterogels swelling. The measured τ values for SMMA1, SMMA4 and SMMA7 heterogels exposed to vapor are plotted versus swelling time, t_s, in Figure 14a, b and c, respectively. It can be seen that, in all samples, the τ values decrease as vapor uptake is increased. The plots of the vapor uptake, W, for the SMMA1, SMMA4 and EMMA7 heterogels swollen in chloroform vapor are shown in Figure 15a, b and c, respectively, which are typical vapor uptake curves, obeying Li-Tanaka Equation (Eq.6). Logarithmic form of the data in Figure 15 are fitted to Eq.(18).

The fits are presented in Figure 16, from where B_1 and τ_c values are produced and are listed in Table IV. The plot of τ_c versus styrene content is shown in Figure 17a where, as the styrene content in heterogels is increased, penetration of vapor molecules slow down, as expected, i.e. decreased polymer-solvent interaction, prevents the gel from swelling. Here, we have to mention that the solubility parameters of PMMA, PS and chloroform are 19, 18.2 and 19 $(MPa)^{1/2}$, respectively, which results in high degree of swelling in MMA rich content heterogels than S rich content gels. The behaviour of D_c in Figure 17b predicts that gel segments move much faster in high polymer-solvent interaction i.e. as MMA content is increased gel segments move much faster in the heterogel during vapor penetration.

Table IV.

symbols	strene Con. (vol.%)	τ_c (min)	B_1	D_c x10^{-5} (cm^2s^{-1})	D_m x10^{-7} (cm^2s^{-1})
S	100	64.52	0.72	0.40	1.24
SMMA8	80	58.68	0.87	0.98	1.44
SMMA7	70	53.55	0.93	1.83	1.70
SMMA5	50	50.25	0.94	2.35	1.30
SMMA4	40	46.52	0.56	0.40	1.22
SMMA3	30	44.64	0.93	3.72	1.20
SMMA2	20	42.66	0.90	4.03	1.49
SMMA1	10	40.66	0.96	4.74	1.47
SMMA2,5	2.5	37.74	0.97	5.60	1.87

The sum of the mutual diffusion coefficients were calculated from Eq.(13) by using the k_q values and found to be around $D_m \cong 10^{-7}$ cm^2s^{-1}, where R is taken as 0,8nm. The observed mutual diffusion coefficients, D_m, are listed in Table IV which are typical for a small molecule diffusing in a swollen rubbery environment[15]. It is interesting to note that D_m values are much smaller than the cooperative diffusion coefficient, D_c. This result is expected, because an element of swollen network moves much faster, due to the restraining forces, than the P and chloroform molecules in the swollen, viscous environment.

Swelling of PS Gels under Organic Vapor at Various Temperature

In order to probe vapor induced swelling at various temperature, the fluorescence decay curves were measured and fitted to the mono exponential functions given Eq.(10). The normalized τ values for the gels swelling at 25, 35 and 45°C, are plotted versus swelling time, t_s in Figure18a. It can be seen that in all samples τ values decrease as vapor uptake is increased. The increase in temperature causes faster decrease in τ values. The plots of the vapor uptake, W for the gels swollen at 25, 35 and 45°C in chloroform vapor are shown in Figure 18b. Logarithmic form of the data in Figure18b are fitted to the Eq.(18) and presented in Figure19, from where B_1 and τ_c values are determined and are listed in Table V, where it is seen that as the temperature is increased penetration of vapor molecules into the gel increase, as expected. By knowing B_1 values one can obtain α_1 values and then from Eq.(7) D_c values are produced and listed in Table V. The behavior of D_c versus temperature predicts that gel segments move much faster at high temperature during the vapor penetration. As seen in Table V, the D_c values increase as the temperature is increased, which predicts that the D_c-T relation may obey the following Arrhenius law:

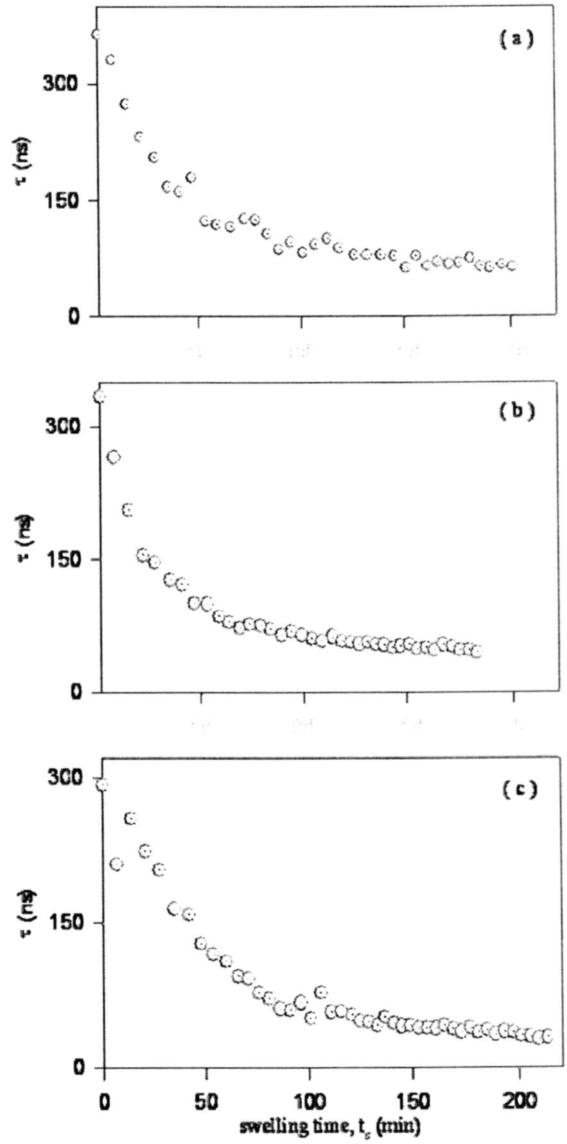

Figure 14. The plots of pyrene lifetimes, τ versus swelling time, t_s for the heterogels a) SMMA1, b) SMMA4 and c) SMMA7 heterogels swollen in chloroform vapor. τ values were obtained by fitting the I(t) data to Eq.(10).

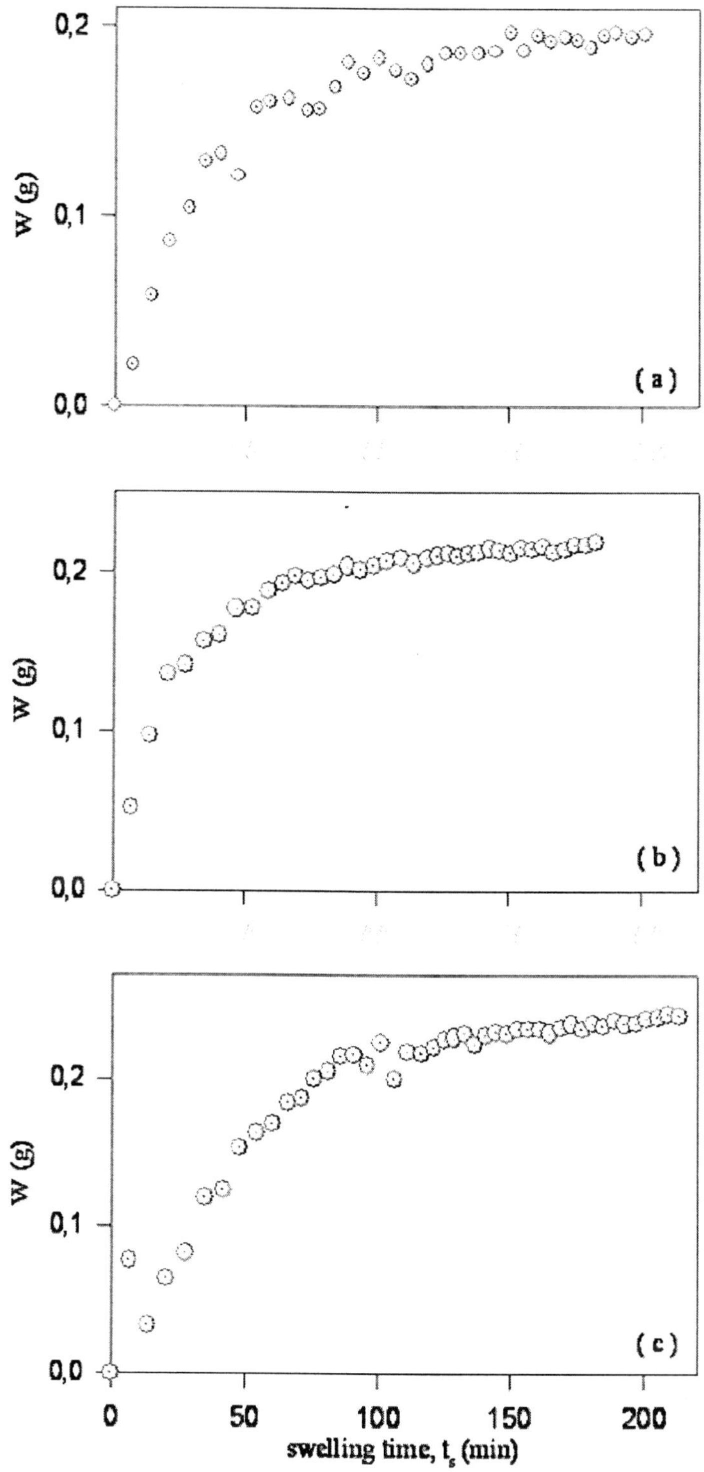

Figure 15. The plots of the vapor uptake, W versus swelling time, t_s for the a) SMMA1, b) SMMA4 and c) SMMA7 heterogels swollen in chloroform vapor.

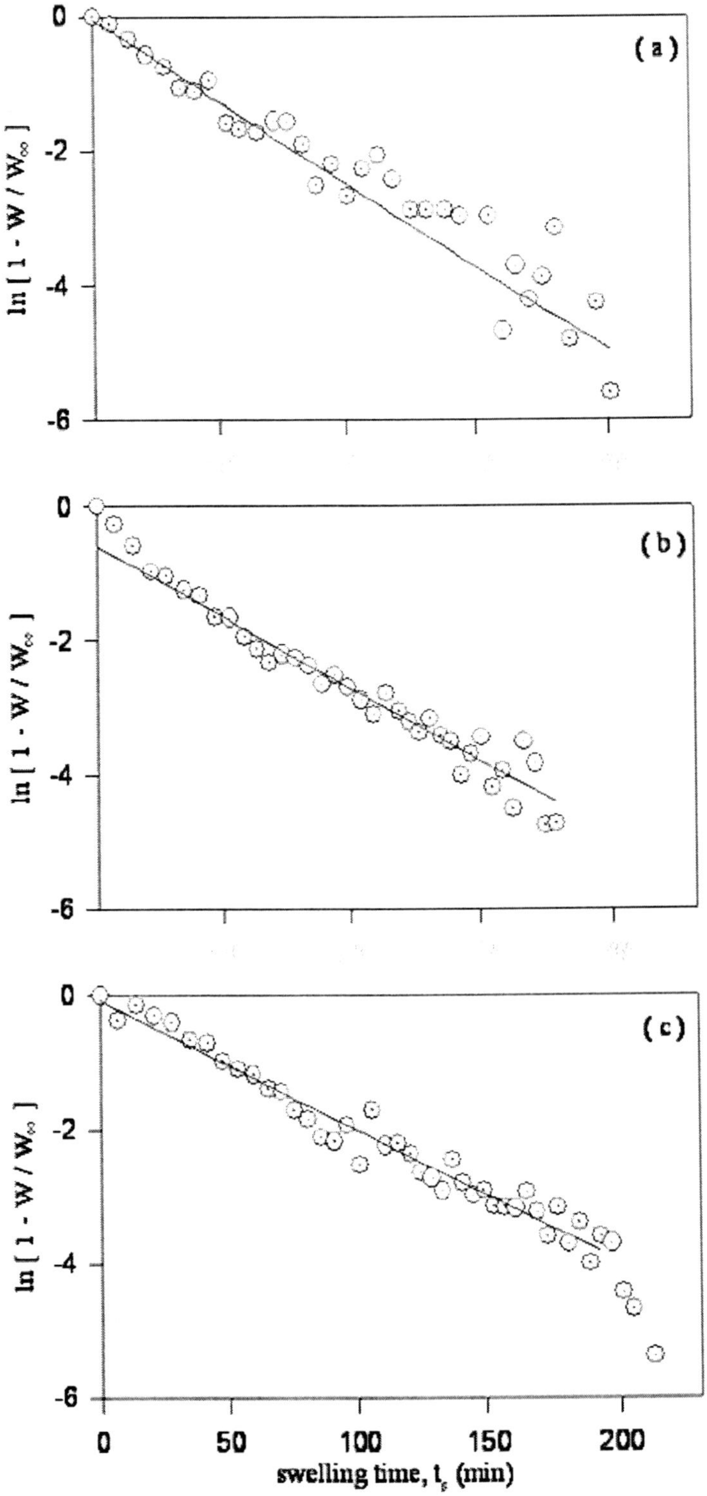

Figure 16. Linear regression of the data in Figure 15 according to Eq.(18) for the a) SMMA1, b) SMMA4 and c) SMMA7 heterogels.

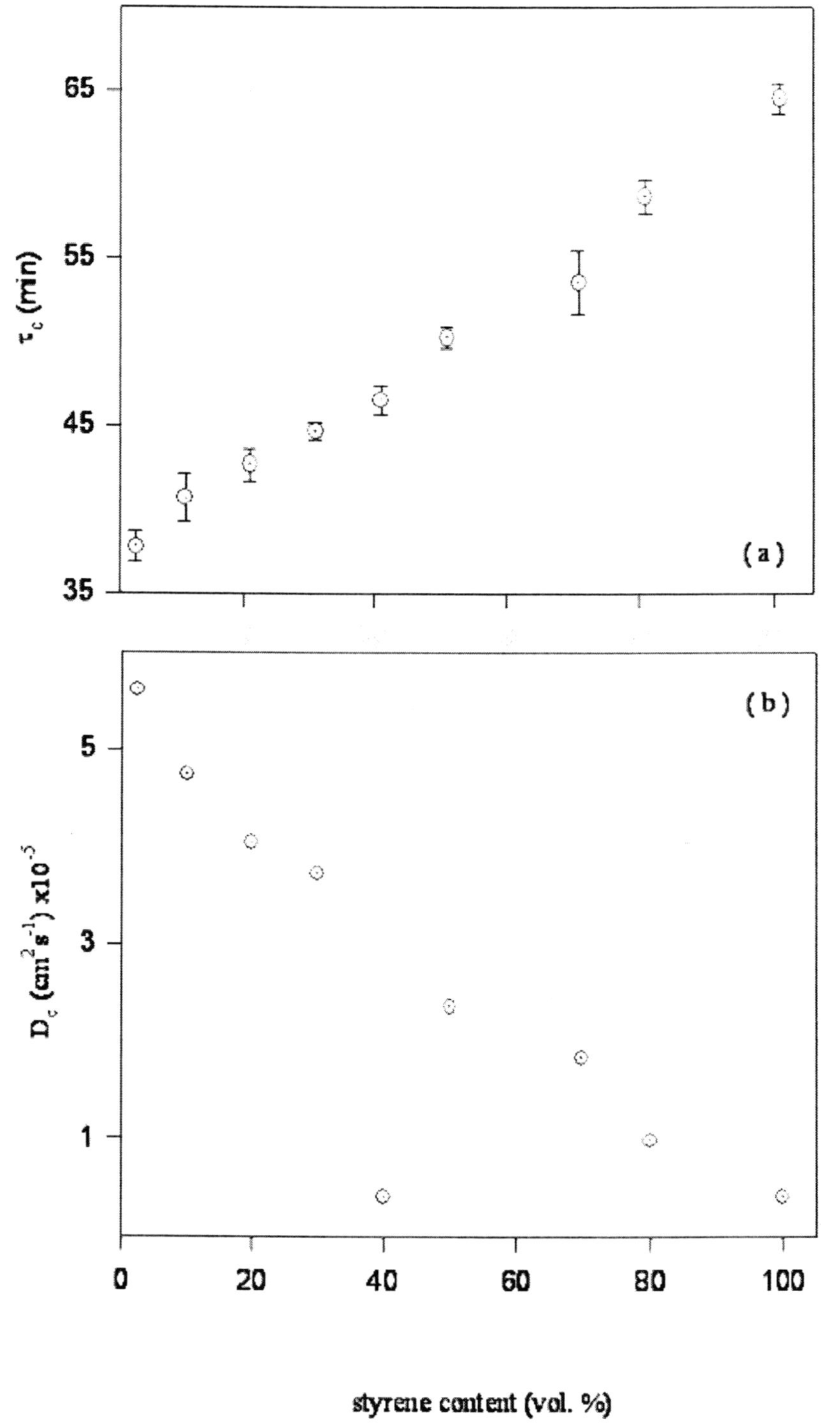

Figure 17. The plots of the a) τ_c and b) D_c values versus styrene content in heterogels.

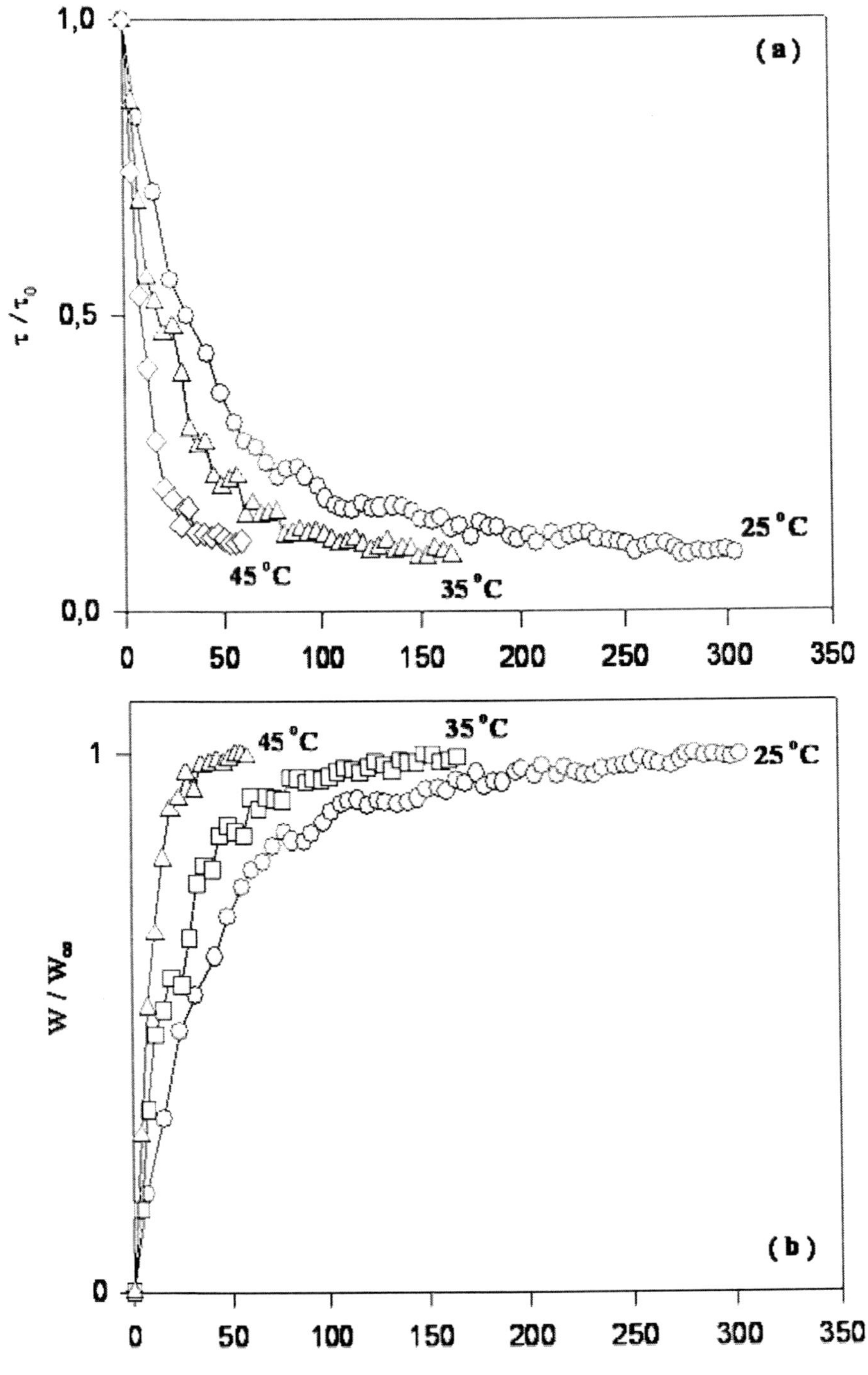

Figure 18. a-) The plots of normalized pyrene lifetimes, τ versus swelling time, t_s for the PS gel swollen at 25, 35 and 45°C. τ values were obtained by fitting the I(t) data to Eq.(10). **b-)** The plots of normalized vapor uptake, W versus swelling time, t_s for the gel swollen at 25, 35 and 45°C.

$$D_c = D_{co} \exp(-\Delta E_s / kT) \tag{19}$$

where the ΔE_s is the activation energy for swelling, k the Boltzmann's constant and D_{co} the cooperative diffusion coefficient at $T=\infty$. The logarithmic form of the D_c data are plotted versus T^{-1} in Figure 20 where the slope of the linear relation produces the activation energy ΔE_s for the swelling gel as 80.0 kJ mol^{-1}.

Table V

Temperature (°C)	m_i (g)	m_f (g)	d_i (cm)	d_f (cm)	B_1	$k_q(M^{-1}s^{-1})$ x10^5	τ_c (min)	$D_c(cm^2s^{-1})$ x10^{-5}	$D_m(cm^2s^{-1})$ x10^{-7}
50	0.062	0.168	0.20	0.26	0.86	1.0	8.9	5.08	1.75
45	0.067	0.166	0.21	0.26	0.85	1.0	10.5	4.24	1.72
40	0.065	0.167	0.20	0.26	0.80	1.1	17.6	1.85	1.85
35	0.067	0.211	0.21	0.28	0.71	0.9	34.7	0.78	1.45
30	0.066	0.178	0.21	0.27	0.60	0.9	47.1	1.15	1.50
25	0.064	0.247	0.20	0.27	0.72	0.7	64.5	0.40	1.24

m_i, d_i ; initial weight and thickness of the PS gels
m_f, d_f ;final weight and thickness of the PS gels
τ_c ; swelling time constant
D_c ; collective diffusion coefficient
D_m ; mutual diffusion coefficient

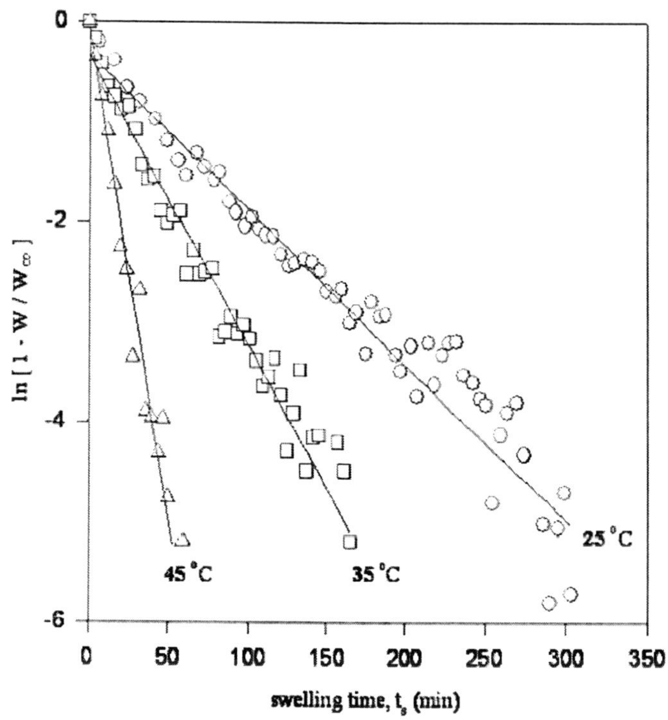

Figure 19. Linear regression of the data in Figure 18b according to Eq.(18) for the PS gels swollen at 25, 35 and 45°C.

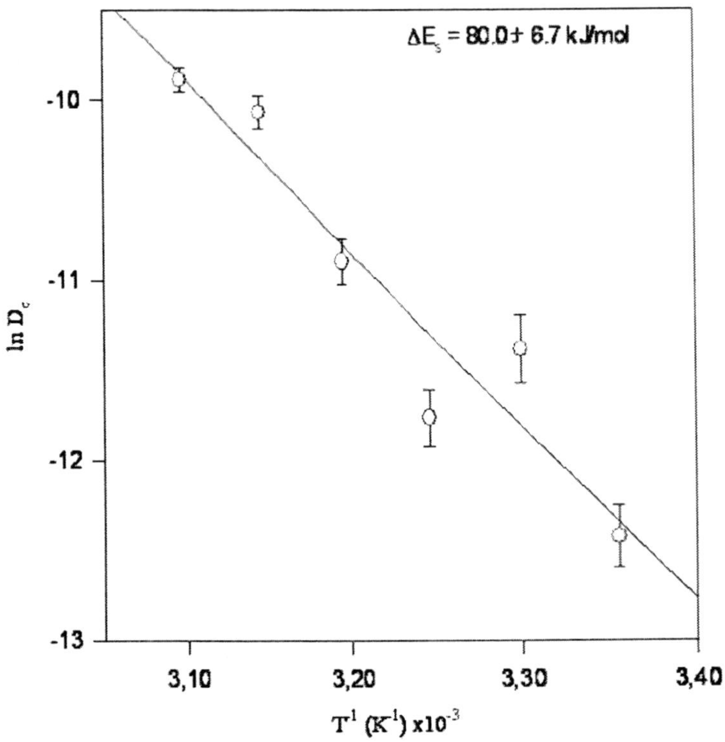

Figure 20. The logarithmic plot of D_c values versus temperature T^{-1} according to Eq.(19). The slope of the linear relation produces the activation energy as ΔE_s= 80.0 kJ mol^{-1} for swelling of the gel.

Conclusion

In this chapter we have showed that FTRF technique can be used to measure time constants, τ_c and cooperative diffusion coefficients, D_c during polymeric gel swelling induced by organic vapor and organic solvent. It is understood that organic molecules penetrate much slower in vapor phase than in solvent into PMMA gels. It is also shown that penetration of organic molecules are much faster in loosely formed gels than densely formed gels in both phases. It is seen in Table I and II that macroscopic measurements such as Δm and Δd are not very sensitive to the environment of the gel i.e. similar Δm and Δd values were produced in both vapor and solvent phases. Δm and Δd values are also independent of the EGDM content in PMMA gels. Solvents of poor solution quality penetrates the gel much more slowly than solvent of good solution quality. Gels swell much faster and to a larger extent in high quality solvent than in bad quality solvents

Swelling of S-MMA heterogels under chloroform vapor using FTRF technique showed that polymer-solvent interaction plays dominant role during swelling processes i.e. MMA-rich heterogels swell much faster than S-rich heterogels. Several studies related to diffusion of aqueous and nonaqueous agents into polymeric networks have been reported in literature where corresponding activation energies are presented. Water sorption of networks based on

2-hydroxyethyl methacrylate-triethylenglycol dimethacrylate copolymers was investigated [41] gravimetrically and the apparent activation energy was found to be independent of composition of the networks, given a value of 15.5 kJ mol^{-1}. Solvent transport aromatic hydrocarbons through both crosslinked and uncrosslinked poly(ethylene-co-vinyl acetate) membranes has been investigated by using gravimetric technique [42]. The activation energies for the process of diffusion, sorption and permeation have been measured and found to be in between 3-20 kJ mol^{-1} depending on the penetrant size. An investigation of sorption kinetics and diffusion of chloroalkanes into tetrafluoro ethylene/propylene copolymer membranes was reported [43], where activation energy of diffusion for chloroform was found to be around 15 kJ mol^{-1}. Diffusion and permeation properties of substituted benzenes through variety of membranes formed from blends of nitrile rubber and natural rubber have been studied in the temperature interval of 28-70 °C and the values of activation energies were measured to be around 10 kJ mol^{-1} [44]. These activation energies all correspond to typical diffusion processes and much lower than our finding (80.0 kJ mol^{-1}). The difference between literature values and our finding most probably originates from the difference between solvent and vapor penetration processes. From here one may conclude that high concentration gradient during solvent penetration lowers the value of the activation energy, however system needs more energy during vapor sorption due to low concentration gradient. Resent work [45] oxygen diffusion in copolymers of ethylene and norbornene support these findings where the diffusion activation energy was found to be around 35-40 kJ mol^{-1} over the range of 5-35 °C temperature. Here it can be concluded that the FTRF technique can be used to study gel swelling in various internal and external conditions with a quite high accuracy.

REFERENCES

[1] Tanaka, T.; Fillmore, D.J.; Sun, S.T.; Nishio, I.; Swislo, G.; Shah, A. *Phys Rev Lett* 1980, 45, 1636.

[2] Dusek, K.; Peterson, D. *J Poly Sci* 1968, *A2*, 1209.

[3] Tanaka, T. *Phys Rev Lett* 1980, 45,1636.

[4] Kajiwara, K; et al. *Polymer Gels*, Plenum Press, New York-London, 1990.

[5] Tobolsky, A.V.; Goobel, J.C. *Macromolecules* 1970, 3, 556.

[6] Schild, H.G. *Prog Polym Sci* 1992, 17,163.

[7] Amiya, T.; Tanaka, T. *Macromolecules* 1987, 20, 1162.

[8] Tanaka, T.; Filmore, D. *Chem Phys* 1979, 70, 1214.

[9] Peters, A.; Candau, S.J. *Macromolecules* 1986, 19, 1952.

[10] Chiarelli, P.; De Rossi, D. *Prog Colloid Polym Sci* 1988, 7, 44.

[11] Dusek, K.; Prins, W. *Adv Polym Sci* 1969, 6, 1.

[12] Candau, S.; Baltide, J.; Delsanti, M. *Adv Polym Sci* 1982, 7, 44.

[13] Li, Y.; Tanaka, T. *J Chem Phys* 1990, 92, 1365.

[14] Horie, K.; Mita, I.; Kawabata, J.; Nakahama, S.; Hirao, A.; Yamazaki, N. *Polym J* 1980, 12, 319.

[15] Pekcan, Ö.; Winnik, M.A.; Egan, L.S.; Croucher, M.D. *Macromolecules* 1983, 16, 669.

[16] Pekcan, Ö.; Winnik, M.A.; Croucher, M.D. *Phys Rev Lett* 1988, 61, 641.

[17] Pekcan, Ö. *Chem Phys Lett* 1992, 20, 198.

[18] Pekcan, Ö. *Trends in Polym Sci* 1994, 2, 236.
[19] Pekcan, Ö. *Chem Phys* 1993, 177, 619.
[20] Pekcan, Ö.; Winnik, M.A.; Croucher, M.D. *Macromolecules* 1990, 23, 2673.
[21] Wang, Y.; Winnik, M.A. *Macromolecules* 1993, 26, 3147.
[22] Pekcan, Ö. *J Appl Polym Sci* 1993, 49, 151.
[23] Pekcan, Ö. *J Appl Polym Sci* 1996, 59, 521.
[24] Peters, A.; Candau, S.J. *Macromolecules* 1998, 21, 2278.
[25] Bastide, J.; Duoplessix, R.; Picot, C.; Candau, S. *Macromolecules* 1984, 17, 83.
[26] Wu, C.; Yan, C.Y. *Macromolecules* 1994, 27, 4516.
[27] Panxviel, J.C.; Dunn, B.; Zink, J.J. *J Phys Chem* 1989, 93, 2134.
[28] Pekcan, Ö.; Yılmaz, Y.; Okay, O. *Chem Phys Lett* 1994, 229, 537.
[29] Pekcan, Ö.; Yılmaz, Y.; Okay, O. *Polymer* 1996, 37, 2049.
[30] Yılmaz, Y.; Pekcan, Ö. *Polymer* 1998, 39, 5351.
[31] Erdoğan, M.; Pekcan, Ö. *J Polym Sci Part B* 2000, 38, 739.
[32] Kropp, L.J.; Dawson, R.W. *Fluorescence and Phosphorescence of Aromatic Hydrocarbons in Polymethylmethacrylate in International Conference on Molecular Luminescence,* E.C. Lim, Ed., New York, 1969.
[33] Bixon, M.; Jortner, J. *J Chem Phys* 1968, 48, 715.
[34] Birks, J.B.; Lumb, M.D.; Munro, I.H. *Proc Roy Soc. A* 1964, 277, 289.
[35] Kamioka, K.; Weber, S.E.; Morishima, Y. *Macromolecules* 1988, 21,972.
[36] Jones, P.F.; Siegel, S. *J Chem Phys* 1964, 50, 1134.
[37] Birks J.B. *Photophysics of Aromatic Molecule* J Wiley Interscience, New York, 1971.
[38] Lakowicz, J. R. *Principles of Fluorescence Spectroscopy* Plenum Press, New York, 1983.
[39] Ware, W.R.; James, D.R.; Siemianczuk, A. *Rev Sci Inst* 1992, 63, 1710.
[40] Birks, J.B.; Lumb M.D.; Mumra J.H. *Proc R Soc Sev A* 1989, 277, 289.
[41] Peniche, C.; Cohen, M.E.; Vazquer, B.; Roman, *J.S. Polymer* 1997, 38, 5977.
[42] Kumar, S.A.; Tomas, S.; Kumaran, M.G. *Polymer* 1997, 38, 4629.
[43] Aminabhari, T.M.; Harlapur, S.F.; Balundgi, R.H.; Ortega, J.D. *Polymer* 1998, 39, 1067.
[44] Mathai, A.E.; Singh, R.P.; Thomas, S. *J. Membrane Sci* 2002, 202, 35.
[45] Poulsen, L.; Zebger, I.; Klinger, M.; Larsen, P.S.; Ogilby, R.P. *Macromolecules* 2003, 36, 7189.

In: Polymer Research Developments
Editor: Robert K. Bregg, pp. 57-73

ISBN 1-59454-742-4

Chapter 3

Cyclic and Linear Oligomers Formation in the Synthesis of Bisamide Tetraazamacrocycles

H. Mouaziz,* J.-C. Soutif and V. Montembault

LCOM - Chimie des Polymères, UCO2M, UMR n°6011 CNRS-Université du Maine, Avenue Olivier Messiaen, Le Mans cedex, France

Abstract

Research concerning polymeric materials containing tetraazamacrocycles is of great interest for many applications because their complexes show exceptional kinetic and thermodynamic properties, and has attracted increasing attention during the last four decades. The elaboration of these polymeric materials implies first the synthesis of bifunctional tetraazamacrocycles. The methods for the formation of these structures have been extensively reviewed. However, their development is limited by the difficulty to obtain these macrocycles with a high yield. Classical strategy for this kind of reactions involves the use of high dilution conditions. Under these conditions the probability of intramolecular reactions is significantly greater than the probability of intermolecular reactions. We were particularly interested in synthetic procedures as short as possible to prepare *N, N'*-difunctionalized tetraazacyclododecanes in order to immobilize the tetraazamacrocyclic structure in a polymer chain by polycondensation with difunctional comonomers such as diols or diamines. Two different procedures in one step using high dilution conditions or using a catalyst system have been investigated for the synthesis of bisamide tetraazamacrocycles. In order to explain the low reaction yield of ring closure, the side-products have been characterized by MALDI-TOF MS to identify their cyclic or linear structure and their size. The influence of experimental parameters on the nature of the structures (cyclic or linear oligomers) has also been discussed. The results have confirmed the competition between cyclization and chain-growth polymerization

* E-mail: hannamouaziz@hotmail.com

occurring during synthesis of tetraazamacrocycles. This explains the presence of cyclic and linear oligomers which prevents a better reaction yield for ring closure. Nevertheless, we report our investigation on the best conditions to favour obtaining of these two macrocycles.

Keywords: Cyclocondensation / Tetraazamacrocycle / Oligomers / MALDI-TOF MS.

INTRODUCTION

The many potential applications of functionalized azamacrocycles in coordination chemistry, for separation processes [1-5], biomedical applications or in the field of radiotherapy [6-9], have made them very attractive targets because of their ability to interact with both metal cations and anionic species [10-13]. The polyaza macrocyclic molecules are known to form metal complexes more stable than the analogous open chains. Numerous synthetic routes have been developed for producing new macrocyclic polyamines but no general method exists for the preparation of these structures and one limitation is the low reaction yield because macrocyclic compounds cannot be synthesized without occurrence of organic polymers as side-products [14]. The preferential formation of only one macrocycle often depends on the conditions of reaction such as template effects, temperature, concentration of starting materials, and kinetic and thermodynamic behaviour [15, 16]. Thermodynamic stability depends on the ring size which, in spite of considerable investigation, has not been formulated quantitatively in general but has been qualitatively studied [16] (Figure 1).

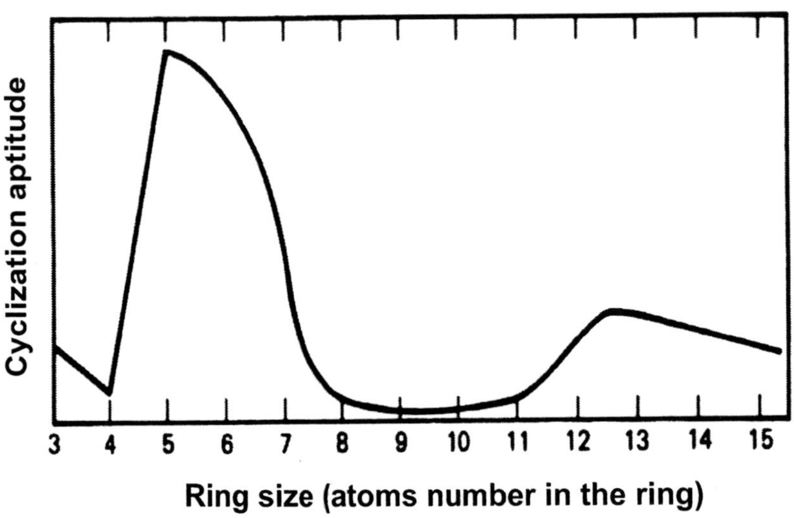

Figure 1. Evolution of the cyclization aptitude versus ring size [16]

Ring strain is managed by both angular and conformational stress for rings of less than 11 atoms and interfere little or not for the macrocyclic structures studied in this work. Ring formation is a kinetically controlled reaction. For cyclization involving equilibrium reaction (Equation 1):

$$P_{m+n} \underset{}{\overset{k_n}{\rightleftharpoons}} C_n + P_m \qquad (1)$$

P_{m+n} and P_m are linear chains and C_n are cycles.

The Jacobsen-Stockmayer theory suggests [17] that $[C_n] \approx n^{-5/2}$.

For irreversible systems [18, 19] (Equations 2 & 3), a kinetic control explains the results:

$$P_{i,j} + P_{m,n} \xrightarrow{k_{(i,m)}} P_{i+m,\,j+n-1} \qquad (2)$$

$$P_{i,1} \xrightarrow{k_{c(i)}} C_i \qquad (3)$$

where $P_{i,j}$, $P_{m,n}$ and C_i are linear and cyclic macromolecules.

Considering the equations of the reaction rates (order 2 and 1 respectively), a different effect of the reaction parameters on the relative reaction rates (temperature, concentration, stoechiometry...) can be expected.

The concentration of ring molecules is proportional to the concentration of linear chains of type $P_{i,1}$ [17]: $[C_i] = P_{i,1} * i^{-5/2}$. Dilution has a double effect [19]: on the one hand, the total rate of the reaction decreases, on the other hand, the relative rate of the reaction of cyclization increases. This fact makes us anticipate an optimal dilution.

All the theoretical approaches carried out in cases of studies of the role of cyclization in step-growth polymerization in particular by H.R. Kricheldorf [20-24] can be applied to explain some low reaction yields obtained for cyclization reactions. Determination of the structure of side-products (by RMN and SEC [25-27]) was difficult because of the lack of an accurate technique, and has received more attention since the development of MALDI-TOF MS (Matrix-assisted laser desorption/ionization time-of-flight mass spectroscopy) but the main examples concern step-growth polymerization [20-24, 28-30]. This competition between cyclization and macromolecular chain formation has also been turned to account in case of cyclodepolymerization of condensation polymers leading much more likely to the formation of cycles at low concentrations [31-33]. For instance, cyclodepolymerization of functional linear polyesters leads to macrocycles which, on equilibrium *via* transesterification, give soluble combinatorial libraries of crown ether-ester analogues which have been characterized by MALDI-TOF MS.

The present work deals with the synthesis of *N, N'*-difunctionalized tetraazacyclododecanes with carboxylate functions (acid, ester) in order to immobilize the tetraazamacrocyclic structure in a polymer chain by step-growth polymerization with difunctional comonomers such as diols or diamines [34]. There has been a few methods to prepare *N, N'*-difunctionalized tetraazacyclododecanes containing carboxylate groups using simple ring closure reactions under high dilution conditions [35, 36]. These examples have shown that reaction yields of ring closure are low because of the competition reaction which takes place between cyclization and chain-growth polymerization. However, up to now, there has been very few studies concerning the identification of the resulting side-products. In order to explain the low reaction yield of ring closure, the identification of the side-products in terms of their size and their cyclic or linear structures has been carried out. MALDI-TOF MS is a powerful analytical technique that is convenient for elucidation for such structures. MALDI-TOF MS as a soft ionization method provides measurements of absolute molecular weights, compositions, and functionality of intact molecules.

We have been particularly interested by the synthesis of two bifunctionalized tetraazamacrocycles obtained by two different ways in order to insert the tetraazamacrocyclic complexing site in a polymer chain. We have studied the synthesis of 1,4,7,10-tetraazacyclododecane-2,6-dione (dioxocyclen) because its synthesis in one step allows to obtain a macrocycle with two protected trans positions inducing a possibility of a direct further functionalization on the two other positions.

We have also studied the synthesis of 6,11-dioxo-1,4,7,10-tetraazacyclododecane-1,4-diacetic acid (DTDA) because its synthesis in one step allows to obtain directly a macrocycle with two protected trans positions and two acetic acid functions, inducing the possibility of a further direct polycondensation with diols or diamines.

We report in this study our investigation of the best conditions to favour obtaining of these two macrocycles and on the identification of each side-product.

SYNTHESIS OF 1,4,7,10-TETRAAZACYCLODODECANE-2,6-DIONE (DIOXOCYCLEN (1))

Synthesis of dioxocyclen **1** (Scheme 1) has been briefly described by Fabbrizzi et al. [35] from diethyl iminodiacetate and diethylenetriamine ($C = 5.10^{-2}$ mol.L^{-1}) in refluxing ethanol for 5 days followed by crystallisation from ethanol. The advantage of this synthesis is the direct protection of two nitrogen atoms through amide functions. However, the yield of only 10% is indeed substantially low because the crude dioxocyclen is highly contaminated with oligomeric substances. ^{1}H NMR analysis has shown that the major side-product of this reaction consists of two diester subunits and of one diamine subunit yielding a linear diamide-diester [37, 38].

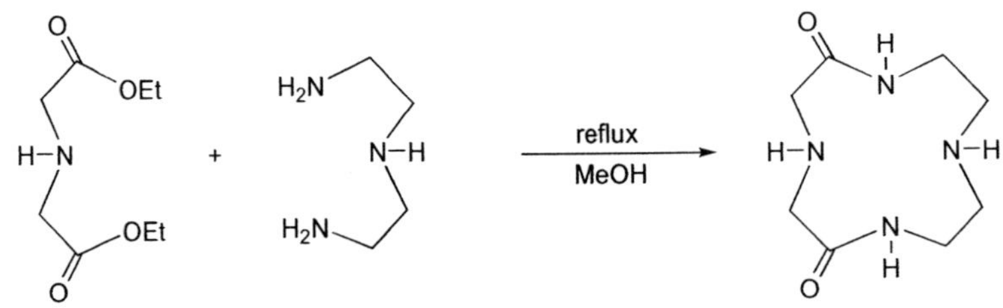

1

Scheme 1. Preparation of dioxocyclen **1**

We have then realized this reaction while varying various parameters in order to study more precisely this synthesis. A kinetic study of the macrocycle formation at 60°C has been carried out by ^1H NMR spectroscopy to follow the disappearance of the starting products. The results have confirmed the necessity of a reaction time of 5 days to obtain a complete conversion of the starting products. The same synthesis has been carried out at higher temperatures (T = 80°C and 100°C, MeOH, t = 5 days, C = 5.10^{-2} mol.L^{-1}). We have also used the method reported by Gryko and al. [39] (rt, MeOH/MeONa, t = 8 h, C = 0.1 mol.L^{-1}). They have developed a simple catalyst system to accelerate the reaction of diesters with diamines. Use of sodium methoxide NaOMe has allowed obtaining the dioxocyclen with a significantly higher yield which can be explained by base-accelerated intramolecular reaction of the intermediate compound possessing one free ester group and one free amino group. The increase of yield is associated with the use of anhydrous methanolates instead of metal hydroxides which prevents the partial hydrolysis of the starting esters by nucleophilic attack of OH^- and is not due to the template effect owing to the Na^+ ion [39]. The compounds obtained using high dilution conditions or a simple catalyst system have been characterized by ^1H and ^{13}C NMR. ^1H NMR analysis was coherent with the expected dioxocyclen structure. However ^{13}C NMR has shown more than four different carbons, which induces a mixture of compounds. Therefore, in order to identify the structure of the compounds, MALDI-TOF MS analysis has been realized on the mixtures of compounds synthesized using high dilution conditions and a simple catalyst system. An example of MALDI-TOF mass spectrum corresponding to the reaction with a simple catalyst system is shown Figure 2.

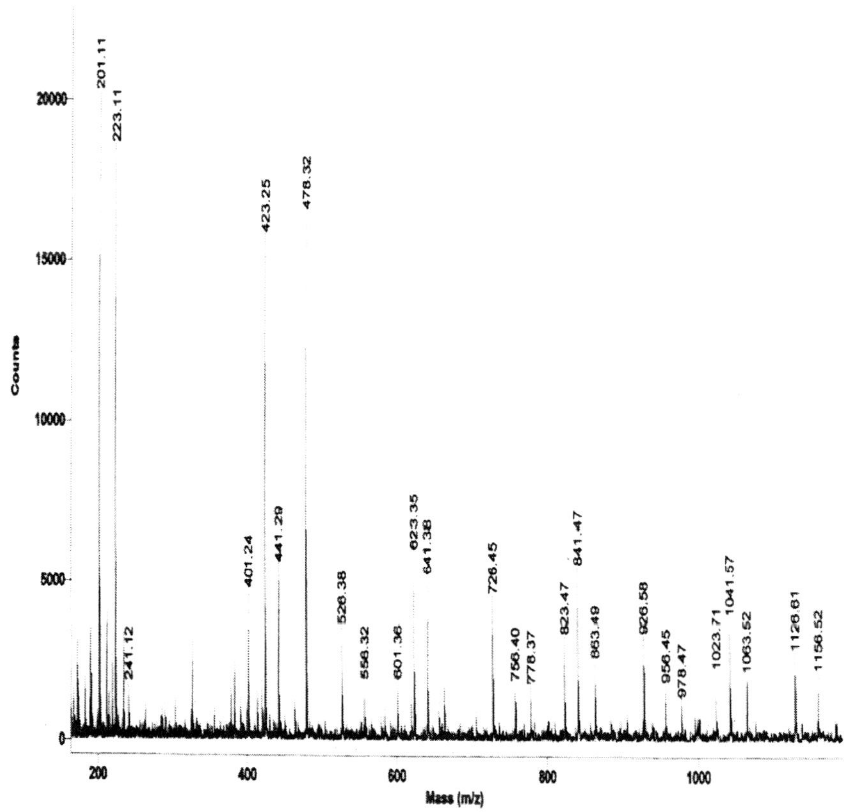

Figure 2. MALDI-TOF mass spectrum of the products obtained by reaction using a catalyst system

Table 1. Identification of the oligomers formed during the dioxocyclen synthesis

Entry	Nature of the oligomers	Structure
I	cyclic oligomers	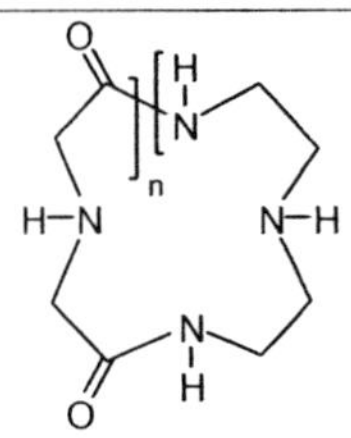
II	linear oligomers with amine ends	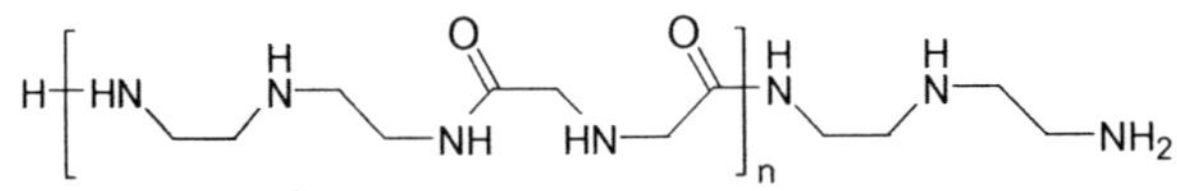
III	linear oligomers with acid and ester ends	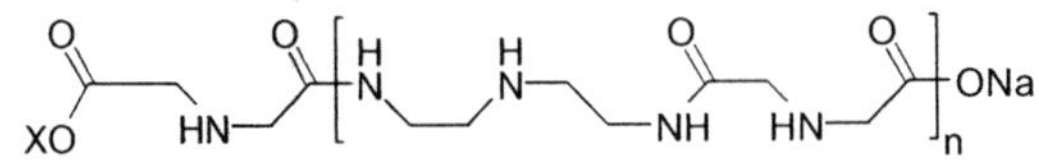
IV	linear oligomers with acid ends	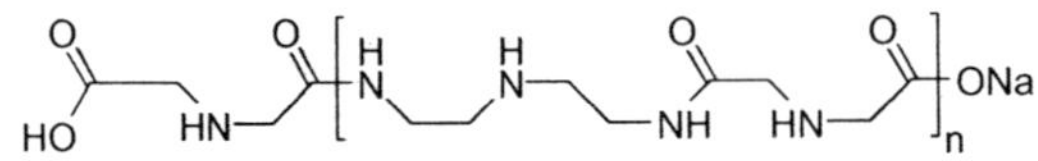
V	linear oligomers with acid and amine ends	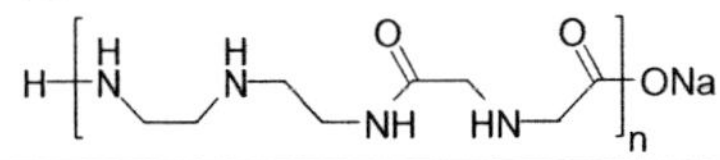

The analysis has shown the formation of five different families (Table 1). The difference between each peak owning to a same family corresponds to the dioxocyclen unit (ΔM = 200.1). The first element of each family corresponds respectively to m/z 200.11, 223.11, 441.29, 526.38, 556.32 and 863.49. The first family is easily identified and is attributed to the cyclic oligomers (Table 1, Entry I) ionised by a proton (201.11) and by sodium (223.11). The series starting at 441.29 and 863.49 could correspond to linear oligomers with sodium carboxylate and amine ends (Table 1, Entry V), ionised by a proton (441.29) and by sodium (863.49). The family starting at 526.38 could correspond to linear oligomers with amine ends (Table 1, Entry II), ionised by sodium. Finally, the family starting at 556.32 could correspond to linear oligomers with sodium carboxylate and acid ends (Table 1, Entry IV) ionised by a proton. The analysis of the mixtures of compounds obtained using high dilution conditions at 60, 80 and 100°C has shown one more family that could be attributed to linear oligomers with sodium carboxylate and ester ends (Table 1, Entry III). In consequence five structures have been identified corresponding to cyclic and linear oligomers obtained during dioxocyclen synthesis.

Table 2. Number of units observed for each family at 60, 80, 100°C using high dilution conditions and at room temperature using a catalyst system

Families	(I)	(II)	(III) and (IV)	(V)
T = 60°C	n = 1 to 3	-	n = 1 to 3 (III)	-
T = 80°C	n = 1 to 4	n = 2 to 4	n = 1 to 3 (III)	-
T = 100°C	n = 1 to 5	n = 1 to 11	n = 1 to 4 (III)	-
T = rt (CH_3ONa/CH_3OH)	n = 1 to 8	n = 1 to 10	n = 2 to 6 (IV)	n = 1 to 10

Table 2 summarizes the number of units observed for each family corresponding to the reactions using high dilution conditions at 60, 80, 100°C and for the reaction at room temperature using a catalyst system. Families I and II are observed whatever the strategy used for the synthesis of dioxocyclen. Families III and IV observed respectively for reaction using high dilution conditions and CH_3ONa as catalyst system are very close. One has a sodium carboxylate end whereas the other has an ester end. The hydrolysis of the ester function may be due to α–cyano-4-hydroxycinnamic acid used as a matrix for MALDI-TOF MS analysis. Family V is only present for the reaction using a catalyst system. Moreover, the maximum number of units increased for each family, as observed when temperature was increased in the case of high dilution conditions. This induced the synthesis of more species, especially higher linear oligomers. These results have shown that the increase of reactivity favours the synthesis of oligomers but not the formation of the smallest macrocycle. Gryko et al. [39] have also noticed that a higher yield using a catalyst system isn't observed systematically for every macrocycle depending on their chemical structure, especially on the positions of the carbonyl groups in relation to nitrogens in the macrocycle. In the case of dioxocyclen, its chemical structure didn't allow an increase of the yield using a catalyst system. Therefore, to limit the formation of linear oligomers and of larger macrocycles the best way is to carry out the synthesis of dioxocyclen under methanol reflux at 60°C during 5 days using high dilution conditions.

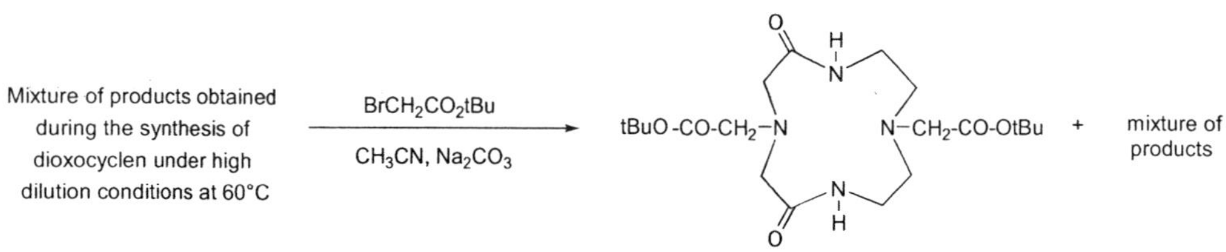

Scheme 2. Chemical modification of the mixture of products obtained during the synthesis of **1**

Different attempts of separation of dioxocyclen from the linear and cyclic oligomers have been realized by HPLC varying the ratio methanol:water. However, the structure of the different compounds seems too close to get a satisfactory separation of the mixture. In order to separate the smallest macrocycle from the other compounds present in the mixture, a chemical modification has then been carried out by reaction of alkylbromoacetates (tertiobutyl bromoacetate, ethyl bromoacetate, benzyl bromoacetate) on the mixture of products obtained during the synthesis of dioxocyclen under high dilution conditions at 60°C. The solubility of the dibutylester and the dibenzylester was too poor to get a good separation whatever the composition of the eluent. However, the ditertiobutylester tetraazamacrocycle **2** (Scheme 2) has been successfully isolated by HPLC (methanol:water, 70/30). The difunctional macrocycle **2** corresponds to 12% of the mixture.

The mixture of products has also been analysed by SEC (Figure 3). Elution volumes for compound **2** and tertiobutyl bromoacetate are respectively 32.65 and 29.07 mL. Elution volume of 30.28 mL would correspond to the monosubstituted macrocycle. Higher oligomers have been observed between 24.52 and 27.81 mL. The percentage of oligomers corresponds to 68% of the mixture and the smallest expected macrocycle corresponds to 27%. This result is higher than the percentage calculated from HPLC. This might be due to the necessary filtration before the analysis in SEC, which removes higher mass compounds. In consequence, the percentage of the compound **2** is more likely 12%. This yield is limited because of the apparent difficulty to form the smallest macrocycle from the intermediate α-amino-ω-ester intermediate as it has been highlighted by the MALDI-TOF MS analyses.

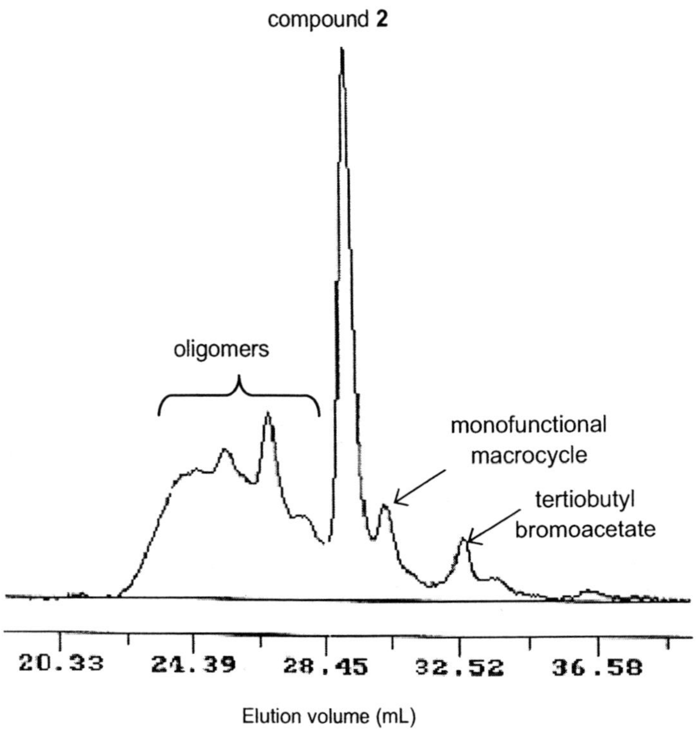

Ve (mL)	Compounds	Percentage (%)
24.52 25.36 26.67 27.81	oligomers	68
29.07	compound **2**	27
30.28	monofunctional macrocycle	3
32.65	tertiobutyl bromoacetate	2

Figure 3. SEC chromatogram in THF of the mixture of products obtained during the synthesis of dioxocyclen at 60°C under high dilution conditions

SYNTHESIS OF 6,11-DIOXO-1,4,7,10-TETRAAZACYCLODODECANE-1,4-DIACETIC ACID (DTDA (3))

Inoue et al. [36] have produced 6,11-dioxo-1,4,7,10-tetraazacyclododecane-1,4-diacetic acid from a condensation reaction with ethylenediaminetetraacetic dianhydride and ethylenediamine in DMF in dilute conditions but with yields of only 24%. This one-step reaction allows obtaining a macrocycle directly with two protected trans position and with two acetic acid functions, inducing the possibility of a direct further polycondensation with diols or diamines. This reaction has resulted in the formation of two macrocycles as shown in Scheme 3 which were isolated by fractional extraction and characterized by X-ray crystal analyses. The others side-products have been identified as linear molecules without identifying their structures.

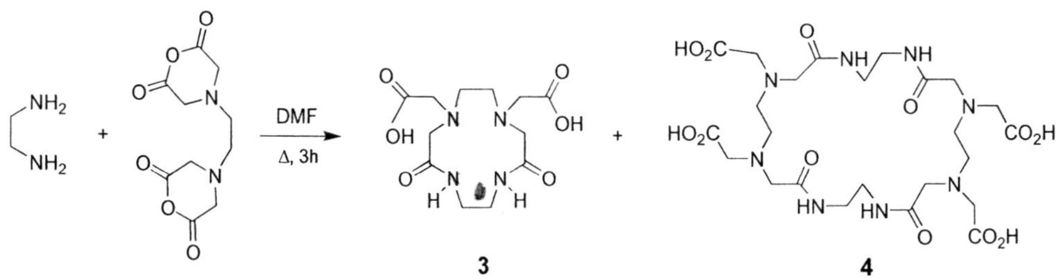

Scheme 3. Preparation of DTDA **3**

We have carried out the same synthesis in DMF using high dilution technique in order to favour the smallest macrocycle formation and MALDI-TOF MS analysis has been used in order to identify the structures of the side-products. The spectrum (Figure 4) doesn't show any formation of linear molecules but the formation of a mixture of macrocycles with a range of sizes between 12 atoms and 120 atoms.

The precipitation of DTDA has been observed with addition of water to the mixture whereas the macrocycles of bigger sizes have been obtained by precipitation in THF. The crude product DTDA is only obtained after several washings with water until complete elimination of larger macrocycles. Among them, the macrocycle **4** (24 atoms) has been isolated after several extractions with THF.

In order to improve the yield reported in the literature [36], we have studied the effect of the concentration on the yield of the reaction (Table 3). The yield of the compound **4** (24 atoms) didn't depend on the concentration whereas the yield of the compound **3** varies with the concentration. The best conditions to obtain DTDA (12 atoms) with the best yield is to work at $C = 2{*}10^{-2}$ $mol.L^{-1}$. Under high dilution conditions ($C = 1{*}10^{-2}$ $mol.L^{-1}$), the yield of the smallest macrocycle didn't increase, maybe due to partial hydrolysis of the anhydride functions before the end of the reaction. Using more concentrated conditions ($C = 4{*}10^{-2}$ $mol.L^{-1}$), the formation of linear chains is favoured inducing a lower yield in macrocycles.

Table 3. Influence of the concentration on the yields of the compounds 3 and 4 (T = 60°C, DMF)

[C] ($mol.L^{-1}$)	**3** (Yield %)	**4** (Yield %)
$4{*}10^{-2}$	27.1	2
$2{*}10^{-2}$	38	3
$1{*}10^{-2}$	10.4	3

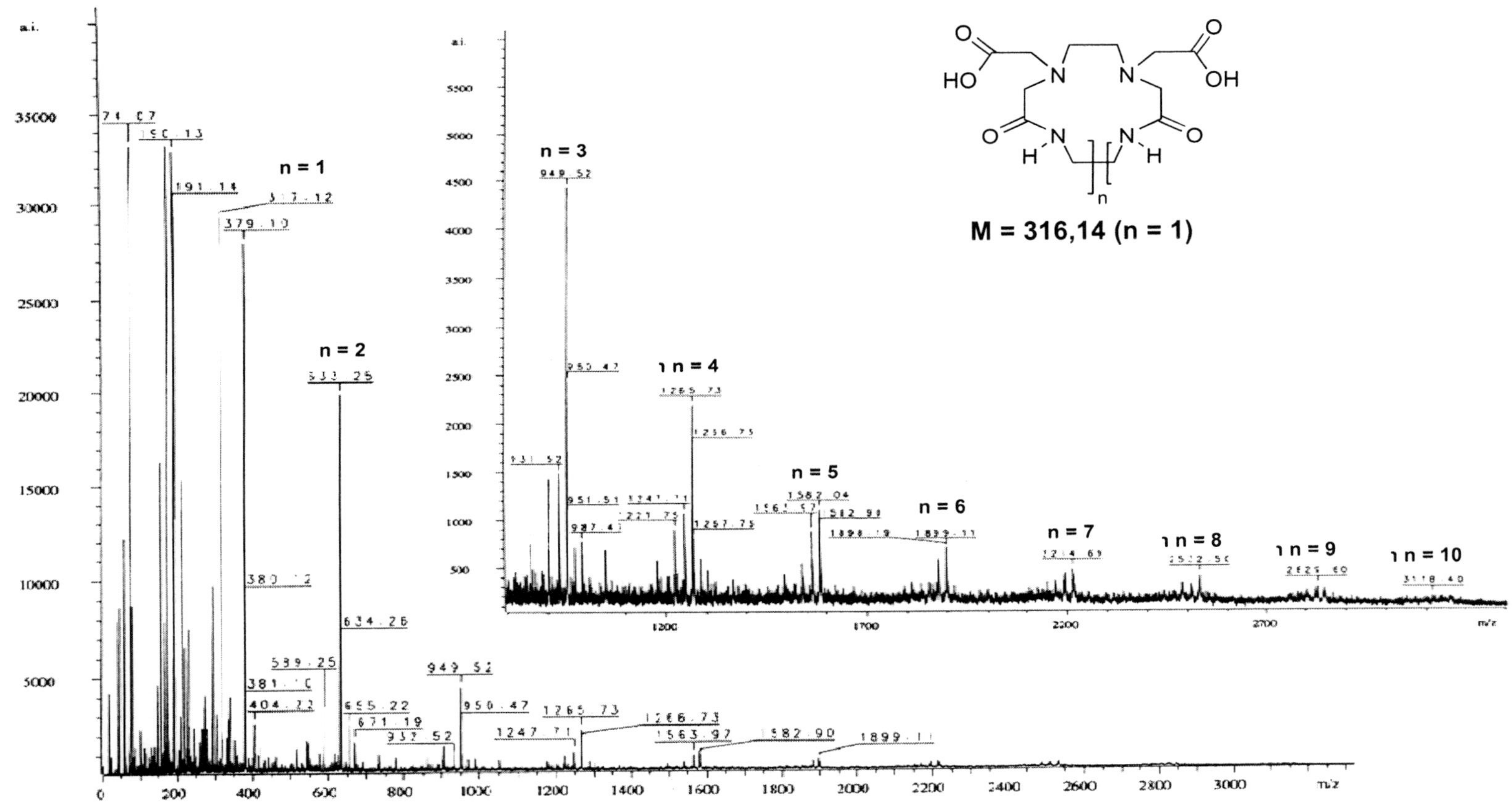

Figure 4. MALDI-TOF mass spectrum of cyclic oligomers

The compounds **3** and **4** have been characterized by ^{1}H and ^{13}C NMR, by infrared spectroscopy and by MALDI-TOF MS. Using theoretical tables [40], the chemical shifts order of the methylene groups present in the macrocycles has been established:

$$\delta\,(\mathbf{CH_2}N) < \delta\,(\mathbf{CH_2}CONH) < \delta\,(\mathbf{CH_2}COOH) < \delta\,(\mathbf{CH_2}NHCO).$$

Therefore, the four peaks present on ^{1}H NMR spectra of the compounds **3** and **4** have been attributed following this order. Then the carbons have been attributed by correlation between ^{1}H and ^{13}C. The compounds **3** and **4** have analogous structures. However, a slight difference has been observed between their chemical shifts in ^{1}H and ^{13}C NMR (Table 4). For the smallest macrocycle **3** (12 atoms), the chemical shifts for $\mathbf{CH_2}CONH$ and $\mathbf{CH_2}COOH$ are similar in ^{1}H NMR but different in ^{13}C NMR whereas for **4** (24 atoms), very close chemical shifts have been noticed in ^{1}H NMR for $\mathbf{CH_2}CONH$, $\mathbf{CH_2}COOH$ and $\mathbf{CH_2}NHCO$ and different values in ^{13}C NMR for $\mathbf{CH_2}CONH$ and $\mathbf{CH_2}COOH$.

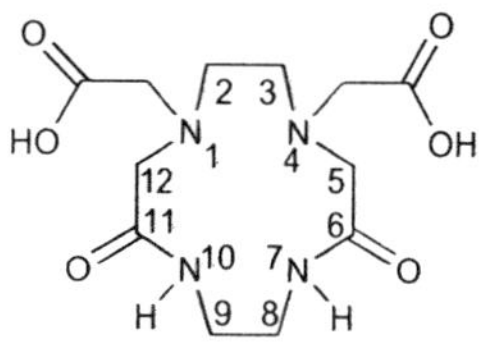

3
(12 atoms)

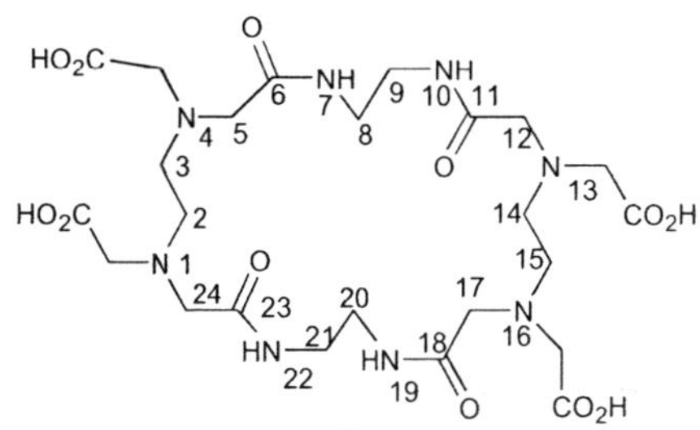

4
(24 atoms)

Table 4. Chemical shifts of the compounds 3 and 4 in 1H and 13C NMR (D2O)

		CH_2N	CH_2COOH	CH_2CONH	CH_2NHCO	COOH	CONH
^{1}H NMR (δ ppm)	**3**	2.62	3.23	3.14	3.46	-	-
	4	2.66	3.24	3.20	3.38	-	-
^{13}C NMR (δ ppm)	**3**	53.7	59.6	59.6	36.5	178.8	174.9
	4	52.2	57.9	58.4	38	178.7	174.7

In order to confirm the formation of macrocycles larger than 12 or 24 atoms, an analysis by MALDI TOF MS has also been realized on:

- the precipitate formed during the reaction,
- the filtrate from which the compounds **3** and **4** have been isolated by precipitation respectively in water and in THF.

The MALDI-TOF spectrum of the precipitate (3.5 % in weight of the mixture) corresponds to the condensation between EDTA bisanhydride and one or two molecules of ethylenediamine (Scheme 4).

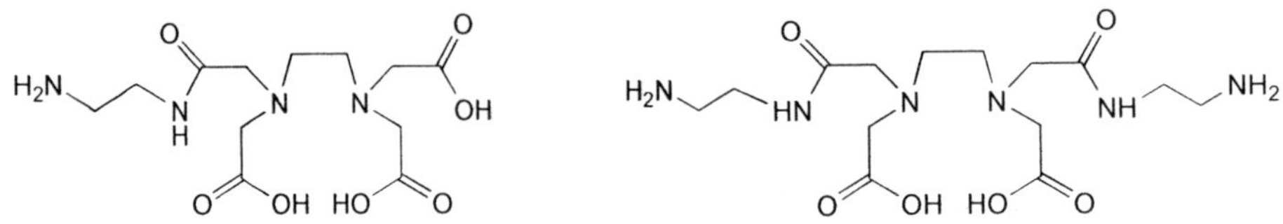

Scheme 4. Compounds obtained by precipitation during the reaction between ethylenediamine and EDTA bisanhydride

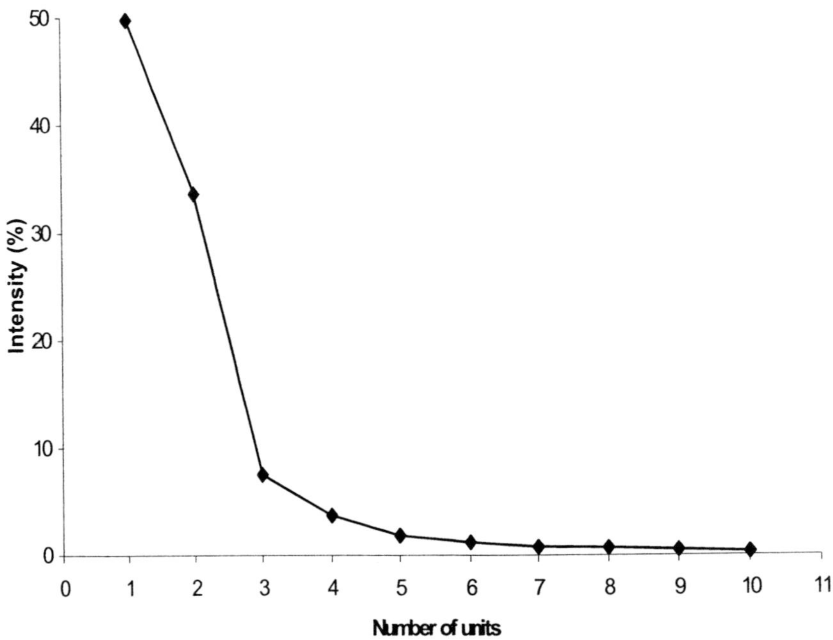

Figure 5. Distribution of cyclic oligomers by sizes with n = number of units. The percentage is calculated from the intensity of each group of peaks

^1H and ^{13}C NMR analyses carried out on the filtrate have shown numerous peaks among which those of **3** and **4** can be distinguished. The analysis by MALDI-TOF MS of the filtrate has shown macrocycles with sizes varying between 12 and 120 atoms. Each group of peaks is formed by a main peak corresponding to the ionisation by a proton $[M+H]^+$ and by two other peaks corresponding respectively to the ionisation by Na^+ and by K^+. The distance between each group of peaks corresponds to $\Delta M = 316.1$, which is the molecular mass of DTDA (compound **3**). The graph representing the relative intensity of the peaks (%) with the molecular weight (Figure 5) shows that **3** (49.8%) and **4** (33.6%) are the main products in the mixture. This confirms the NMR analyses which have shown the preponderant formation of the rings with 12 atoms (n = 1) and 24 atoms (n = 2).

It is well known that MALDI-TOF MS analysis of MW distribution becomes less reliable as the molecular weight and polydispersity of samples increase [41-45], but in the range studied here the fraction of each macrocycle is assumed to be equivalent to the intensity of the signal (I_i), and so to the quantity of each product (n_i). Then, from the spectrum (Figure 4), the number-average molecular weight ($\overline{M_n}$), the weight-average molecular weight ($\overline{M_w}$) and the polydispersity index PDI ($\overline{M_w}/\overline{M_n}$) have been calculated following these equations:

$$\overline{M_n} = (\Sigma n_i \cdot M_i)/(\Sigma n_i) \text{ et } \overline{M_w} = (\Sigma n_i \cdot M_i^2)/(\Sigma n_i \cdot M_i) \text{ with } n_i = k^*I_i$$

thus :

$$\overline{M_n} \approx (\Sigma I_i \cdot M_i)/(\Sigma I_i) \text{ and}$$

$$\overline{M_w} \approx (\Sigma I_i \cdot M_i^2)/(\Sigma I_i \cdot M_i)$$

Thus, $\overline{M_n} = 600\ g.mol^{-1}$, $\overline{M_w} = 930\ g.mol^{-1}$ and PDI = ($\overline{M_w}/\overline{M_n}$) = 1.5

These values are absolute masses values when those obtained by SEC are alcanes equivalents or polystyrene equivalents. Anyway, this mixture couldn't be analysed by SEC because of the chemical structure of the macrocycles (acid and amide functions), which could interfere with the chromatographic columns.

The crude macrocycle DTDA with 12 atoms has then be obtained with acetic functions on the positions 1 and 4 in one step and can then be used as a difunctional monomer in polycondensation reactions [34].

CONCLUSION

Bisamide tetraazacyclododecanes have been directly obtained by condensation of linear precursors in one step by use of high dilution technique or a simple catalyst system. Evidence has been given that macrocycle syntheses whatever the used conditions have led to the expected macrocycle with the simultaneous formation of cyclic and linear oligomers which explains the difficulty to reach high yield for ring closure.

Determination of the structure of each obtained oligomer as well as their size has been carried out by MALDI-TOF MS and has confirmed the competition occurring between cyclization and chain-growth polymerization.

Influence of temperature and sodium methoxide as catalyst has been highlighted by these analyses. This study has confirmed that the currently available method using the high dilution technique has led to the best reaction yield in ring closure for 1,4,7,10-tetraazacyclododecane-2,6-dione or dioxocyclen (under methanol reflux) and for 6,11-dioxo-1,4,7,10-tetraazacyclododecane-1,4-diacetic acid (DTDA). Their syntheses in one step provide bifunctional macrocycles. These compounds are of interest since they show very strong binding for metal transition cations and furthermore they are expected to have enhanced selectivity for metal cations relative to related acyclic compounds. Moreover the cyclic oligomers obtained during the syntheses of dioxocyclen and DTDA could be expected to be valuable building blocks in the synthesis of supramolecular structures [46].

EXPERIMENTAL SECTION

Materials

All the chemicals used were analytical grade chemical products: diethylenetriamine, ethyl iminodiacetate, tertiobutyl bromoacetate, ethylenediaminetetraacetic dianhydride, ethylenediamine, sodium, Na_2CO_3, anhydrous methanol, acetonitrile, DMF, THF. ^{1}H-NMR and ^{13}C-NMR spectra were recorded on a Bruker spectrometer in $CDCl_3$, in D_2O+NaOD and in CD_3OD solvents. FT-IR spectra were taken by using Bruker IFS 66, with KBr discs. High Performance Liquid Chromatography (HPLC) analyses were performed on a WATERS modular chromatograph equipped with two detectors (UV-vis spectrometer and differential refractometer), two Model 510 pumps and a Model U6 K injector. Various columns were used in relation with the applications: a Waters RCM Radialpak, Nova-pak C18 column (100*5 mm i.d., particle size of 4 μm) in analytical mode with methanol as eluent (flow 1 mL/min); a μ Bonda-pak C18 column (150*19 mm i.d., particle size of 10 μm) in semi-preparative mode with methanol (flow 2 mL/min).

MALDI-TOF Mass Spectrometry

The measurements were conducted with a Bruker Biflex III MALDI-TOF mass spectrometer. All measurements of this paper were recorded in the linear mode. For all the experiments, the matrix used was α-cyano-4-hydroxycinnamic acid (HCCA). The polymer was dissolved in methanol at a concentration of 10 $mg.mL^{-1}$ and was mixed in a ratio 1/10 by vol. with a 10 $mg.mL^{-1}$ matrix solution in the same solvent as used to dissolve the polymer. 1 μL of the mixture was actually applied to the probe surface.

Synthesis of 1,4,7,10-tetraazacyclododecane-2,6-dione (dioxocyclen 1)

By High Dilution Technique

A three-neck flask was assembled with a condenser and two dropping funnels containing solutions (5.10^{-2} $mol.L^{-1}$) of diethylenetriamine (2.58 g: $2.5.10^{-2}$ $mol.L^{-1}$ in 500 mL of methanol) and of ethyl iminodiacetate (4.73 g: 2.5. 10^{-2} $mol.L^{-1}$ in 500 mL of methanol). These solutions were simultaneously added dropwise in the three-neck flask heated to reflux. After addition (time: 5 h), the resulting solution was refluxed for 5 days. The solution was then evaporated to lead to a yellow solid (4.5 g). *^{1}H NMR (D_2O) δppm: 2.65 (4H, m, CH_2NH), 3.24 (4H, s, CH_2CONH), 3.26 (4H, m, CH_2NHCO). IR (KBr) ν(cm^{-1}): 3285 (2νCO), 2928 (CH_2), 1643 (C=O), 1550 (∂NH), 1455 (∂CH_2).*

By a Catalyst System

Sodium (0.23 g: 10 mmol) is added to anhydrous methanol (100 mL). The mixture was cooled down to 5°C and diethylenetriamine (1.03 g: 10 mmol) and ethyl iminodiacetate (1.8921 g: 10 mmol) were added. The mixture was stirred at room temperature during 8 hours and the solvent was then evaporated to lead to an orange powder (2.5 g). *^{1}H NMR (D_2O)*

δppm: 2.65 (4H, m, CH_2NH), 3.24 (4H, s, CH_2CONH), 3.26 (4H, m, CH_2NHCO). IR (KBr) ν(cm^{-1}): 3285 (2 νCO), 2928 (CH_2), 1643 (C=O), 1550 (∂NH), 1455 (∂CH_2).

Synthesis of 1,7-di(tertiobutylacetate)-3,11-dioxo-1,4,7,10-tetraazacyclododecane (2)

Na_2CO_3 (1.06 g: 10 mmol) and tertiobutyl bromoacetate (1.98 g: 10 mmol) were added to a solution of the mixture of products obtained during the synthesis of dioxocyclen under high dilution conditions at 60°C (1 g: 5 mmol) in acetonitrile (50 mL). The mixture was refluxed during 18 h. The white precipitate was filtered off and the orange filtrate was evaporated leading to a brown mixture. After HPLC (CH_3OH:H_2O, 70/30), the compound **2** was obtained (12%). *^1H NMR ($CDCl_3$) δ(ppm): 1.5 (18H, s, $C(CH_3)_3$), 2.7 (4H, s, CH_2N), 3.2 (4H+4H+4H, m, NCH_2CO+-CH_2NHCO+$CH_2COOtBu$). ^{13}C NMR ($CDCl_3$) δ(ppm): 28.5 (CH_3), 37.6 (CH_2N), 54.3 (CH_2NHCO), 60.3 (NCH_2CO), 62.1 ($CH_2COOtBu$), 82.7 ($C(CH_3)_3$), 168.7 (COOtBu), 172.5 (CONH). IR (KBr) ν(cm^{-1}): 3374 (2 νCO), 1732 (C=O), 1662 (CONH). MS (APCI) (CH_3CN:H_2O, 70/30): m/z = 428.3 (MH^+).*

Synthesis of 6,11-dioxo-1,4,7,10-tetraazacyclododecane-1,4-diacetic Acid (3)

In a three-neck flask assembled with a condenser, a dropping funnel, and a nitrogen-inlet tube, ethylenediaminetetraacetic dianhydride (2.5 g: 8.0 mmol) was suspended in dry DMF (350 mL) with stirring under a nitrogen atmosphere at 50°C. To the suspension was added dropwise a DMF solution (50 mL) containing ethylenediamine (0.48 g: 8.0 mmol) through the dropping funnel in a period of 2 h. The resulting reaction mixture was heated at 60°C overnight and left to stand at room temperature. The precipitate formed during the reaction was removed by filtration. The filtrate was concentrated by the use of rotary evaporator at a temperature below 60°C and water (10 mL) was added to the resulting viscous liquid. The finely divided solid that was formed was separated from the mixture by filtration, washed with THF, and dried in vacuum, this product is the 12-membered macrocycle *3* (Yield: 0.96 g ; 38%). Mp 254°C. *^1H NMR (D_2O+NaOD) δ(ppm): 2.62 (4H, s, (CH_2N), 3.46 (4H, s, (CH_2NHCO), 3.14 (4H, s, CH_2CONH), 3.23 (4H, s, CH_2COOH). ^{13}C NMR (D_2O+NaOD) δ(ppm): 53.7 (CH_2N), 36.5 (CH_2NHCO), 59.6 (CH_2CONH+CH_2COOH), 174.9 (CONH), 178.8 (COOH). IR (KBr) ν(cm^{-1}): 3305 (2 νCO), 1720 (COOH), 1645 (CONH).*

Synthesis of 6,11,18,23-tetraoxo-1,4,7,10,13,16,19,22-octaazacyclotetracosane-1,4,13,16-tetraacetic Acid (4)

The filtrate obtained at the preceding step was concentrated with a rotary evaporator, and the resulting viscous solution was mixed with a large amount of THF. The colorless solid that precipitated was found to be the 24-membered macrocycle **4**. The product was collected on a sintered glass filter, washed with THF, and dried in vacuum overnight (Yield: 0.15 g ; 3%). Mp 212°C. *^1H NMR (D_2O) δ(ppm): 2.66 (4H, s, CH_2N), 3.38 (4H, s, (CH_2NHCO), 3.20 (4H,* *s, CH_2CONH), 3.24 (4H, s, CH_2COOH). ^{13}C NMR (D_2O) δ(ppm): 52.2 (CH_2N), 38 (CH_2NHCO), 58.4 (CH_2CONH), 57.9 (CH_2COOH), 174.7 (CONH), 178.7 (COOH). IR (KBr) ν(cm^{-1}): 3305 (2νCO), 1700 (COOH), 1649 (CONH).*

References

[1] Blain, S.; Appriou, P.; Chaumiel, H.; Handel, H. *Anal.Chim.Acta 1990*, *232*, 331.

[2] Louvet, V.; Handel, H.; Appriou, P.; Guglielmetti, R. *Eur.Polym.J. 1987*, *23(8)*, 585.

[3] Puranik, D.B.; David, V.A.; Morris, R.E.; Chang, E.L. *Energy Fuel 1997*, *11*, 1311.

[4] Chaumeil, H.; Handel, H. *Eur.Polym.J. 1991*, *27*, 269.

[5] Fujiwara, M.; Nakajima, Y.; Matsushita, T.; Shono, T. *Polyhedron 1985*, *4*, 1859.

[6] Aime, S.; Botta, M.; Crich, S.G.; Giovenzanna, G.; Palmisano, G.; Sisti, M. *Bioconjugate Chem. 1999*, *10*, 192.

[7] Lewis, M.R.; Raubitschek, A.; Shively, J.E. *Bioconjugate Chem. 1994*, *5*, 565.

[8] Alexander, V. *Chem.Rev. 1995*, *95*, 273.

[9] Parker, D., *Chem.Britain 1994*, 818.

[10] Bradshaw, J.S. *Aza-crown Macrocycles*; Wiley: New-York, USA, *1993*.

[11] Izatt, R.M.; Pawlak, K.; Bradshaw, J.S.; Bruening, R.L. *Chem.Rev. 1991*, *91*, 1721.

[12] Bradshaw, J.S.; Krakiowak, K.E.; Izatt, R.M. *Tetrahedron 1992*, *48*, 4475.

[13] Lehn, J.-M. *Angew.Chem., Int.Ed.Engl. 1988*, *27*, 89.

[14] Krakowiak, K.E.; Izatt, R.M.; Bradshaw, J.S. *J.Heterocyclic Chem. 2001*, *38*, 1239.

[15] Fradet, A. *Initiation à la Chimie et à la Physico-Chimie Macromoléculaire*; GFP: Strasbourg, FR, *1997*, vol. 11, 267.

[16] Odian, G. *La polymérisation : Principes et Applications*; Polytechnica: Paris, FR, *1994*.

[17] Jacobsen, H.; Stockmayer, W.H. *J.Chem.Phys. 1950*, *18(12)*, 1600.

[18] Irzhak, T.F.; Peregudov, N.I.; Irzhak, V.I.; Rozenberg, B.A. *Polym.Sci. 1993*, *35(9)*, 1290.

[19] Irzhak, T.F.; Peregudov, N.I.; Irzhak, V.I.; Rozenberg, B.A. *Polym.Sci. 1993*, *35(7)*, 905.

[20] Kricheldorf, H.R. *Macromol.Rapid Commun. 2000*, *21*, 528.

[21] Kricheldorf, H.R. *Macromol.Symp. 2003*, *199*, 15.

[22] Kricheldorf, H.R.; Rabenstein, M.; Maskos, M.; Schmidt, M. *Macromolecules 2001*, 34(4), 713.

[23] Kricheldorf, H.R.; Böhme, S.; Schwarz, G. *Macromolecules*, *2001*, *34(26)*, 8879.

[24] Kricheldorf, H.R. *Macromolecules 2003*, *36(7)*, 2302.

[25] Djonlagic, J.; Sépulchre, M.-O.; Sépulchre, M.; Spassky, N.; Dunjic, B.; Jacovic, M.S. *Macromol.Chem. 1990*, *191*, 1529.

[26] Mertens, I.J.A.; Jenneskens, L.W.; Vlietstra, E.J.; Van der Kerk-van Hoof, A.C.; Zwikker, J.W.; Smeets, W.J.J; Spek, A.L. *J.Chem.Soc., Chem.Commun. 1995*, 1621.

[27] Hamilton, S.C.; Semlyen, J.A. *Polym. 1997*, *38(7)*, 1685.

[28] Miola, C.; Delolme, F.; Zanella-Cléon, I.; Dessalces, G.; Hamaide, T. *Polym.Bull. 1998*, *40*, 541.

[29] Mezoul, G.; Lalot, T.; Brigodiot, M.; Maréchal, E. *Macromol.Rapid Commun. 1995*, *16*, 613.

[30] Chevallier, P.; Soutif, J.-C.; Brosse, J.-C.; Brunelle, A. *Rapid Commun.Mass Spectrom. 2002, 16*, 1476.

[31] Wood, B.R.; Semlyen, J.A.; Hodge, P. *Polymer 1997, 38*, 2287.

[32] Ben-Haida, A.; Baxter, I.; Colquhoun, H.M.; Hodge, P.; Kohnke, F.H.; Williams, D.J. *Chem.Commun. 1997*, 1533.

[33] Monvisade, P.; Hodge, P.; Ruddick, C.L. *Chem.Commun. 1999*, 1987.

[34] Mouaziz, H.; Soutif, J.-C.; Montembault, V. *Eur.Polym.J. 2003, 39*, 1773.

[35] Fabbrizzi, L.; Kaden, T.A.; Perotti, A.; Seghi, B.; Siegfried, L. *Inorg.Chem. 1986, 25*, 321.

[36] Inoue, M.B.; Villegas, C.A.; Asano, K.; Nakamura, M.; Inoue, M.; Fernando, Q. *Inorg.Chem. 1992, 31*, 2480.

[37] Jurczak, J.; Stankiewicz, T.; Salanski, P.; Kasprzyk, S.; Lipkowski, P. *Tetrahedron 1993, 49(7)*, 1478.

[38] Sharghi, H.; Eshghi, H. *Tetrahedron 1995, 51(3)*, 913.

[39] Gryko, D.T.; Gryko, D.; Jurczak, J. *Synlett 1999, 8*, 1310.

[40] Pretsch, E.; Clerc, T.; Seibl, J.; Simon, W. *Spectral Data for Determination of Organic Compounds*; Springler Verlag: Berlin-Heidelberg, GE, *1983*.

[41] Belu, A.M.; Hunt, M.O.; De Simone, J.M.; Linton, R.W. *Macromolecules 1994, 27*, 1905.

[42] Räder, H.J.; Spickermann, J.; Müllen, K. *Macromol. Chem. Phys. 1995, 196*, 3967.

[43] Danis, P.O.; Karr, D. E. *Macromolecules 1995, 28*, 8548.

[44] Montaudo, G.; Montaudo, M. S.; Puglisi, C.; Samperi, F. *Macromolecules 1995, 28*, 4562.

[45] Whittal, R.; Li, L.; Lee, S.; Winnick, M. A. *Macromol. Rapid Commun. 1996, 17*, 59.

[46] Mouaziz, H.; Montembault, V.; Soutif, J.-C.; Brosse, J.-C. *Synth. Commun. 2002, 32(24)*, 3763.

In: Polymer Research Developments
Editor: Robert K. Bregg, pp. 75-90

ISBN 1-59454-742-4

Chapter 4

THE SYNTHESIS AND CHARACTERIZATION OF OLIGO-2-[(PYRIDINE-2-YL-METHYLENE) AMINO] PYRIDINE-3-OL

İsmet Kaya* and Süleyman Çulhaoğlu
Department of Chemistry, Faculty of Sciences and Arts,
Çanakkale Onsekiz Mart University, Çanakkale, Turkey

ABSTRACT

The oxidative polycondenzation reaction conditions and optimum parameters of 2-[(pyridine-2-yl-methylene) amino] pyridine-3-ol were determined by oxidants such as air oxygen, H_2O_2, and NaOCl in an aqueous alkaline solution between 40 and 90 °C. Synthesized 2-[(pyridine-2-yl-methylene) amino] pyridine-3-ol (2-PMAP) and oligo-2-[(pyridine-2-yl-methylene) amino] pyridine-3-ol (O-2-PMAP) were characterized by ^{1}H-NMR, ^{13}C-NMR, FT-IR, UV-Vis, size exclusion chromatography (SEC) and elemental analysis techniques. At the optimum reaction conditions, the yield of oligo-2-[(pyridine-2-yl-methylene) amino] pyridine-3-ol was found to be 47.00% (for air O_2 oxidant), 43.00% (for H_2O_2 oxidant) and 73.50% (for NaOCl oxidant). The number-average molecular weight (M_n), weight-average molecular weight (M_w) and polydispersity index (PDI) values of (O-2-PMAP) were found to be 9708 g mol^{-1}and 12235 g mol^{-1} and 1.260, using NaOCl, 3476 g mol^{-1} and 4126 g mol^{-1} and 1.187, using air O_2, and 18466 g mol^{-1}and 22318 g mol^{-1} and 1.209, using H_2O_2, respectively. DTA and TG analyses were shown to be stable of oligo-2-[(pyridine-2-yl-methylene) amino] pyridine-3-ol through to temperature and thermo-oxidative decomposition.

Keywords: oxidative polycondenzation, air oxygen, NaOCl, H_2O_2, oligo-2-[(pyridine-2-yl-methylene) amino] pyridine-3-ol.

---

* To whom all correspondence should be addressed. E mail: kayaismet@hotmail.com; Fax: +90 286 218 05 33

INTRODUCTION

The oligophenols and their Schiff base derivatives, which included –OH and –CH=N groups, have been used in various fields. They have useful properties such as paramagnetizm, semi conductivity, electrochemical cell and resisting to high energy. Because of these properties, they were used to prepare of composites with resistance at high temperature, thermostabilisations, graphite materials, and epoxy oligomer and block copolymers, adhesives, photo resists and antistatic materials. The halogen and sulphur derivatives of oligophenols were used to prepare the composite materials enduring to flame of the specific detergents such as lead storing battery cathodes [1-9]. These compounds were profited to new properties to their structures adding to other functional groups. More functions oligophenols may be used to cleaning of poisonous heavy metals in the industrial waste waters. Therefore, the synthesis of oligomer-metal complexes is very important at analytic and environmental chemistry. Coordination compounds of salicylaldehyde Schiff base have proven to be an excellent model system to study the enzymatic reactions of pyridoxal phosphate, because both of these systems effectively catalyze transamination and racemization reaction [10].

In this paper, we have investigated the effects of different parameters such as temperature, times and beginning concentration of 2-PMAP, NaOCl, H_2O_2 and alkaline for the 2-[(pyridine-2-yl-methylene) amino] pyridine-3-ol. 2-PMAP and O-2-PMAP were characterized by using FT-IR, UV-Vis, ^{1}H-NMR, ^{13}C-NMR, elemental analysis and TGA, DTA. Also, the molecular weight distribution of O-2-PMAP was determined by SEC technique.

MATERIALS AND METHODS

Materials

2-amino-3-hydroxy pyridine, 2-pyridinecarbaldehyde, 1, 4-dioxane, methanol, ethanol, benzene, toluene, acetonitrile, ethyl acetate, $CHCl_3$, CCl_4, acetone, n-hexane, n-heptane, tetrahydrofurane, THF, dimethylformamide, DMF, dimethylsulfoxide, DMSO, H_2SO_4 (98%), KOH, H_2O_2 (30% aqueous solution) and hydrochloric acid (HCl, 37%) were supplied from Merck Chemical Co. as chromatographic grade and they were used as received. Sodium hypo chloride (NaOCl), (30% aqueous solution) was supplied from Paksoy Chem. Co. (Turkey).

Preparation of 2-[(pyridine-2-yl-methylene) amino] pyridine-3-ol (2-PMAP)

2-[(pyridine-2-yl-methylene) amino] pyridine-3-ol was prepared by the condensation of 2-pyridinecarbaldehyde (0.56 g, 0.0052 mol) and 2-amino-3-hydroxypyridine (0.572 g, 0.0052 mol) in methanol 15 ml achieved by boiling the mixture under reflux for 3 h (Scheme 1). The precipitated 2-[(pyridine-2-yl-methylene) amino] pyridine-3-ol was filtered and recrystalizied from methanol and dried in vacuum desiccators (yield, 82 %).

Anal. calcd. for 2-PMAP: C, 66.33; H, 4.52; N, 21.11. Found: C, 65.50; H, 5.00; N, 20.25. UV-Vis (λ_{max}): 218, 233, 244, 261, 310 and 372 nm. FT-IR (KBr, cm^{-1}): ν (O-H) 3400 s, ν (C-H Phenyl) 3070 m, ν (C=N) 1616 s, ν (ring-C=N) 1596 s, ν (C=C phenyl) 1500, 1476 s, ν (C-O) 1302 s. ^{1}H-NMR (DMSO): δ ppm, 9.18 (s, 1H, -C<u>H</u>=N-); 10.00 (s, 1H, -OH), 6.59 (d, 1H, C2-H), 7.40 (m, 1H, C3-H), 8.05 (d, 1H, C4-H), 8.00(d, 1H, C8-H), 7.95 (m, 1H, C9-H), 7.59 (m, 1H, C10-H), 8.60 (d, 1H, C11-H). ^{13}C-NMR (DMSO): ppm, 1137.52 (C1-ipso), 119.05 (C2-H), 112.89 (C3-H), 128.93 (C4-H), 194.24 (C5-ipso), 149.81 (C6-H), 150.86 (C7-ipso), 125.04(C8-H), 127.33 (C9-H), 122.21 (C10-H), 139.75 (C11-H).

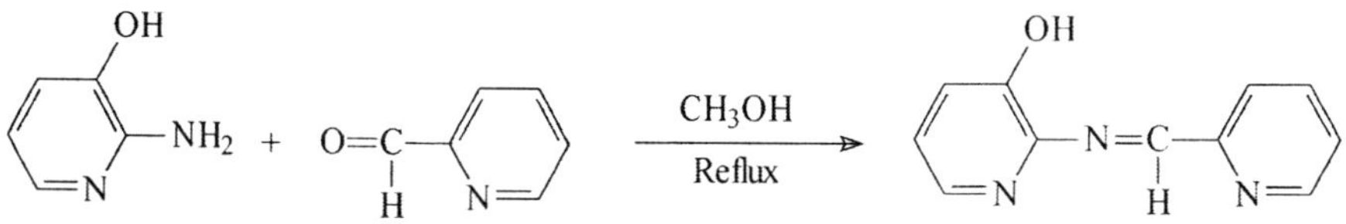

Scheme 1.

Synthesis of oligo-2-[(pyridine-2-yl-methylene) amino] pyridine-3-ol with Air Oxygen[11]

2-PMAP (0.199 g, 0.001 mol) was dissolved in an aqueous solution of KOH (10%) (0.056 g, 0.001 mol) and placed into a 100 ml three-necked round-bottomed flask. It was fitted with a condenser, thermometer, stirrer and addition to placed glass tubing over condenser for sending air oxygen. The reaction mixture was heated at the various temperatures and times (Table 1). The air was passed into rate 8.5 L/h during the course of the reaction. In order not to have water loss in the reaction mixture and to unneutralize to CO_2 of air oxygen to KOH, air oxygen passed into an aqueous solution of KOH (20%) before sending reaction tubing (Scheme 2). It was cooled to room temperature and then 0.087 ml HCl (37%) was added to the reaction mixture. The mixture was filtered and washed with hot water (3 x 25 ml) for separating from mineral salts and dried in the oven at 110°C.

Anal. calcd. for O-2-PMAP: C, 67.01; H, 3.55; N, 21.32. Found: C, 66.48; H, 4.40; N, 20.70. UV-Vis (λ_{max}): 221, 239, 262 and 322. FT-IR (KBr, cm^{-1}): ν (O-H) 3428 s, ν (C-H Phenyl) 3068 m, ν (C=N) 1618 s, ν (C=C phenyl) 1550, 1589 s, ν (C-O) 1280 s. ^{1}H-NMR (DMSO): δ ppm, 8.72 (s, 1H, -C<u>H</u>=N-); 10.60 (s, 1H, -OH), 8.05 (s, 1H, C3-H), 6.88 (d, 1H, C8-H), 6.40 (m, 1H, C9-H), 7.40 (m, 1H, C10-H), 8.68 (d, 1H, C11-H). ^{13}C-NMR (DMSO): ppm, 1137.52 (C1-ipso), 119.05 (C2-H), 112.89 (C3-H), 128.93 (C4-H), 194.24 (C5-ipso), 149.81 (C6-H), 150.86 (C7-ipso), 125.04(C8-H), 127.33 (C9-H), 122.21 (C10-H), 139.75 (C11-H).

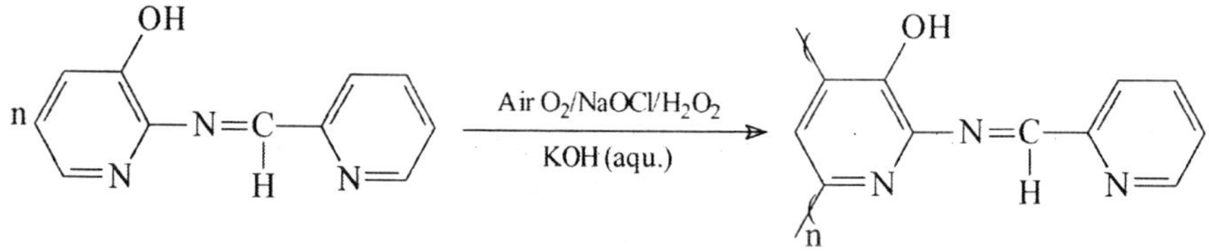

Scheme 2.

Synthesis of oligo-2-[(pyridine-2-yl-methylene) amino] pyridine-3-ol with NaOCl[11]

O-2-PMAP was synthesized through oxidative polycondensation of 2-PMAP with solution in water of NaOCl (30%). The 2-PMAP (0.199 g, 0.001mol) was dissolved in an aqueous solution of KOH (10%, 0.056 g, 0.001mol) and placed into a 100 ml three-necked round-bottomed flask. It was fitted with a condenser, thermometer, stirrer and an addition to funnel containing NaOCl. After heating at 40°C for 30 min., NaOCl was added drop wise over about 20 min. The reaction mixture was heated at the various temperatures and times (Table 2). The mixture was neutralized with 0.087 ml HCl (37%) at room temperature (Scheme 2) and then was filtered and washed with hot water (3 x 25 ml) for separating from mineral salts and dried in the oven at 110°C.

Synthesis of oligo-2-[(pyridine-2-yl-methylene) amino] pyridine-3-ol with H_2O_2[11]

O-2-PMAP was synthesized through oxidative polycondenzation of 2-PMAP with solution in water of H_2O_2 (30%). The 2-PMAP (0.199 g, 0.001 mol) was dissolved in an aqueous solution of KOH (10%, 0.056 g, 0.001 mol) and placed into a 100 ml three-necked round-bottomed flask. It was fitted with a condenser, thermometer, stirrer and an addition to funnel containing H_2O_2. After heating at 40 °C for 30 min., H_2O_2 was added drop wise over about 20 min. The reaction mixture was heated at the various temperatures and times (Table 3). The mixture was neutralized with 0.087 ml HCl (37%) at room temperature (Scheme 2). The mixture was filtered and washed with hot water (3 x 25 ml) for separating from mineral salts and dried in the oven at 110°C.

Characterization Techniques

The infrared and ultraviolet-visible spectra were measured by Perkin Elmer FT-IR BX II and UV-1208 Shimadzu, respectively. Elemental analysis was carried out with a Carlo Erba 1106. The FT-IR spectra were recorded using KBr disc (4000-350 cm^{-1}). UV-Vis spectra of 2-PMAP and O-2-PMAP were determined by using DMSO. 2-PMAP and O-2-PMAP were characterized by using ^{1}H-NMR and ^{13}C-NMR spectra (Bruker Avance DPX-400 and 100.6 MHz, respectively) recorded at 25°C by using deuteriated DMSO as solvent. TMS was used as internal standard. Thermal data were obtained by using Seteram Labsys TG-DTA 16 Thermal Analysis. TG-DTA measurements were made between 20-1000°C (in N_2, rate 10°C/min). The number-average molecular weight (M_n), weight-average molecular weight (M_w) and polydispersity index (PDI) were determined by size exclusion chromatography (SEC) techniques of Shimadzu Co. For SEC investigations were used a SGX (100 Å and 7 nm diameter loading material) 3.3 mm i.d. x 300 mm columns; eluent: DMF (0.4 ml/min), polystyrene standards. A refractive index detector (at 25°C) was used to analyze the product.

RESULTS AND DISCUSSION

The Investigation of Synthesis Conditions of O-2-PMAP

Because of electron-donor effect of phenylazomethine, oligo-2-[(pyridine-2-yl-methylene) amino]pyridine-3-ol have not oxidized at neutral medium by air O_2, H_2O_2 and NaOCl. But, 2-PMAP only oxidized by these oxidants, in an aqueous alkaline medium. In this reaction, medium was brown, because of phenylazomethine-phenoxy radicals. The oxidative polycondenzation of oligo-2-[(pyridine-2-yl-methylene) amino]pyridine-3-ol was formed in an aqueous alkaline medium between 40 and 90°C by effect of air O_2, H_2O_2 and NaOCl.

The conditions of oxidative polycondenzation reaction of 2-PMAP with air O_2 in aqueous alkaline medium are given in Table 1. The conversion of 2-PMAP was 47.0% at optimum conditions such as $[2\text{-PMAP}]_0=[KOH]_0=0.033$ mol/L, at 80°C for 5h. At the same conditions, when reaction temperature increased in 50°C to 80 °C, the conversion of 2-PMAP changed from 11.0% to 41.0%. At the same conditions, while reaction time 25 h, the yield of 2-PMAP decreased from 47.0% to 38.5%. These values demonstrated the product to negative effect of the reaction time. That is, as thought the oxidative polycondenzation reaction time increased, the yield decreased.

Table 1. The oxidative polycondenzation of 2-[(pyridine-2-yl-methylene) amino] pyridine-3-ol with air oxygen in an aqueous KOH

Sample No	$[2\text{-PMAP}]_0$ mol/L	$[KOH]_0$ mol/L	Air O_2 (L/h)	Temp. °C	Time (h)	Conversion of 2-PMAP %
1	0.033	0.033	8.5	50	10	11.0
2	0.033	0.033	8.5	60	10	37.5
3	0.033	0.033	8.5	70	10	39.0
4	0.033	0.033	8.5	80	10	41.0
5	0.033	0.033	8.5	90	10	40.5
6	0.066	0.066	8.5	50	10	33.5
7	0.066	0.066	8.5	60	10	19.5
8	0.066	0.066	8.5	70	10	16.5
9	0.066	0.066	8.5	80	10	15.5
10	0.066	0.066	8.5	90	10	4.8
11	0.033	0.033	8.5	80	5	47.0
12	0.033	0.033	8.5	80	15	40.5
13	0.033	0.033	8.5	80	25	38.5
14	0.066	0.066	8.5	50	5	36.5
15	0.066	0.066	8.5	50	15	19.5
16	0.066	0.066	8.5	50	25	18.0

The conditions of oxidative polycondenzation reaction of 2-PMAP with NaOCl in aqueous alkaline medium are given in Table 2. The conversion of 2-PMAP was 73.5% at optimum conditions such as $[2\text{-PMAP}]_0=[KOH]_0=0.033$ mol/L and $=[NaOCl]_0=0.084$ mol/L, at 90°C for 3h. At the same conditions, when molar amount of NaOCl increased in two coating dependence on 2-PMAP, total yield was 45.0% (see Table 2).

Table 2. The oxidative polycondenzation of 2-[(pyridine-2-yl-methylene) amino] pyridine-3-ol with NaOCl in an aqueous KOH

Sample No	$[2\text{-PMAP}]_0$ mol/L	$[KOH]_0$ mol/L	$[NaOCl]_0$ mol/L	Temp. °C	Time (h)	Conversion of 2-PMAP %
1	0.033	0.033	0.084	50	3	62.5
2	0.033	0.033	0.084	60	3	53.5
3	0.033	0.033	0.084	70	3	66.0
4	0.033	0.033	0.084	80	3	60.0
5	0.033	0.033	0.084	90	3	73.5
6	0.033	0.066	0.084	50	3	56.0
7	0.033	0.066	0.084	60	3	54.0
8	0.033	0.066	0.084	70	3	53.5
9	0.033	0.066	0.084	80	3	36.5
10	0.033	0.066	0.084	90	3	25.5
11	0.033	0.033	0.168	50	3	62.5
12	0.033	0.033	0.168	60	3	70.0
13	0.033	0.033	0.168	70	3	54.0
14	0.033	0.033	0.168	80	3	48.0
15	0.033	0.033	0.168	90	3	45.0
16	0.033	0.033	0.084	90	5	47.5
17	0.033	0.033	0.084	90	10	40.5
18	0.033	0.033	0.084	90	15	32.3
19	0.033	0.033	0.084	90	25	28.0
20	0.033	0.066	0.084	50	5	35.0
21	0.033	0.066	0.084	50	10	24.3
22	0.033	0.066	0.084	50	15	24.0
23	0.033	0.066	0.084	50	25	20.5
24	0.033	0.033	0.168	60	5	47.8
25	0.033	0.033	0.168	60	10	34.5
26	0.033	0.033	0.168	60	15	25.0
27	0.033	0.033	0.168	60	25	20.0

The conditions of oxidative polycondenzation reaction of 2-PMAP with 30% H_2O_2 solution in aqueous alkaline medium are given in Table 3. The conversion of 2-PMAP was 43.0% at optimum conditions such as $[2\text{-PMAP}]_0=[KOH]_0=0.033$ mol/L and $[H_2O_2]_0=0.326$ mol/L, at 50°C for 3h. The conversion of 2-PMAP was 30.0% at the reaction conditions such as $[2\text{-PMAP}]_0=[KOH]_0=0.033$ mol/L and $[H_2O_2]_0=0.326$ mol/L at 50°C for 15h. At the same conditions, when molar amount of H_2O_2 and reaction time increased, the yield of 2-PMAP decreased (see Table 3).

As is seen from Table 1, 2 and 3, in these reactions the yield of the 2-PMAP was dependent upon temperature, beginning concentrations of oxidants and time factors. At the same conditions, at the conversion of 2-PMAP, NaOCl was more active than air O_2 and H_2O_2. The reason of this may be depolymerization to monomer of oligomer at high concentration of KOH. According to temperatures and times, % conversions of 2-PMAP are given in Figures 1-6 for air O_2, NaOCl and H_2O_2 oxidants.

Table 3. The oxidative polycondenzation of 2-[(pyridine-2-yl-methylene) amino] pyridine-3-ol with H_2O_2 in an aqueous KOH

Sample No	$[2\text{-PMAP}]_0$ mol/L	$[KOH]_0$ mol/L	$[H_2O_2]_0$ mol/L	Temp. °C	Time (h)	Conversion of 2-PMAP %
1	0.033	0.033	0.326	50	3	43.0
2	0.033	0.033	0.326	60	3	42.0
3	0.033	0.033	0.326	70	3	36.5
4	0.033	0.033	0.326	80	3	9.0
5	0.033	0.033	0.326	90	3	1.5
6	0.033	0.066	0.326	50	3	12.5
7	0.033	0.066	0.326	60	3	6.5
8	0.033	0.066	0.326	70	3	3.0
9	0.033	0.066	0.326	80	3	1.5
10	0.033	0.066	0.326	90	3	0.9
11	0.033	0.033	0.652	50	3	30.5
12	0.033	0.033	0.652	60	3	27.5
13	0.033	0.033	0.652	70	3	26.0
14	0.033	0.033	0.652	80	3	10.5
15	0.033	0.033	0.652	90	3	3.5
16	0.033	0.033	0.326	50	5	39.5
17	0.033	0.033	0.326	50	10	32.3
18	0.033	0.033	0.326	50	15	30.0
19	0.033	0.033	0.326	50	25	10.5
20	0.033	0.033	0.652	50	5	25.0
21	0.033	0.033	0.652	50	10	20.5
22	0.033	0.033	0.652	50	15	18.0
23	0.033	0.033	0.652	50	25	10.2
24	0.033	0.033	0.326	40	3	16.2
25	0.033	0.033	0.652	40	3	21.3

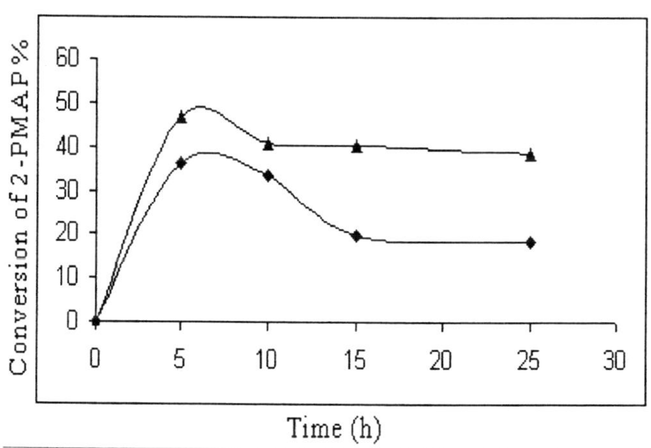

Figure 1. Change of yield % of 2-PMAP, according to temperature and times (air O_2= 8.5 L/h and ▲: 80 °C ♦: 50 °C)

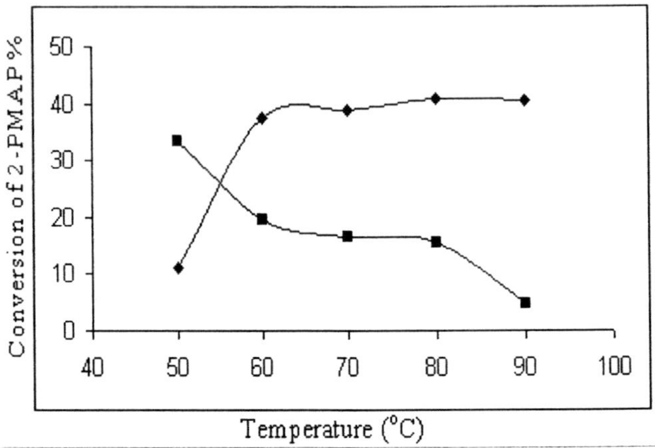

Figure 2. Change of yield % of 2-PMAP, according to temperature (air O_2: 8.5 L/h and ♦: $[KOH]_o$=0.033 mol/L ■: $[KOH]_0$=0.066 mol/L)

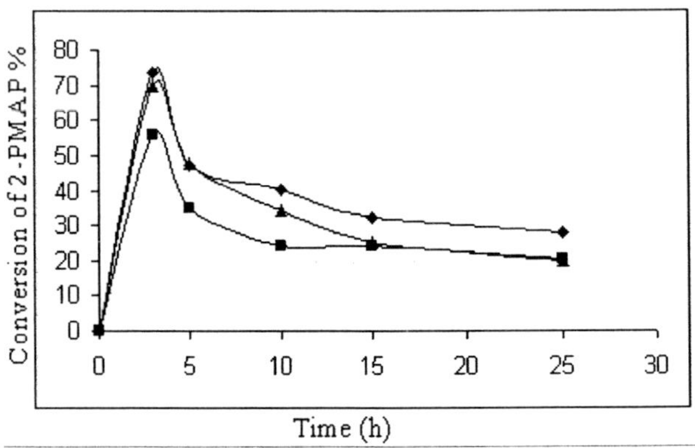

Figure 3. Change of yield % of 2-PMAP, according to times ($[NaOCl]_0$: 0.084 mol/L and ♦= 50 ▲ = 60 and ■ = 90 °C)

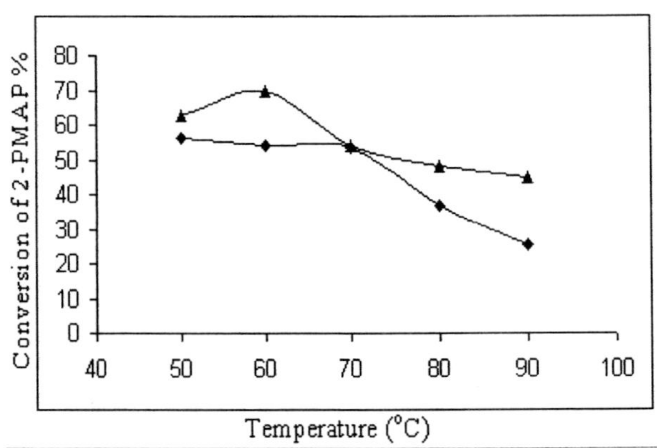

Figure 4. Change of yield % of 2-PMAP, according to temperatures (♦: $[NaOCl]_0$ = 0.084 mol/L and $[KOH]_o$=0.066mol/L ▲: $[NaOCl]_0$ = 0.168 mol/L and $[KOH]_o$=0.033 mol/L)

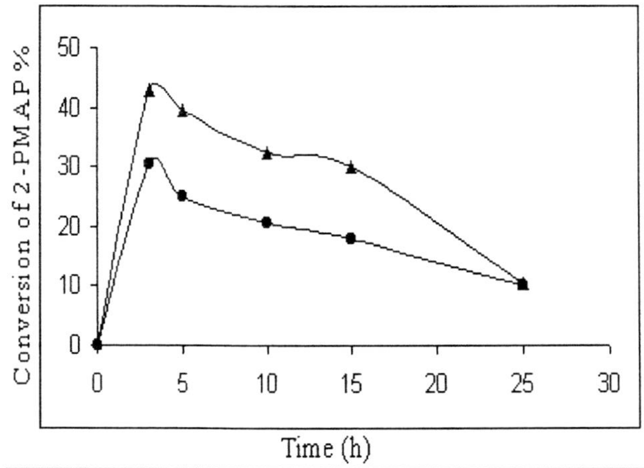

Figure 5. Change of yield % of 2-PMAP, according to times (•:$[H_2O_2]_0$= 0.326 mol/L and $[KOH]_0$=0.033 mol/L ▲:$[H_2O_2]_0$= 0.652 mol/L and $[KOH]_0$=0.033 mol/L)

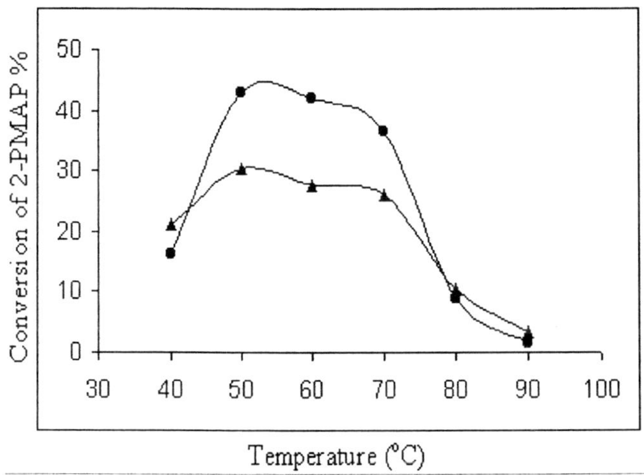

Figure 6. Change of yield % of 2-PMAP, according to temperature (•:$[H_2O_2]_0$= 0.326 mol/L and ($[KOH]_0$=0.033 mol/L ▲:$[H_2O_2]_0$= 0.652 mol/L and $[KOH]_0$=0.033 mol/L)

Solubility

O-2-PMAP was dark brown at powder forms and it was completely soluble in organic solvents such as DMF, THF, DMSO, acetone, aqueous alkaline and conc.H_2SO_4 but it was partly soluble in methanol and ethanol. O-2-PMAP was any insoluble in n-heptane, n-hexane, acetonitrile, benzene, toluene, ethyl acetate, $CHCl_3$, CCl_4 and 1, 4-dioxane.

Structure of O-2-PMAP

SEC analysis chromatograms of O-2-PMAP are given in Figure 7. When the air O_2 was used as oxidant, M_n, M_w and PDI values of O-2-PMAP were found to be 3476 g mol^{-1}, 4126 g mol^{-1}, and 1.187, respectively, (Figure 7 a). Two peaks were observed in the chromatogram of O-2-PMAP. 48% of the mass of O-2-PMAP was low molecular weight (M_n= 4044 g mol^{-1}, M_w= 4146 g mol^{-1}, PDI= 1.025) but 52% of its weight was high molecular weight (M_n= 9635 g mol^{-1}, M_w= 12041 g mol^{-1}, PDI= 1.250). According to oxidant H_2O_2, M_n, M_w and PDI values of O-2-PMAP was found to be 18466 g mol^{-1}, 22318 g mol^{-1} and 1.209, respectively (Figure 7 b). Two peaks were observed in the chromatogram of O-2-PMAP. 40% of the mass of O-2-PMAP was low molecular weight (M_n= 1485 g mol^{-1}, M_w= 3007 g mol^{-1}, PDI= 2.025) but 60% of its weight was high molecular weight (M_n= 60562 g mol^{-1}, M_w= 65036 g mol^{-1}, PDI= 1.074). According to NaOCl oxidant, M_n, M_w and PDI values of O-2-PMAP was found to be 9708 g mol^{-1}, 12235 g mol^{-1} and 1.260, respectively (Figure 7 c). Two peaks were observed in the chromatogram of O-2-PMAP. 42% of the mass of O-2-PMAP was low molecular weight (M_n= 2065 g mol^{-1}, M_w= 4408 g mol^{-1}, PDI= 2.135) but 58% of its weight was high molecular weight (M_n= 31980 g mol^{-1}, M_w= 32015 g mol^{-1}, PDI= 1.001).

The UV-Vis spectra of O-2-PMAP and 4-PMAP were similar to one together. However, at the spectrum of 2-PMAP, K bands of phenol and C_5H_4-N= were observed in 218 nm and 233 nm, respectively. Benzene band of 2-PMAP and strength R band of -CH=N- groups were observed in 261 and 310 nm, respectively. At the UV-Vis spectrum of O-2-PMAP, K bands of phenol and NC_5H_4-N= and benzene band and strength R band of -CH=N- groups were observed in 221 and 239, 262 and 322 nm, respectively.

At the FT-IR spectra (Figure 8) of the oxidative polycondenzation product of 2-[(pyridine-2-yl-methylene) amino] pyridine-3-ol are only different by reduction of band strength and numbers. At the FT-IR spectra of O-2-PMAP, bands of -OH and –CH=N groups were observed in 3400 and 1616 cm^{-1}, respectively. At the FT-IR spectra of O-2-PMAP, bands of -OH and –CH=N groups were observed in 3428 and 1618 cm^{-1}, respectively. In order to identify the structures of monomer and polymer, the ^{1}H-NMR spectra were recorded in DMSO-d_6. ^{1}H-NMR and ^{13}C-NMR spectra of the monomer are given in Figure 9 and 10, respectively. At the ^{1}H-NMR spectra of 2-PMAP and O-2-PMAP (Figure 11), the signals of –OH and –CH=N groups were observed in 10.00 and 9.18 ppm and 10.60 and 8.72 ppm, respectively. The FT-IR spectral data and the results of the ^{1}H-NMR spectra of the O-2-PMAP confirm together. According to ^{13}C-NMR, 2-PMAP has eleven signals [δ=1137.52 (C1-ipso), 119.05 (C2-H), 112.89 (C3-H), 128.93 (C4-H), 194.24 (C5-ipso), 149.81 (C6-H), 150.86 (C7-ipso), 125.04 (C8-H), 127.33 (C9-H), 122.21 (C10-H), 139.75 (C11-H)] which show that the structures in solution are symmetrical (Scheme 3).

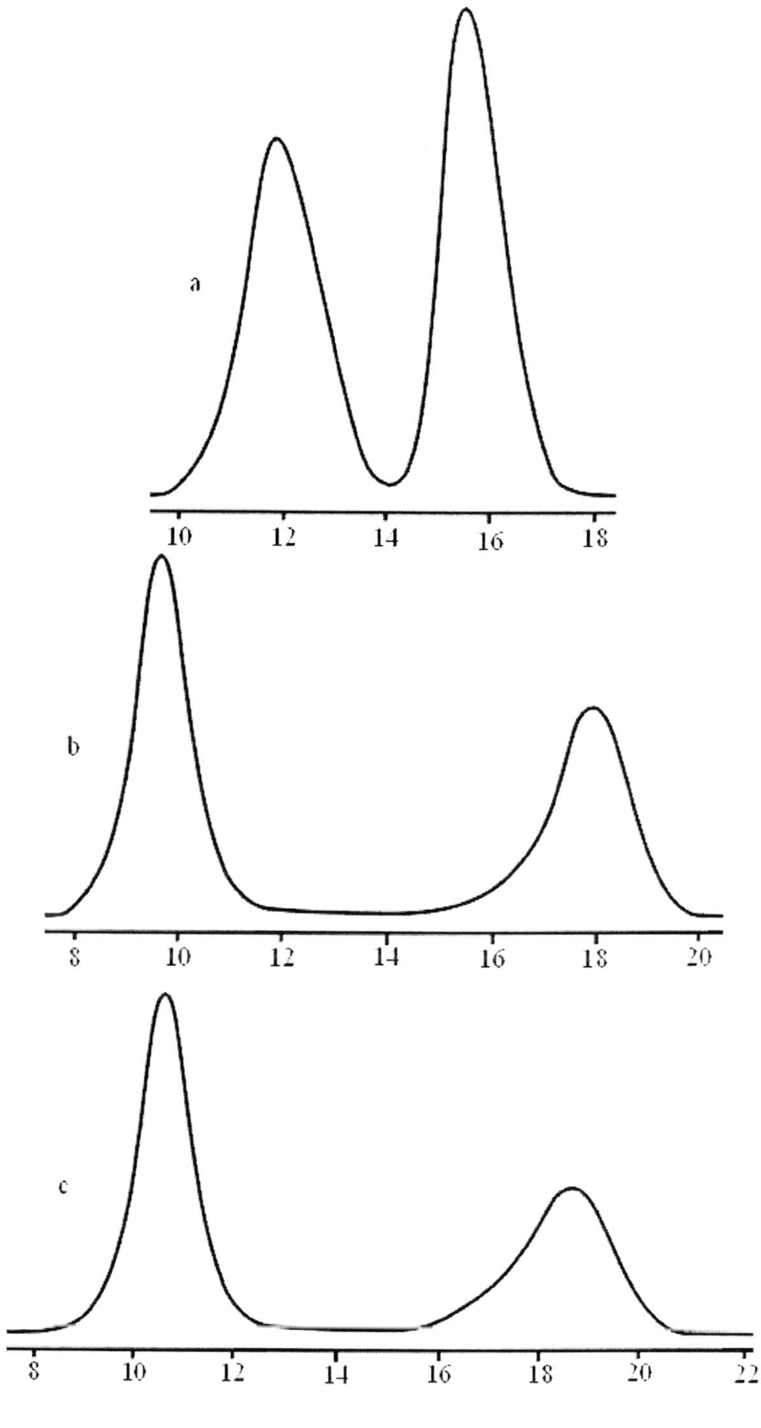

Figure 7. The chromatographic curves of the oxidative polycondenzation product of 4-PMAP a- (air O_2, as oxidant) b- (H_2O_2, as oxidant) c- (NaOCl, as oxidant)

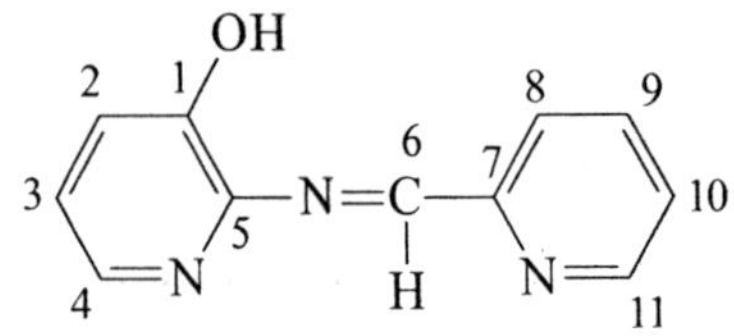

Scheme 3

According to spectral and SEC analyses, O-2-PMAP's formula is given as follows (Scheme 4):

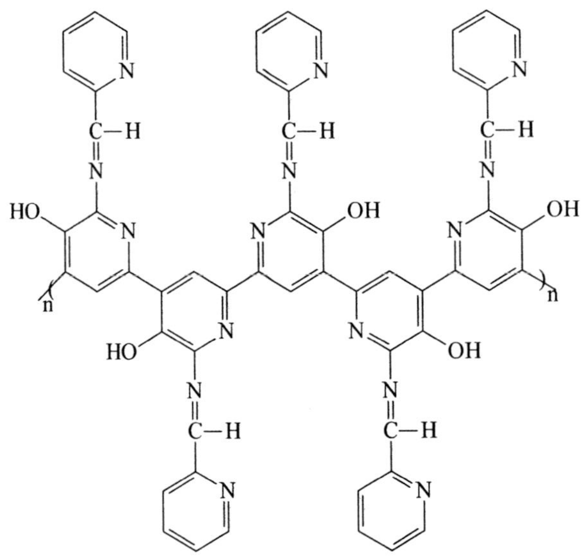

Scheme 4. The structures of 2-[(pyridine-2-yl-methylene) amino] pyridine-3-ol

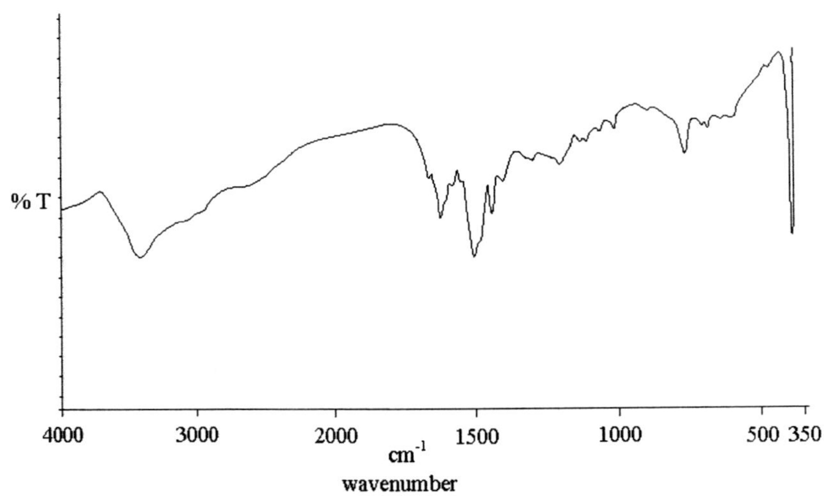

Figure 8. FT-IR spectrum of oligo-2-[(pyridine-2-yl-methylene) amino] pyridine-3-ol

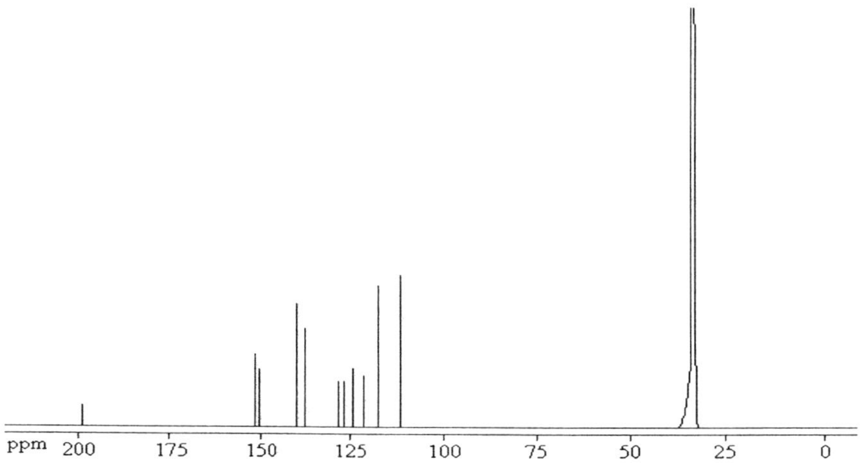

Figure 9. ^{13}C-NMR spectrum of 2-[(pyridine-2-yl-methylene) amino] pyridine-3-ol

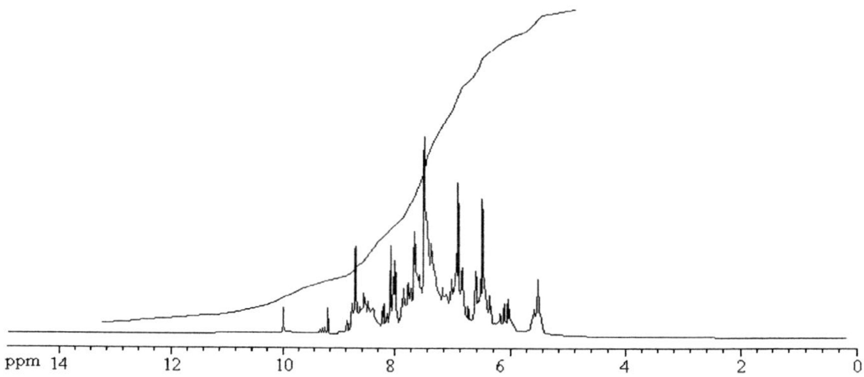

Figure 10. ^{1}H-NMR spectrum of 2-[(pyridine-2-yl-methylene) amino] pyridine-3-ol

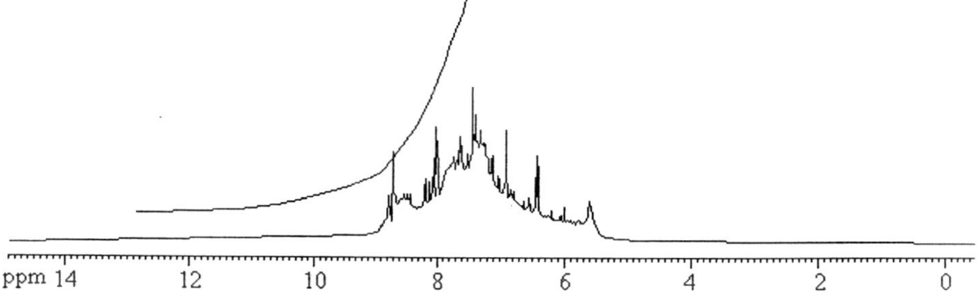

Figure 11. ^{1}H-NMR spectrum of oligo-2-[(pyridine-2-yl-methylene) amino] pyridine-3-ol

İ. Kaya at al. oligo-N-4-aminopyridine, oligo-2-[(pyridine-4-yl-imino) methyl] phenol[12], oligo-2-amino-3-hydroxypyridine and Schiff base oligomer[13] had been synthesized and characterized by UV-Vis, FT-IR, ^{1}H-^{13}C-NMR, elemental analysis, DTA-TG and size exclusion chromatography. According to SEC analysis, for air O_2 oxidant, M_n, M_w and PDI values of oligo-2-amino-3-hydroxypyridine and oligo-2-[benzilydeneimino] pyridine-3-ol had been 1433 g mol^{-1}, 1912 g mol^{-1}, 1.33 and 2637 g mol^{-1}, 5106 g mol^{-1} and 1.94 respectively[13].

Thermal Analysis of 2-PMAP and O-2-PMAP

The thermal degradations of 2-PMAP and O-2-PMAP were studied by DTA-TG analyses at the air medium Figure 12 and 13). 5% and 50% of O-2-PMAP's mass was lost at 170°C and 263°C, respectively. The total mass of it was lost between 300-350°C. Thermal degradation lasted to a high speed at 281°C (1.28 mg/min). O-2-PMAP demonstrated more resistance than 2-PMAP against to temperature at the same condition. 50% of O-2-PMAP's mass was lost at 541°C. Thermal degradation was lasted to a weak speed (0.90 mg/min). According to DTA, stretch exothermic process of O-2-PMAP started at 468 °C and T_{max} was at 575 °C.

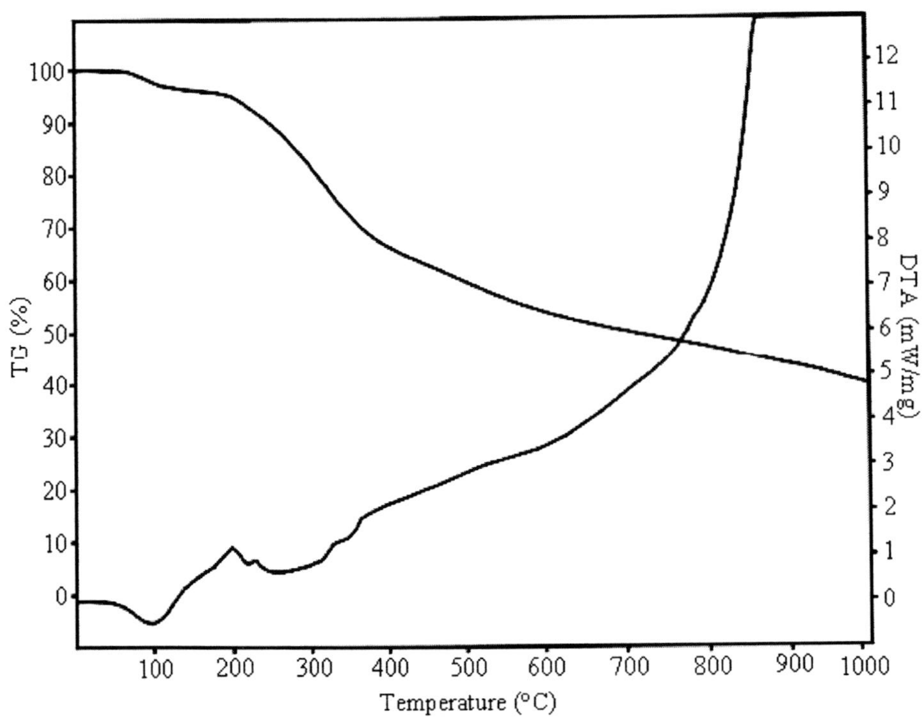

Figure 12. DTA-TGA curves of 2-[(pyridine-2-yl-methylene) amino] pyridine-3-ol

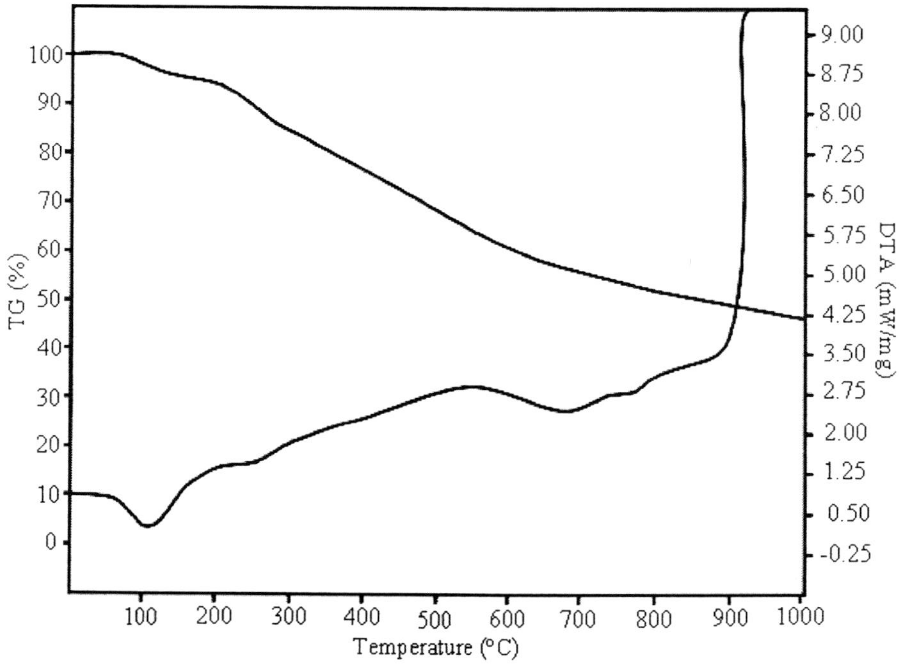

Figure 13. DTA-TGA curves of oligo-2-[(pyridine-2-yl-methylene) amino] pyridine-3-ol

CONCLUSION

According to spectral analyses, synthesized oligo-2-[(pyridine-2-yl-methylene) amino] pyridine-3-ol macromolecules were appointed to formation from units bonding with *o*-and *p*-carbons of azomethinephenol. At the optimum reaction conditions, the yield of oligo-2-[(pyridine-2-yl-methylene) amino] pyridine-3-ol (O-2-PMAP) was found to be 47.00% (for air O_2 oxidant), 43.00% (for H_2O_2 oxidant) and 73.50% (for NaOCl oxidant). The number-average molecular weight (M_n) and weight-average molecular weight (M_w) values of O-2-PMAP were found to be 9708 g $mol^{-1}$and 12235 g mol^{-1} and 1.260, using NaOCl, 3476 g mol^{-1} and 4126 g mol^{-1} and 1.187, using air O_2, and 18466 g $mol^{-1}$and 22318 g mol^{-1} and 1.209, using H_2O_2, respectively. According to TGA and DTA curves, polymer demonstrated high stability against thermo-oxidative degradation.

REFERENCES

[1] Suh, S.C. and Shim, S.C. 2000. Synthesis and properties of a novel polyazomethine, the polymer with high photoconductivity and second-order optical nonlinearity. *Synthetic Metals*, 114 (1): 91-95.

[2] Mamedov, B.A., Vidadi. Yu.A., Alieva, D.N. and Ragimov, A.V. 1997. Paramagnetism and electrical conductivity centre relation in semiconductor oligoresorcinol. *PolymerInternational*. 43 (2): 126-128.

[3] Feast, W. J., 1985. Synthesis and properties of some conjugated, potentially conductive, polymers. *Chemistry and Industry*. 15: 263-68.

[4] Ragimov, A.V., Mamedov, B.A. and Gasanova, S.G. 1997. New efficient dielectric and antistatic materials based on oligoaminophenol. *Polymer International*, 43 (4): 343-346.

[5] Baughman, R.H., Bredas, J.L., Chance, R.R., Elsenbaumer, R.L. and Shacklette, L.W. 1982. Structural basis for semi conducting and metallic polimer/dopant systems. *Chem.Rev*. 82:209-222.

[6] Kaya, İ., and Şenol, D., 2003. The synthesis and characterization of oligo-2-hydroxy- 1-naftaldehyde and some of its Schiff-base oligomers, *J Appl.Polym. Sci.*, 90 (2): 442-450.

[7] Kaya, İ., and Cihangiroğlu, N, 2004. Synthesis, characterization and anti-microbial activity of oligo-N-2-aminopyridinylsalicylaldimine and some oligomer-metal complexes, *J Polymer Research*, 11 (1) 37-42.

[8] Diaz, F.R., Moreno, J., Tagle, L.H., East, G.A., Radic, D., 1999. Synthesis, characterization and electrical properties of polyimines derived from selenophene, *Synthetic Metals*, 100 (2):187-94.

[9] Aly, K.I. and Khalaf, A.A., New polymer syntheses. 2000. IX. Synthesis and properties of new conducting polyazomethine polymers containing main chain cycloalkanone and pyridine moieties, *J Appl. Polym. Sci.*, 77 (6): 1218-229.

[10] Kaya, İ., Vilayetoğlu, A.R., and Mart, H. 2001. The Synthesis and Properties of Oligosalicylaldehyde and its Schiff Base Oligomers. *Polymer*. 42 (11): 4859-4865.

[11] Kaya, İ. and Cihangiroğlu N., 2004.Synthesis, characterization and anti-microbial activity of oligo-N-2-aminopyridinylsalicylaldimine and some oligomer-metal complexes, *J Polymer Research*, 11 (1) 37-42.

[12] Kaya, İ. and Koyuncu S., 2003.The synthesis, characterization of oligo-N-4-aminopyridine, oligo-2-[(pyridine-4-yl-imino) methyl] phenol and its some oligomer-metal complexes *Polymer*, 44 (24) 7299-7309.

[13] Kaya, İ. and Koça S., 2004. Synthesis, characterization and optimum reaction conditions of oligo-2-amino-3-hydroxypyridine and its Schiff base oligomer, *Polymer*, 45 (6) 1743-1753.

In: Polymer Research Developments
Editor: Robert K. Bregg, pp. 91-106
ISBN 1-59454-742-4

Chapter 5

THE PROTEIN-IMPRINTED POLYMER MATERIALS

Xingshou Pang and Guoxiang Cheng*
School of Materials Science and Engineering, Tianjin University, Tianjin 300072, China

ABSTRACT

Molecular imprinting is a rapidly developing technique for designing and preparing host compounds for molecular recognition. Although molecular imprinting technique has demonstrated great potential for specific recognition of small molecules, the reports on the imprinting of biological macromolecules such as proteins, whole cells and viruses have been scarce. However, there is still a strong incentive to prepare bio-imprinted polymers for the use of diagnostics, sensors, bioseparation, etc. In this paper we analyzed the principle and characteristic of protein molecular imprinting technique and reviewed the methods of preparing the protein-imprinted polymer that have been reported. The difficulties of protein molecular imprinting technique were also presented. Moreover, in our research team, we have done some work on how to prepare protein-imprinted polymer, especially how to prepare protein-imprinted soft-wet gel beads which is fit to adsorb and recognize template protein. So we introduced the preparation method, the recognition mechanism of protein imprinted soft-wet gel beads. In the end, the prospect of protein-imprinted technique was also given.

INTRODUCTION

Molecular imprinting is a method for preparing polymers of predetermined selectivity for separation and analysis of a vast variety of biologically active or harmful substances [1,2]. Molecular imprinting, shown schematically in Figure 1, is based on the preparation of a highly cross-linked polymer around a template in the presence of a suitable monomer. The template and monomers are first mixed in order to form a stable prepolymerization complex

---

* E-mail: gxcheng@tju.edu.cn

in a selected solvent. Subsequently, the polymerization is initiated in the presence of a suitable crosslinker. After polymerization, traditionally, bulk polymerization, the polymer is ground and sieved to an appropriate particle size, and the template is removed by some ways, leaving cavities complementary in shape, size and functionality. These cavities are able to selectively rebind, in given conditions, the template from a complex mixture. Chemically and mechanically stable molecularly imprinted polymers (MIPs) able to recognize specific substances may successfully serve as substitutes for antibodies, enzymes or other native biological structures in fundamental investigations of molecular recognition and may have numerous other applications in biotechnology, medicine, environment control, etc.

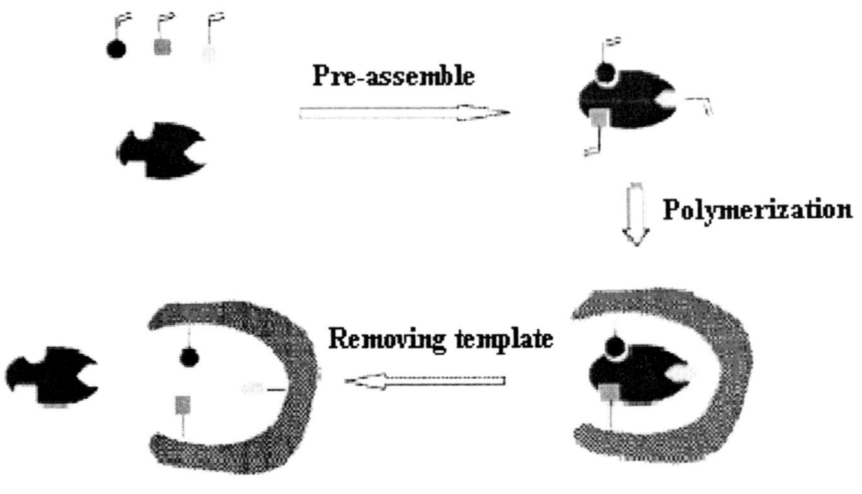

Figure 1 Schematic diagram of the principle of molecular imprinting

Molecular imprinting has demonstrated great potential for specific recognition of small molecules [3-7]. However, the reports on the imprinting of biological macromolecules such as proteins, cells and viruses have been scarce. As we all know, proteins are the components of life and have close relationship with the life activity. MIPs with protein as template can be used as substitutes of antibody, enzyme or natural bio-structure and have wide application prospects in biotechnology, life science, medicine, etc. Scientists have done many meaningful researches on the fields.

Characteristic of Protein Molecular Imprinting

Most proteins are bulky, fragile and have complex structure and chemical properties, so it is not convenient for imprinting protein compared small molecule imprinting. To date, at least two effects have been reported to hider the synthesis of MIPs selective to proteins: steric and thermodynamic [8]. The steric effect is based on the notion that bulky protein templates can not slip in and out of a polymer network. Thermodynamic considerations indicate that the use of non-rigid templates, such as proteins, yields less well-defined recognition sites in MIPs. Therefore, in order to succeed in imprinting protein, the following conditions must be fulfilled [9]: designing a proper polymerization reaction to keep the integrity of protein structure and conformation; after the polymerization and remove of protein, the imprinting nano-cavity should be kept; MIPs should have minimum non-specific binding.

Two different strategies were proposed for the design of MIPs, specific for high molecular mass compounds such as proteins. The first approach is based on the proper placement of few functional groups able to form strong interaction with the template proteins. Predominantly metal-chelating or strong electrostatic interactions were explored for template recognition. The second approach is based on MIPs ability to recognize a template by using a combination of shape complementary and multipoint weak interaction provided by the monomers able to form hydrophobic interactions or hydrogen bonds. These two strategy can be used as the foundation of designing protein-imprinted polymers.

Functional Monomer and Crosslinker for Protein Imprinting Technology

The polymerization of functional monomer and crosslinker can form the framework of MIPs. As to protein imprinting, the generally used functional monomers include 3-aminopropyltriethoxysilane, acrylamide and the derivative of acrylic acid. The crosslinkers include tetraethylorthosilicate and N,N'-methylene diacrylamide. Acrylamide is a kind of inert gel monomer and it is widely used in chromatogram and electrophoresis, which is useful to separate and purify biomaterials. The amido group of acrylamide can form strong hydrogen bond with the peptide bonds in the protein even in polar solvent, and it can not be ionized in aqueous solution. When proteins are used as template, acrylamide has great application potential. In addition, some natural polymer such as protein can also be used as matrix to imprint other molecules. Some natural amylose such as chitosan is used as matrix to imprint some protein and MIPs have specific recognition ability for template protein.

Molecular Imprinting Method of Protein

Protein Entrapment

The general used method is entrapment method, that is bulk polymerization method (see in Figure 2 a). Firstly, proteins pre-assemble with the functional monomers by covalent or non-covalent, after polymerization, proteins are buried in the block polymer. Then, these monoliths are ground and sieved to produce particles of the desired dimensions and after the remove of the exposed proteins, the protein imprinted polymer can be obtained. These protein-imprinted polymers have imprinted cavities that have complementary shape and chemical groups with the template.

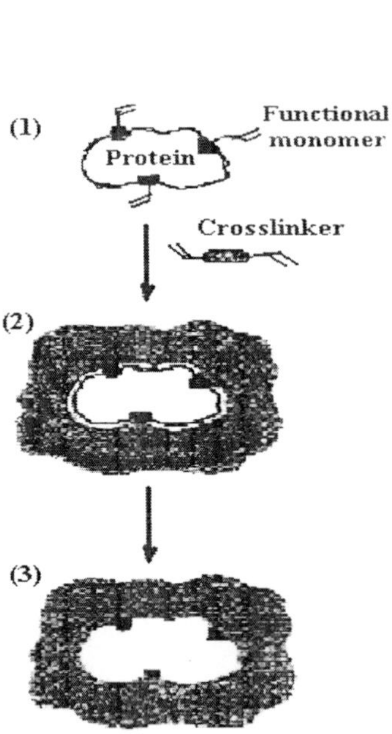

a. Protein entrapment

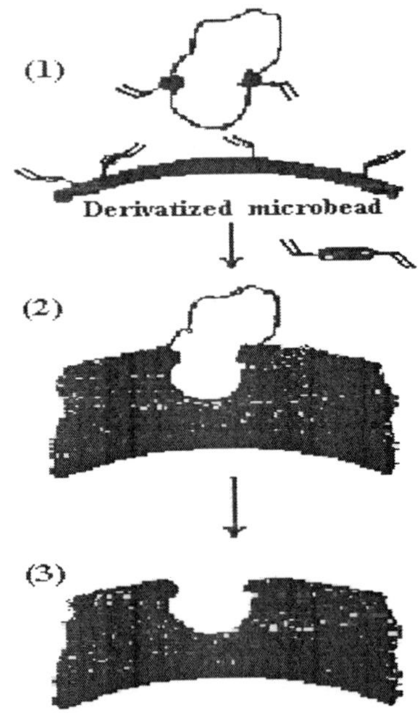

b. Microbead surface imprinting

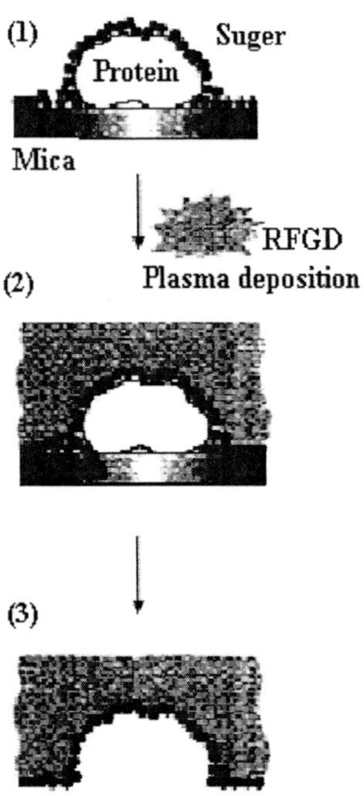

c. Flat surface imprinting

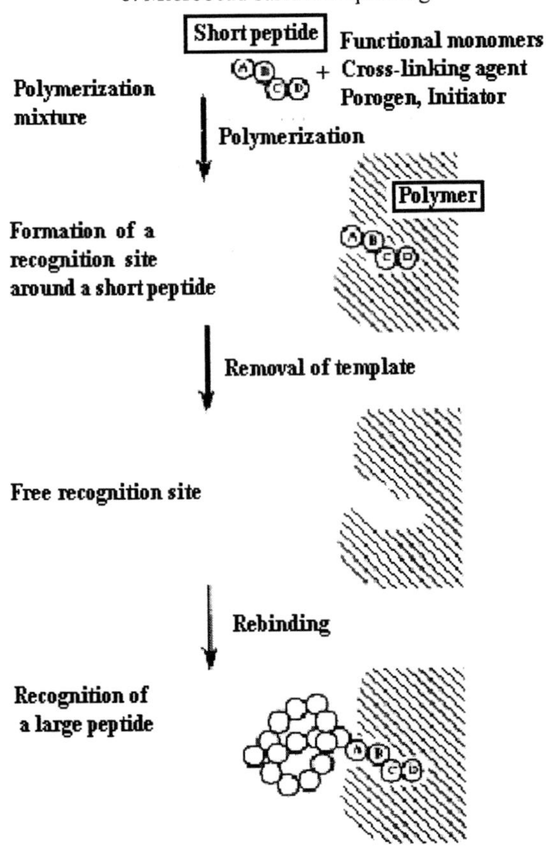

d. Epitope approach

Figure 2 Molecular imprinting method of protein

Venton[10] prepared polysiloxane polymers from 3-aminopropyltriethoxysilane and tetraethylorthosilicate (1:3) in the presence of two proteins: urease and BSA. The entrapped proteins were removed by pronase digestion and washing and the resulting polymers evaluated for their ability to again bind the two proteins. It was found that urease preferentially bound to the polymer made in the presence of urease, and BSA preferentially bound to the polymer made in the presence of BSA. The absolute preferential binding excess was greater (30%)for urease binding relative to that observed for BSA(3%). However, in both cases the same relative binding ratio of 1.5 or 50% excess was found. A similar study using the closely related heimoglobin and myoglobin proteins failed to show comparable excess binding in the presence of the predetermined protein. In the latter case, it was demonstrated that the rebound proteins did not equilibrate with labeled solution proteins, indicating a very tight association with polymer surface possibly masking any specificity that existed. However, it was possible to show that urease release of rebound hemoglobin from the polymer made in the presence of hemoglobin was less than for myoglobin bound to the same polymer and visa versa, again suggesting induced properties unique to the polymer prepared with the predetermined protein. To the extent that this notion of induced complementary order is correct, it may have applications in the development of protein specific adsorbents and in our understanding of polymer surface adhesion and the molding of template fine structure.

Another protein entrapment technology realized the specific recognition of proteins. Hjerten SJ[11] and Liao JL[12]prepared chromatographic gel by poymerizing acrylamide and N,N'-methylenebisacrylamide in the presence of a protein. The gel was pressed through a sieve and obtained gel particles. The gel particles were packed into a chromatographic tube to study the recognition properties. The results indicated that the gel particles could recognize selectively some particular protein. A gel with specific adsorption of three model proteins was also synthesized in order to demonstrate that the beds could be employed to remove (traces of) several proteins contaminating a sample. The degree of selective recognition is high, to judge from the fact that myoglobin from horse, but not that from whale, was adsorbed onto a column designed to bind specifically the former protein. |This selectivity is noteworthy since these two proteins have similar amino acid sequence and 3-D structures. The recognition mechanism is discussed in terms of a new classification system for chromatographic beds based on the number of bonds between the solute and the matrix and the strength of each bond and "non-specific bonds". The selectivity recognition is characterized by a large number of weak bonds formed between proteins and the complementary imprinted cavities.

Microbead Surface Imprinting

Bulk polymerization yields polymer monoliths. Some proteins are permanently buried in the monoliths, which is a waste for expensive proteins. Meanwhile, the monoliths have to be ground and sieved. This process is not only wasteful and time-consuming, but also produces irregularly shaped fragments, which are unfavorable for applications. Microbead surface imprinting is a good way to solve these problems(see in Figure 2 b). Firstly, proteins pre-assemble with the functional monomers on the surface of the derivative microbeads, then add cross-linker to form polymer layer with partly buried proteins. After removing the protein, the binding cavities can be gained. The prepared microbeads are fit to use as chromatography bed.

During the early attempt to imprinting protein by Glad M [13], the synthesis of boronate-silane was carried out. The compound was included n the monomer mixture used for the preparation of a polysiloxane-coated silica showing affinity for the glycoprotein transferring. Another way to imprinted monomer is based on the metal chelating interaction [14]. At the presence of metal ions and the enzyme Rnase A, a metal chelating monomer polymerized onto metharylate-derivatized silica particles and the protein imprinted polymer was treated with EDTA to remove the proteins. The imprinting principle is the enzyme Rnase A contains exposed histidine residues which have chelating effect with metal ions. The results show the Rnase A-imprinted polymer had higher affinity for Rnase A than BSA. However, this way can only adapt to imprint proteins with exposed hisidine residues [15]. Some researchers prepared MIPs with specific recognition sites for glucose oxidase via polymerization of methacrylic acid (MAA) and acrylamide (AM) on silicon dioxide particles with microbead surface imprinting technology [16].

Flat Surface Imprinting

This imprinting method is based on radio-frequency glow-discharge (RFGD) plasma deposition technology (see in Figure 2 c)[17]. Firstly, template proteins were adsorbed onto a hydrophilic, atomically flat mica. Then disaccharids were coated onto proteins and multiple hydrogen bonds would be formed between the hydroxyl groups on sugars and the surface polar residues of proteins during dehydration. By optimizing the conditon of RFGD plasma deposition, the coated sugars were incorporated into the plasma film while minimizing the damage to the underlying proteins. After peeling off the mica and solve off the proteins, protein-imprinted nanocavities were thus created that had a shape complementary to the protein and also had precisely positioned hydroxyl groups from the immobilized sugares.

The template-imprinted proteins included albumin, immunoglobulin, fibrinogen, lysozyme, ribonuclease A and glutamine synthetase. Transmission electron microscopy showed that nanometer-sized "pits" in the shape of the imprinted protein were formed on the surfaces of template-imprinted polymer films. Electron spectroscopy for chemical analysis and time-of-light secondary ion mass spectrometry indicated the saccharide covering of imprint surfaces and the removal of template proteins. ^{125}I-labeled protein adsorption from single solutions showed a similar amount of protein was adsorbed to its own imprint as to the imprint of another protein. However, more protein remained on the former surface than on the latter following elution with detergents Tween20 or sodium dodecyl sulfate. Competitive adsorption of a binary protein mixture showed a highly preferential adsorption of template protein to the corresponding imprint. This template recognition diminished as the number of protein-imprinted pits decreased. Structurally unstable proteins exhibited weaker template recognition that "robust" proteins. The hypothesis that protein recognition is due to the complementarity between the protein and its imprinted nanopit was supported by protein turnover experiments that showed template protein adsorbed to its own imprint was less readily displaced by a nontemplate protein.

Epitope Approach

AS we know, in nature, in recognizing an antigen, an antibody interacts only with a small part of it, the epitope (the antigenic site of macromolecule). It should be noted that epitopes may be described as surface domains composed of three to six amino acid residues. Hence, if a short peptide representing only the small exposed fragment of a protein structure is used as a template, then the resulting macroporous, MIPs will also be able to retain the whole protein molecule. The schematic representation of the eepitope approach is shown in Figure 2 d. with Tyr-Pre-Leu-Gly-NH_2 (YPLG) as template, A.Rachkov [8] synthesized MIPs that were able to recognize oxytocin, a larger pepide possessing the same structural fragment as YPLG. A HPLC study demonstrated the important role of ionic interaction and the N-terminal amino group of oxytocin and oxytocin-related peptides in the process of their recognition in the aqueous-rich media[18]. It was also shown that the selectivity of MIPs could be attenuated by water content, ionic strength and pH of the chromatographic mobile phase.

This method for the development of MIPs selective to proteins is also attractive from an economic viewpoint; a small peptide is usually less expensive, and the quantity necessary for polymer preparation is more readily available than that of the corresponding protein. But the approach has the disadvantage that the template protein must have the known exposed fragments.

The Elution Methods of Template Proteins

The elution effect can determine the cycle time and adsorption capacity, and on the other hand, the elution process should not destroy the imprinting cavities in the MIPs. To small molecules, the template can be recycled after they are removed from the MIPs. Some reports say the recycle rate can reach 99%. However, at present, the recycle of template protein is difficult to realize. Most proteins are very expensive, if the problem of recycle cannot be solve, it will limit the applications of protein imprinting technology.

Brow used buffer solution to wash MIPs in order to remove the templates and thought the proteins on the surface of MIPs were removed, but the adsorption capacity of MIPs was very low. The practical adsorption capacity was only 0.557μg GOD/100mg (dry MIPs). The template proteins are usually removed by EDTA and urea when using mental chelating interaction to imprint proteins.

Now people usually take some extreme methods to remove the template proteins, such as SDS+HAC. Through these methods, the MIPs can be re-used by many times, but the expensive proteins can not be recycled. In order to realize the application of protein imprinting technology, some mild elution approach must be developed.

Recognition Mechanism of Protein Imprinting Technology

At the early research on molecular imprinting, people have begun to explore the molecular recognition mechanism. In HPLC, the imprinting molecules have strong interaction with the medium, so people usually study the recognition mechanism by varying polymer, flow phase and the components of sample. But at far there are few studies on the recognition mechanism of macromolecule imprinting.

Generally, because small molecules have less interactive groups, the strong interaction between template and medium can lead to imprinting effect. In small molecule imprinting, people have successfully separated enantiomorph and other compounds with small molecular weight. In these reports, the generally used functional monomer can form strong ionic interaction, such as MAA or 4-vinyl pydine. As to macromolecule, the recognition interaction force should be a combination of shape complementary and multipoint weak interactions, such as hydrogen bond, hydrophobic interaction, etc.

Protein has many charges or nonpolar groups and these groups can form ionic or hydrophobic interaction, and due to protein has many binding sites, the interaction between MIPs and protein can be increased. Therefore, the monomers used for protein imprinting should be inerter than that for small molecule imprinting. To date, protein-imprinted polymers that have better selectivity were synthesized by inert monomer, such as acrylamide[11,12]. In contrast, when using MAA or MMA as functional monomer, the MIPs usually have bad selectivity [10]. These functional monomers are often charged or hydrophobic, accordingly, give rise to strong electrostatic interaction or hydrophobic interaciton, i.e. the MIPs behave to some extent as ion-exchangers or hydrophobic chromatogram. This would result in losing the specific recognition ability of MIPs.

The degree of selective recognition can be high, judging from the fact that MIPs can separate myoglobin from horse and myoglobin from whale. This also reflects that the arrangement of functional monomer plays an important role in recognition precedure. In PAM, the main recognition interaction force is the existence of a lot of hydrogen bonds.

According to some literatures, some template protein cannot be removed in MIPs, so we can consider that the binding force between the template and the imprinting cavities is very strong. On the other hand, because proteins have big volume, it is difficult for them to diffuse in the MIPs and reach the imprinting sites and the effective imprinting sites mainly lie on the surface or shadow part of MIPs. The template proteins existing in the deep part of MIPs can't be eluted. Therefore, the ideal MIPs should have enough interaction with template and also have proper pore distribution to assure effective imprinting sites and less diffusion resistance. Furthermore, during the process of preparation of MIPs, some template proteins can't be imprinted. So we speculate that the existence of proteins in MIPs has the following forms: effective imprinting, entrapment and physical adsorption. After elution, the parts of effective imprinting and physical adsorption are removed, so only a small part of template proteins have contribution for adsorption capacity.

THE WORK OF OUR RESEARCH TEAM ON THE PROTEIN-IMPRINTED POLYMER

Protein-imprinted Soft-wet Gel Beads

As mentioned above, so far most MIPs have been prepared by bulk polymerization, which yields polymer monoliths [19,20]. These monoliths have to be ground and sieved to produce particles of the desired dimensions fore further use. This process is not only wasteful and time-consuming, but also produces irregularly shaped fragments, which are unfavorable for applications, and therefore spherical MIPs are more desirable.

Polyacrylamide(PAM)gel are known to be very inert and biocompatible and mild cross-linked polyacrylamide gel has soft and macroporous structure, which permits large proteins to pass in and out. So in our research team, bovine serum albumin (BSA) molecularly imprinted soft-wet PAM gel beads were prepared via inverse suspension polymerization with acrylamide and N,N'-methylene diacrylamide as polymerization monomers and bovine serum albumin(BSA) as template, respectively. The adsorption property and selectivity of the BSA-imprinted PAM gel beads was studied and imprinting mechanism was also discussed.

Size and Morphology of the Gel Beads

The diameter of the soft-wet gel beads was ranged from 80μm to 280μm. From Figure 4a, it is shown at the top of gel beads the surface was smooth and pores could not be observed because the water evaporation led to the close of the pores. However, in our measurement, the side surface was coarse and we believe that the side surface was still kept wet and it could be close to the initial state. In the amplified ESEM photograph of the side surface (see in Figure 4b), a large quantity of well-distributed macropores could be observed and they were netlike. It is well knows that, in order to facilitate the transfer of macromolecule, such as protein, the gel beads should have macroporous structure. The soft-wet PAM gel beads prepared in our system had this characteristic and these macropores would increase the specific surface area and adsorption capacity of protein-imprinted gel beads and the mass transfer rate of releasing and rebinding template protein.

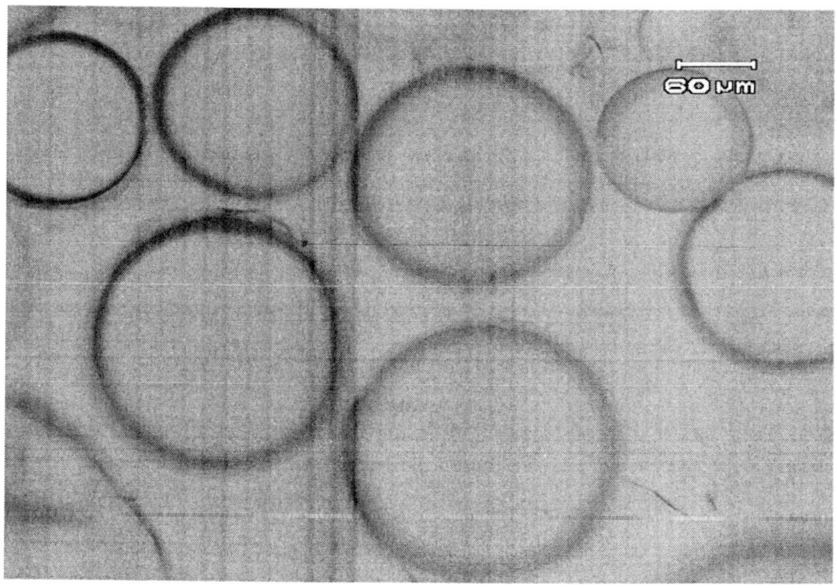

Figure 3 Optical micrograph of the soft-wet gel beads

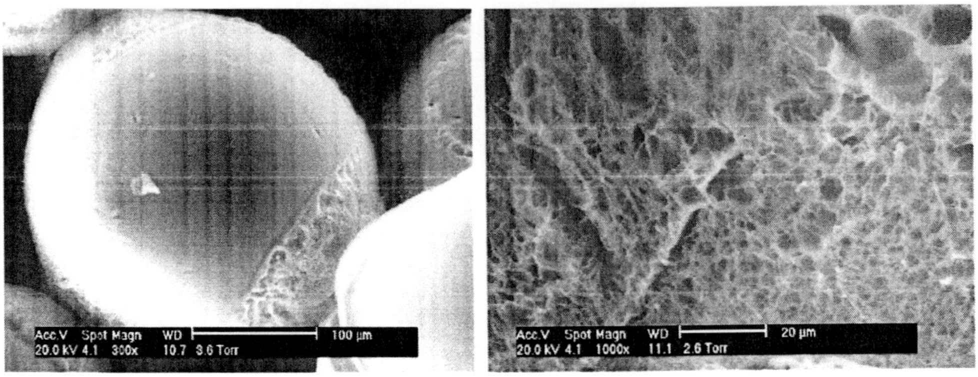

a. Single particle b. Surface magnification

Figure 4 ESEM photographs of soft-wet gel beads

On the other hand, we can see in Figure 5a that the dry gel beads could remain spherical and the surface was smooth. From the photograph of surface magnification it can indicate that pores on the surface were completely closed.

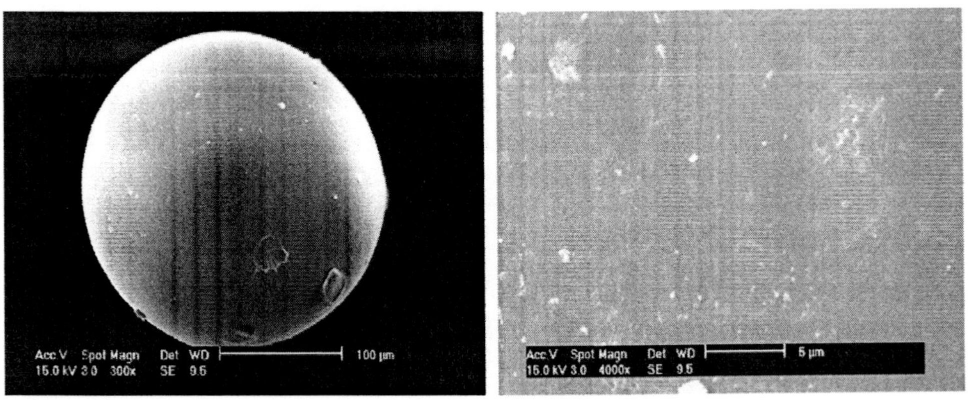

a. Single particle b. Surface magnification

Figure 5 SEM photographs of dry gel beads

Adsorption Kinetics

Figure 6 shows the adsorption kinetics of 20μmol/l BSA solution onto BSA-imprinted and non-imprinted soft-wet gel beads, respectively. We can find both kinds of gel beads had quick adsorption rate. In the early 50minutes, the adsorption capacity increased quickly, but after 300minuntes, the adsorption capacity almost reached equilibrium. At the early time, BSA molecules were easy to reach the surface cavities of the gel beads. With the saturation of the surface cavities, BSA begun to diffuse towards the deep cavities, but BSA has big molecular, so the diffusion of BSA met great resistance and led to the decrease of the adsorption rate.

At the same time, Figure 5 shows the final adsorption capacities of the two gel beads were very different. The BSA-imprinted beads had much higher adsorption capacity than the non-imprinted beads. We considered it caused by the template imprinting effect.

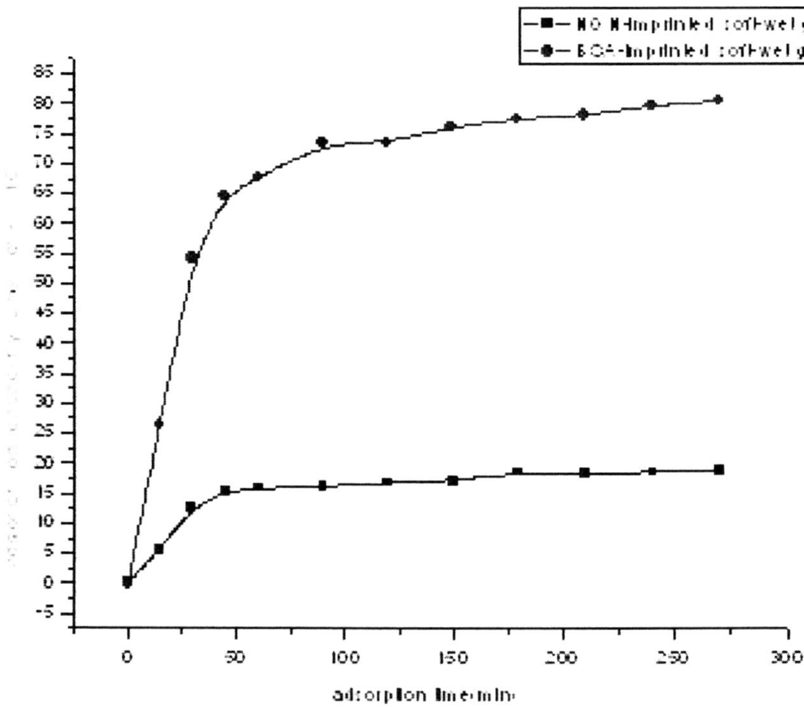

Figure 6 Adsorption kinetics of BSA-imprinted and non-imprinted soft-wet gel beads

Influencing Factors on the Static Adsorption Capacity of BSA-imprinted Gel Beads

Crosslinking Density of Gel Beads

It is indicated in Table1 that the Q decreased with the increase of crosslinking density. The decrease in adsorption capacity at higher crosslinking density is almost certainly due to the inaccessibility of some binding sites. By increasing the crosslinking density, smaller cavities in the gel beads were formed, so the proteins met greater diffusion resistance to reach the imprinted sites and led to the decrease of adsorption capacity.

Table1. Static adsorption capacity of BSA-imprinted gel beads under different crosslinking density

Degree of crosslinking(BisAM/AM, g/g)	0.5/9.5	1/9	1.5/8.5	2/8
SR	4.59	2.28	1.96	1.52
Q(μmol/g) $\times 10^3$	83.03	80.54	56.56	47.29

Static adsorption capacity $Q=(C_0-C_s)\times V/m$, where C_0 is the initial concentration of protein solution; C_s is the protein concentration of the supernatant; V is the volume of the initial solution; m is the mass of the wet gel beads. In our experiment, C_0 is 20μmol/l (BSA).

BSA Amount

From Table2, We can see that Q increased by adding more template protein of BSA. More BSA could produce more binding sites, which resulted in the increase of Q. On the other hand, we also find the increase of Q was not proportional to the BSA amount. In our experimental system, some template proteins were deeply buried in the inner of the gel beads and they were difficult to be removed, so it was wasteful for expensive protein and it was not an optical choose to increase the amount of template in order to gain large adsorption capacity.

Table 2.Static adsorption capacity of BSA-imprinted gel beads with different BSA amount as template

BSA (μmol)	0.5	1
Q(μmol/g)×10^3	80.54	95.02

Ionic Strength

It is demonstrated in Table3 Q decreased when the NaCl content in the BSA solution increased, especially at higher NaCl content. The possible reasons may be as follows. On one hand, the increase of NaCl made the proteins surrounded by more ions, which was a steric resistance for the protein to reach the binding sites; on the other hand, higher ionic strength caused the volume of the gel beads to extract, which made the pore size become smaller, then it was more difficult for BSA to diffuse into the deep parts of the gel beads.

Table 3. Effect of ionic strength on the static adsorption capacity of gel beads

NaCl concentration(mol/l)	0.05	0.1	0.15	0.2	0.25	0.5	1
Q(μmol/g) X 10^3	80.99	81.45	80.09	82.81	78.24	61.20	48.54

Selectivity

To investigate the selectivity of BSA-imprinted soft-wet gel beads, we use lysozyme, ovalbumin, immunoglobulin as control molecules. The results demonstrated the gel beads had higher adsorption capacity for BSA than for all the control proteins□which showed that the gel beads had preferentially selective adsorption for the template protein. We consider multiple-point hydrogen bond interaction between the cavities and template proteins were the main reason leading to imprinting effect.

Another research indicated that the selectivity increased with the increase of molecular weight of control proteins. This phenomenon showed that sterics was also important role in recognizing pr0cess. The larger of the protein, the more difficulty for them to enter the imprinting cavities.

Protein-imprinted Soft-wet Gel Beads with Magnetic Susceptibility

When magnetically susceptible components, such as, Fe, Co, or Ni or their oxides, are encapsulated inside the polymeric beads, the composite beads have magnetically susceptible characteristics. As such, the beads can easily and conveniently be separated from the system in which they are located by external magnetic fields, and hence they are called "DynaBeads".When MIPs encapsulate the magnetically susceptible components just mentioned, they should be easily separated by external magnetic fields after they finish their "active" adsorption and recognition.

In this part, bovine serum albumin (BSA) imprinted soft-gel composite beads with magnetic susceptibility (Figure 7) were prepared by inverse suspension polymerization, using Fe_3O_4 particles (about 0.5μm) as magnetically susceptible component, acrylamide (AM) and N,N'-methylenebisacrylamide (BisAM) as polymeric matrix components. Their magnetic responsibility (Figure 8) was measured by vibrating sample magnetometer. Their recognition specificity (see Table4) was investigated using lysozyme (Lyz) as control molecules.

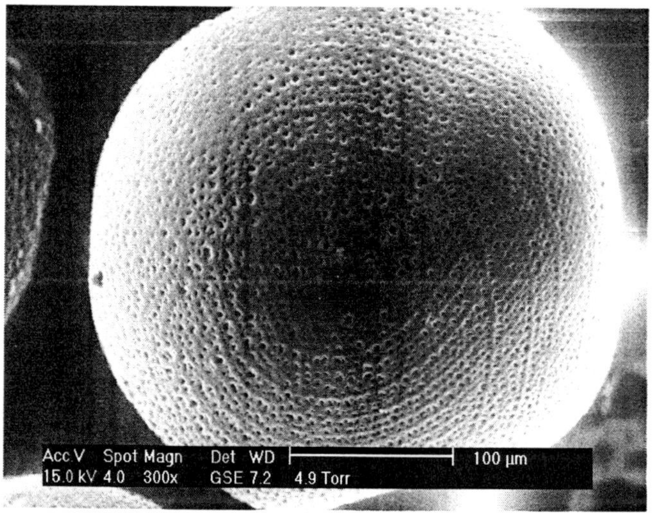

Figure 7 ESEM photographs of BSA-MS-PIGMs in wet condition

Table 4. The results of molecular recognition specificity of BSA-MS-PIGMs

Polymeric Gel microspheres	Analyte	Cs×10³ (μ mol/ml)	Cp×10³ (μ mol/g)	K_D	α	β
BSA-MS-PIGMs	BSA	13.87	30.65	2.21	4.42	4.75
	Lyz	18.17	9.15	0.50		
Non-MS-PIGMs	BSA	16.57	17.15	1.04	0.93	--
	Lyz	16.35	18.25	1.12	1.08	

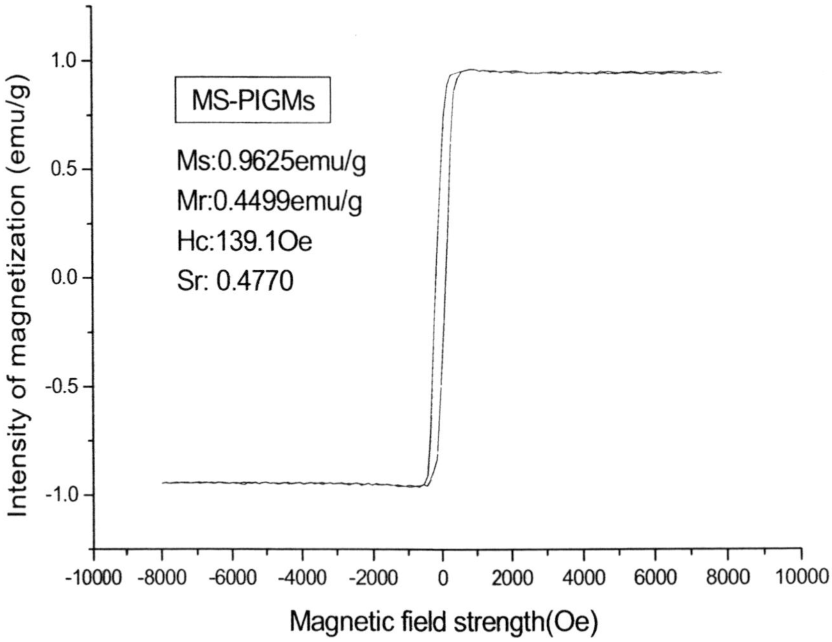

Figure 8 The magnetic hysteresis loop of BSA-MS-PIGMs

Recognition Mechanism

Through the above study, the authors speculated that interaction of proteins with AM and BisAM was synergic action of a large number of hydrogen bond resulted from peptide bond (amido bond) of proteins and amido bond of AM and BisAM. Hydrogen bond is a kind of short distance force, so synergic action could be generated only when PAM gel chains were very close to surface of proteins. When monomers polymerized around proteins, this kind of binding interaction achieved, and complementary protein–polymer surface formed. In other words, synergic action of a large quantity of hydrogen binding was caused by complementation of outer surface of proteins with inner surface of “imprinting cavities”. It was difficult for Non-template proteins to be adsorbed into the ‘inert’ PAM gel because there are too few binding interactions to form synergic action of a large quantity of hydrogen bond.

Problems and Prospect

In the past 2 decades, molecular imprinting technology has made great improvements, however, the macromolecule (including protein) imprinting still has many problems. The mechanism of the selective recognition is but one of many questions which importunately insist upon an answer. The interaction between protein and the functional group has not been systemically studied. The deep investigation in this field would promote the preparation of artificial receptor. Others are how to increase the protein capacity, to improve the reproducibility; to elute the protein adsorbed specifically in a small volume and in a gentle way; to modify the nature and the composition of the gel so that all types of particles (for instance viruses) and molecules, including all proteins, can be adsorbed specifically.

Although it is difficult and there are so many problems in protein imprinting, the imprinting technology of macromolecules has many potential applications in different fields, such as membrane separation and capillary chromatogram. The high specificity of MIPs can also be applied to biosensor field.

ACKNOWLEDGEMENTS

The authors would like to thank the National Natural Science Foundation of China (No. 50373032) and the Teaching and Research Award Program for Outstanding Young Teachers in Higher Education Institutions of MOE, P.R.C.(2002-123)for supporting this research work.

REFERENCES

[1] P.A.G. Cormack, K. Mosbach, Molecular imprinting: developments and the road ahead, *React. Funct. Polym.* 41, 115-124, 1999.

[2] G. Wulff, Molecular imprinting in cross-linked materials with the aid of molecular templates-a way towards artificial antibodies, *Angew. Chem. Int. Ed. Engl.* 31(47), 1812-1832, 1995.

[3] R.J. Ansell, K. Mosbach, Molecular imprinted polymers by suspension polymerization in perfluorocarbon liquids, with emphasis on the influence of the porogenic solvent, *J. Chromatogr. A.* 787, 55-66, 1997.

[4] W. Chen, F. Liu, Y. Xu, K. A. Li, Molecular recognition of procainamide-imprinted polymer, *Anal. Chim. Acta.*432, 277-282, 2001.

[5] S.H. Cheong, S. Mcniven, K. Uezu, M. Goto, S. Furusaki, Testosterone receptor binding mimic constructed using molecular imprinting, *Macromolecules* 30(5), 1317-1322, 1997.

[6] S. L. Lu, G. X. Cheng, X. S. Pang, Preparation of molecularly imprinted Fe3O4/P(St-DVB) composite beads with magnetic susceptibility and their characteristics of molecular recognition for amino *Acid, J. Appl. Polym. Sci.* 89, 3790-3796, 2003.

[7] L. Y. Zhang, G. X. Cheng, C. Fu, Synthesis and characteristics of tyrosine imprinted beads via suspension polymerization, *React. Funct. Polym.* 56, 167-173, 2003.

[8] Rachkov A, Minoura N, Towards molecularly imprinted polymers selective to peptides and proteins. The epitope approach, *Biochimica et Biophysica Acta*, 2001, 1544: 255~266

[9] Ratner B D, Shi H, Recognition templates for biomaterials with engineered bioreactivity, *Current Opinion in Solid State and Materials Science*, 1999, 4: 395~402

[10] Venton DL, Gudipati E, Influence of protein on polysiloxane polymer formation, evidence for induction of complementary protein-polymer interactions, *Biochimica et Biophysica Acta,* 1995, 1250: 126~136

[11] Hjerten S J, Liao J L, Nakazato K, et al, Gels mimicking antibodies in their selective recognition of protein, *Chromatography*, 1997, 44: 227~234

[12] Liao J L, Wang Y, Hjerten S, A novel support with artificially created recognition for the selective removal of proteins and for affinity chromatography, *Chromatography*, 1996, 42: 259~262

[13] Glad M, Narrlow O, Sellergren B, et al, Use of silane monomers for molecular imprinting and enzyme entrapment in polysiloxane-coated porous silica, *J Chromatogr*, 1985, 347:11~23

[14] Kempe M, Glad M, Mosbach K, An approach towards surface imprinting using the enzyme ribonuclease A biorecognition and affinity technology, *Journal of Molecular Recognition,* 1995, 8(1-2): 35~38

[15] Mallik S J, Plunkett S D, Dhal P K, et al, Towards materials for the specific recognition and separation of proteins, *New J Chem*, 1994, 18(3): 299~304

[16] Hirayama K, Burow M, Synthesis of polymer-coated silica particles with specfic recognition sites for glucose oxidase by the molecular imprinting technique, *Chem Lett*, 1998, 8: 731~732

[17] Shi H, Tsai W-B, Ferrari S, et al, Template imprinted nanostructural surfaces for protein recognition, *Nature*, 1999, 398: 593~597

[18] Rachkov A, Minoura N, Recognition of oxytocin and oxytocin-related peptides in aqueous media using a molecularly imprinted polymer synthesized the epitope approach, *Journal of Chromatography*, 2000, 889:111~118

[19] K. Sreenivasan, Effect of the type of monomers of molecularly imprinted polymers on the interaction with steroids, *J. Appl. Polym. Sci.* 68(11), 1863-1866, 1998.

[20] S. S. Milojkovic, K. Dusan, J.J. Comor, J.M. Nedeljkovic, Radiation induced synthesis of molecularly imprinted polymers, *Polymer* 38(11), 2853-2855, 1997.

In: Polymer Research Developments
Editor: Robert K. Bregg, pp. 107-126

ISBN 1-59454-742-4

Chapter 6

CONDUCTIVE POLYMER FILMS: PROPERTIES AND APPLICATIONS TO CHEMICAL SENSING

Luminita Grigore
Optical & Semiconductor Devices Group, Electrical & Electronic Engineering Department, Imperial College London, UK

ABSTRACT

A brief up-to-date review on recent developments regarding structural explanations of conduction mechanism in polymers, kinetic and optical properties, methods for preparation thin conductive polymer films and some of their applications referred to one of the main direction in this field, chemical sensing are presented.

1. INTRODUCTION

First syntheses of polymers were developed in the late 19-th century, but their extensively industrial use started in the 1940s-50s especially in the electronics industry for inactive packaging and insulating material. They were regarded as insulators because they had high electrical resistance, but their range of applications was widening because of their properties, low cost and ease of fabrication.

In 1958, Natta et al. synthesised a black powder of polyacetylene with conductivities similar to semiconductors (between $7x10^{-11}$ and $7x10^{-3}$ Sm^{-1} depending on the fabrication process). Then later, Shirakawa produced by mistake (adding more catalyst than normal) a silvery thin film of polyacetylene with very close value of conductivity as the black powder. Further investigations on polyacetylene black powder showed that its exposure to halogens (e.g. iodine) increased the conductivity value to 10^4 Sm^{-1} [1].

At the beginning of 1960s, research priorities in Germany focused on structural studies of polymers made by oxidative coupling in terms of thermoelectric power and photo- and dark conductivities. To their surprise, measured conductivities for polyphenylene and polythiophene were up to 0.1 $S.cm^{-1}$[2].

Later in the 1980s first polyheterocycles were developed that were more air stable, although their conductivities were not so high (about 10^3 Sm^{-1}). By adding various side groups to the polymer backbone (doping), derivatives soluble in different solvents were prepared, that also had increased values for the conductivities [3].

A logarithmic chart for conductivities of some of the polymers synthesised is shown in figure 1 [4].

All these polymers have a highly π-conjugated electronic state that make them susceptible to chemical/ electrochemical oxidation and reduction, which would alter their electrical and optical properties. Since most of these reactions are reversible and can be controlled, it is possible to develop conducting polymers with good stability and acceptable processing qualities. They are generally homogenous, strongly adherent to the substrate and chemically stable. An exact extent of doping is necessary for charge compensation during their oxidation; as their conductivity increases with the oxidation level, they may behave electrically as an insulator, a semiconductor or a conductor.

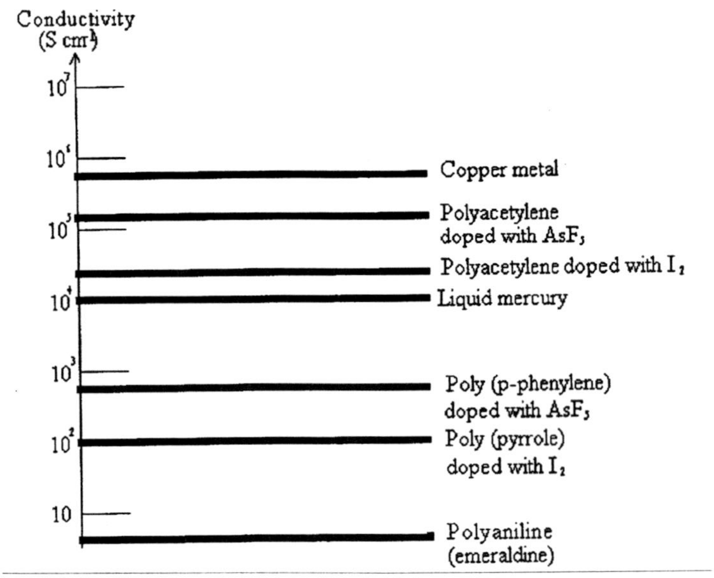

Fig. 1. Logarithmic conductivity ladder placing conductive polymers in comparison with some metals

Extrinsic stability of a conducting polymer is related to its vulnerability to external agents such as oxygen, water, peroxides. However, many of them degrade over time even in dry oxygen free environment. This intrinsic instability is probably caused either by irreversible reaction between charged sites of polymer and either the dopant ion or the p-system of an adjacent neutral chain that produces an sp^3 carbon breaking the conjugation or by a thermally driven mechanism that causes conformational changes in the polymer backbone and hence the loose of the dopant.

There are various techniques to fabricate conductive polymers, such are: cationic, anionic, radical chain growth, co-ordination polymerisation, step growth polymerisation and electrochemical polymerisation. There are two main possibilities to overcome the fact that most of the polymers are intractable, insoluble or difficult to melt: either to modify the polymer as to be more easily processed or to obtain the polymer in its desired shape and form.

Some of the characteristics of conducting polymers regarding conductivity, stability and processing are presented in table1.

Table 1. Stability and processing attributes of some conducting polymers

Polymer	Conductivity ($\Omega^{-1}cm^{-1}$)	Stability (doped state)	Processing possibilities
Polyacetylene	10^3 - 10^5	poor	limited
Polyphenylene	1000	poor	limited
PPS	100	poor	excellent
PPV	1000	poor	limited
Polypyrroles (PP)	100	good	good
Polythiophenes	100	good	excellent
Polyaniline (PAN)	10	good	good

There are two main groups of applications for these conducting polymers. The first one of them uses their conductivity as its main property, and the second, their electroactivity. All of them are shown in table 2.

Table 2. The two groups of applications of conducting polymers

Group 1	Group 2
Electrostatic materials	Molecular electronics
Conductive adhesives	Electrical display
Electromagnetic shielding	Chemical/ Biochemical/ Thermal sensors
Printed circuit boards	Rechargeable batteries and solid electrolytes
Artificial nerves	Drug release systems
Antistatic clothing	Optical computers
Piezoceramics	Ion exchange membranes
Active electronics (diodes, transistors)	Electromechanical actuators
Aircraft structures	"Smart" structures
	Switches

2. STRUCTURE AND PROPERTIES OF CONDUCTIVE POLYMER

2.1 Conduction Mechanism in Polymers

The first attempt to explain the conduction in polymers used the band theory: ideally from a continuous delocalised π-system, a half filled valence bond would be formed. However, further studies showed that the polymers can efficiently lower their energy by bond alteration (alternating short and long bonds) which introduces a band width of 1.5 eV, very much like a high energy gap semiconductor. Moreover, electron spin spectroscopy of doped polymers pointed out that the free spins concentration in polymers is limited to a maximum level no matter of further increase in dopant concentration.

To understand this, it is necessary to examine the two ways in which charge is stored along the polymer chain and its effect. In either an oxidation or a reductive process, the polymer could (1) lose an electron from one of the bands or (2) it could localise the charge over a small section of chain. Localising the charge results in a local distortion due to change in geometry, which decreases the ionisation energy of the chain and increases its electron affinity making it more suitable to accommodate newly formed charges. This method increases polymer energy less than it would if the charge was delocalised and also is consistent with the increase in disorder detected after doping by Raman spectroscopy.

Typical oxidising dopants include: iodine, arsenic, pentachloride, iron (III) chloride and $NOPF_6$ and one of the reductive dopants is sodium naphthalide. Their main ability is either to oxidise/ reduce the polymer without lowering its stability or to initialise side reactions that inhibit polymer capability to conduct electricity (e.g. bromine or $NOPF_6$ are very powerful dopants which add across the double bonds to form sp^3 carbons).

Recent theoretical studies of conjugated polymers structure and properties showed that electronic states involved in the conduction process appear in the gap during doping [5, 6]. Oxidation of polymer chains (p-type doping) leads first to formation of radical sites corresponding to polaron states in the gap [7, 8]: the lowest polaron level (bonding cation level-BCL) is positively charged (spin S=1/2), the upper polaron level (antibonding cation level-ACL) corresponds to a neutral empty state. For higher doping concentrations the two bonding and antibonding levels combined with the new diamagnetic dications formed in the gap band (BDL/ADL) can enlarge to two bands, thus creating the bipolaron state.

In the case of polypyrrole (PP), the oxidative doping (illustrated by a sequence of quinoid-like rings in figure 2) proceeds with the removal of an electron from the backbone p-system that produces a free radical and a spinless positive charge that are coupled to each other via local resonance. The creation and separation of these defects costs an important energy and produces distortion of the lattice that limits the number of quinoid-like rings that might link these two species together (for PP it probably extends over four pyrrole rings). The combination of a charge site and a radical is called polaron, which could be a radical cation or a radical anion. This creates a new localised electronic state in the gap, having the lower energy state occupied by single unpaired electrons. The polaron state of PP is symmetrically located about 0.5 eV from the band edges.

Further oxidation will remove the free radical of the polaron and creates a new spinless defect called bipolaron of lower energy than the creation of two distinct polarons. This might be possible at higher doping levels as polarons are replaced by bipolarons. For PP the bipolarons are located symmetrically with a band gap of 0.75 eV. With continued doping, new bipolarons form and their band gap increases at the expense of the band edges, so as for heavily doped polymers the upper and the lower bipolaron bands will merge with the conduction and the valence bands respectively to produce partially filled bands and metallic like conductivity (as shown in figure 2).

Speaking now about conjugated polymers with degenerate ground state, the mechanism is slightly different as the ground state is twofold degenerated (see figure 3). So, the charged cations are not bound to each other by high energy and they can freely separate along the chain. In this way these defects, called solitons, are independent of one another and domain walls might be formed to separate two phases of opposite orientation and identical energy.

For example in polyacetylene (PA), the solitons are probably delocalised over 12 CH units with maximum charge density next to the dopant ion. Soliton formation creates a new localised electronic state that appears in the middle of the energy gap. At increased doping levels, the charged solitons interact with each other to form a soliton band that may merge with the band edges to create metallic like conductivity (as in figure 3).

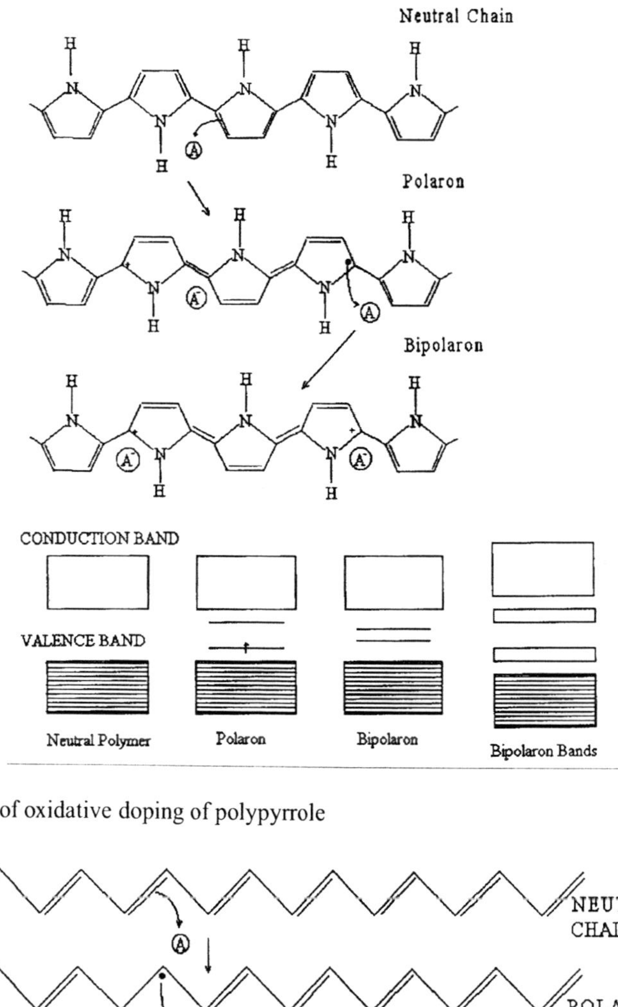

Fig.2. Mechanism of oxidative doping of polypyrrole

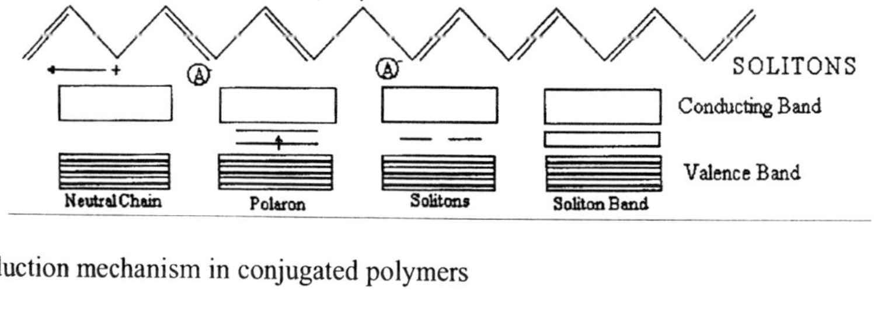

Fig. 3. Conduction mechanism in conjugated polymers

In the case of polyaniline (PAN), considering a model with electroactive sites over two aniline rings [9], because of doping there will be at least two other aniline rings involved in the delocalisation charges and spins. The ideal structure would be a long linear chain arising from head to tail coupling of aniline cation radicals. The complex scheme of redox reaction of PAN (figure 4) starts with the formation of a cation radical (polaron); further oxidation at almost the same potential produces the dication (dipolaron).

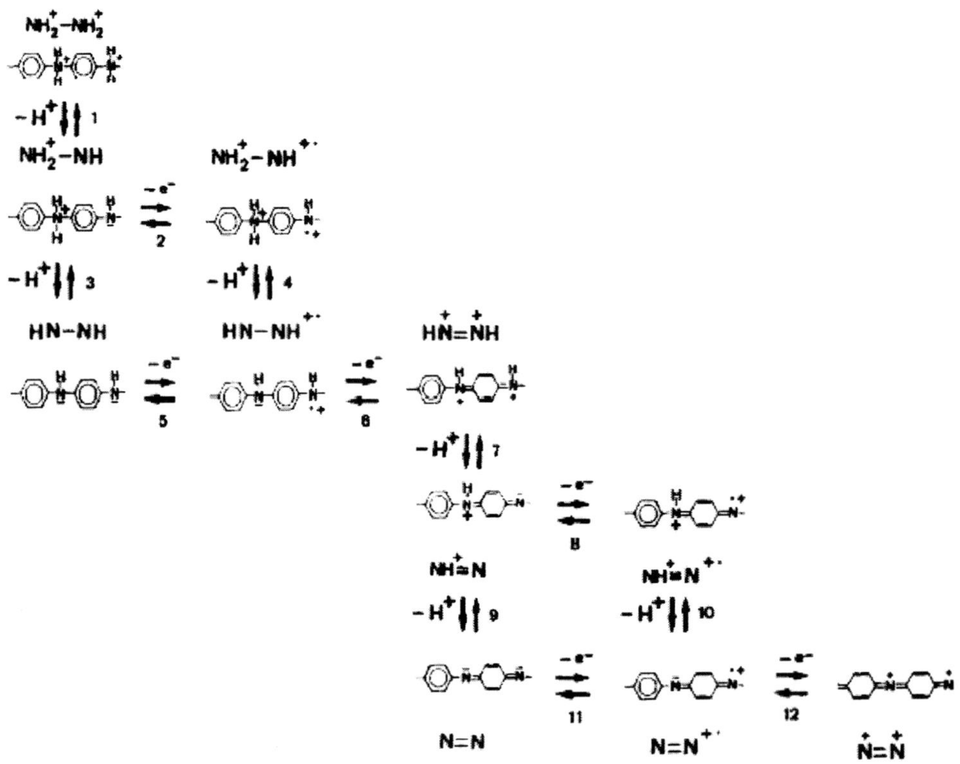

Fig. 4. Redox mechanism of polyaniline

This second oxidation takes place following a mechanism similar to classical anion insertion from PP. Studies in changes of geometry and distribution of doped PAN using the Modified Neglect of Differential Overlap (MNDO) method found that polaron is localised to one six-membered carbon ring [10]; the ring has a character between a phenyl and a quinoid one. Also, doping PAN form polyemeraldine base, an increase in the conjugation of the π-system in the polymer chain takes place that explains the high conductivity in the resulted polymer salt.

Cyclic voltammetry of PAN (figure 5) indicates that the first oxidation peak is attributed to a combination of reactions 1+2+3+4+5+6, the middle peak stands for different structures of PAN resulting from coupling processes of cation radicals of aniline, including probably ortho coupling that will take into account reactions 7+8+9+10+11+12. The third oxidation peak is related to reactions 7+8+9+10+11+12 on the linear polymer coupled in the para position.

The reductive component of the third redox system appears more reversible, because of proton transfers rate and because of the low energy needed for this structural modification.

The reduction of the middle system indicates the presence of several secondary structures of the polymer, which made the reduction of the first system look like a shoulder.

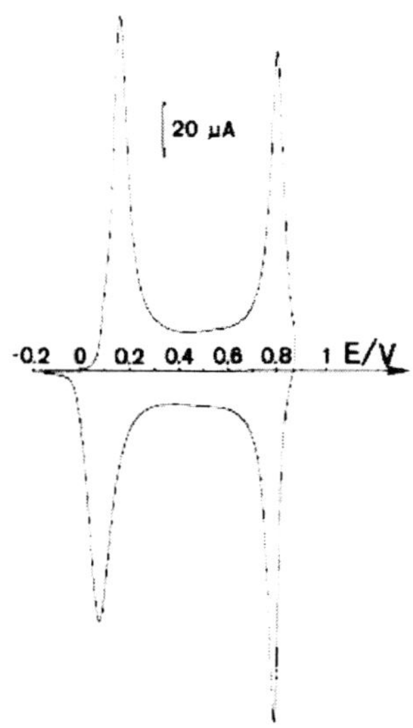

Fig. 5. Cyclic voltammogram of polyaniline film (0.2 μm) in NH4F.2.3 HF at 10 mVs-1prepared by oxidation of aniline 0.5M at 0.7 V(Cu/CuF2) using transparent gold electrode on polyester

2.1.1 Charge Transport

Solitons and bipolarons are possibly the main source of charge carriers in the conduction mechanism, but because polymers are highly disordered (a mixture of crystalline and amorphous phases; for PP it was shown that films are poorly crystalline, arranged in sheets made of planar chains without structural order [5]) it is important to consider both the transport along and between the chains and also at the boundaries of the complex phases. Measurements of doping, temperature, magnetism and current frequency effects in polymers demonstrated that conduction might take place by: (a) movement of charge carriers between localised sites, or between soliton, polaron and bipolaron states; (b) movement of charge carriers between highly conductive domains (created by thermally activated tunnelling or inhomogeneous doping, respectively).

2.2 Kinetics and Optical Properties of Polymers

For the particular case of PP, the redox mechanism is described by the following simplified scheme [5]:

$PP \Leftrightarrow PP^{\cdot +} + 1\,e^-$	E_1	(I): first electron transfer step
$PP^{\cdot +} \Leftrightarrow PP^{2+} + 1\,e^-$	E_2	(II): second electron transfer step
$2\,PP^{\cdot +} \Leftrightarrow PP + PP^{2+}$	K	(III): equilibrium reaction

where: PP = neutral species, associated with six pyrrole units; $PP^{\cdot +}$ = radical cation (polaron), one positive charge located upon six pyrrole cycles; PP^{2+} = dication species (bipolaron), defect of two electrons per 6 cycles.

Assuming that reactions I and II characterised by apparent standard potentials E_1 and E_2 are only restricted by ionic diffusion, and reaction III is characterised by the equilibrium constant: $K = k_1/k_2 = [PP].[PP^{2+}]/[PP^{\cdot +}]^2$ with $\log K = (E_2-E_1)\ nF/RT$ and achieving cycling voltammograms between – 0.5 and + 0.4 V (figure 6), it was pointed out that the second peak is assigned to the oxidation of residual monomer catalysed by the film, the relative broadness of the first system of peaks being attributed to ion-diffusion limitation.

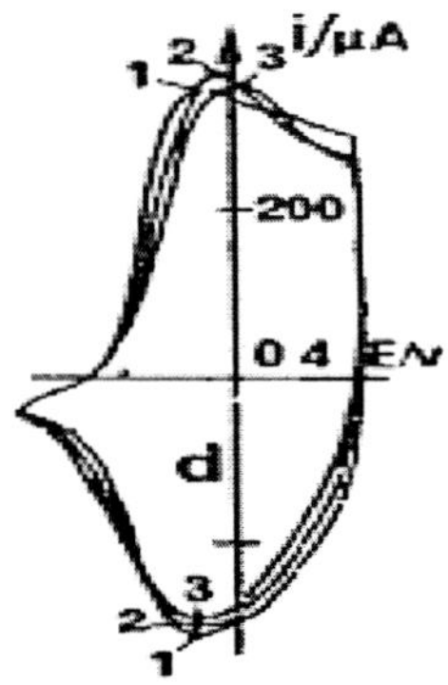

Fig. 6. Cyclic voltammetry of a polypyrrole film (0.24 μm) in CH3CN/ LiClO4

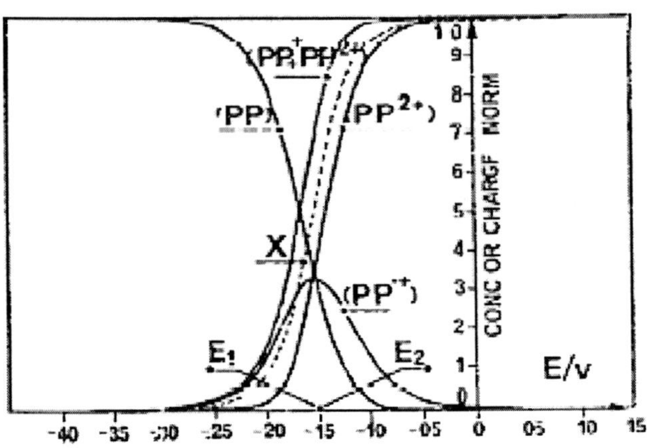

Fig. 7. Concentration profiles of polypyrrole species vs. assigned potential. Curves are estimated from ideal Nernstian model involving two monoelectronic transfers with E_1=E_2= - 0.15V

Two peaks frequently appeared in the reduction part in the range – 0.4 to – 0.2 V, which suggests the redox reaction of PP could involve two-electron transfer steps. In conclusion, PP presents two redox couples involved in a system of peaks localised between – 0.5 and 0 V. EPR experiments showed that the number of PP spins at equilibrium has a maximum of 0.3 spin per 6 pyrrole cycles at – 1.15 V (corresponding for half charge of polymer or x =15%, x being the doping level of the polymer), followed by a decrease at higher potentials. The data are satisfied by a Nernstian process with respect to the species concentration and the charge assuming $E_1=E_2= -0.15V$ (see figure 7).

This condition of potentials equality causes the charge transfer to occur only from – 0.3 to 0 V; the experimental voltammograms indicate that charge appears in a larger potential region (at least 0.8 V). This is probably due to electrostatic interactions between active sites, so that the potentials are probably close but not equal. For these reasons, the two redox couples are not easily separated by voltammetry even at very low scan rates.

Trying to clarify these observations, further absorbance experiments were carried out, knowing that PP strongly absorbs in the visible region (see spectra in figure 8).

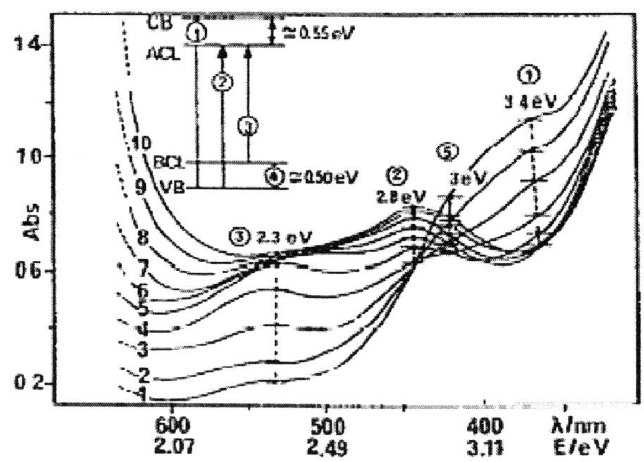

Fig. 8. Absorbance diagram of polypyrrole film (0.24 µm) in CH_3CN/ $LiClO_4$ at increasing levels of doping ranging from – 0.5 V (1) to + 0.4 V (10) by an increment 0.1 V. The single-headed vertical arrows on the energy chart indicate allowed transitions VB=valence band, BCL/ACL=bonding/antibonding cation level, CB=conduction band

Absorption depends on allowed optical transitions in the energy chart. During doping, the two states appear by bringing the molecular orbitals π and π^* closer by adding conjugated double bonds to the system. This change corresponds to the transition from a structure involving only non-conjugated aromatic units (a) to a structure with aromatic units (radical cations) and in the end to a conjugated bond structure containing quinonic units corresponding to dications (b).

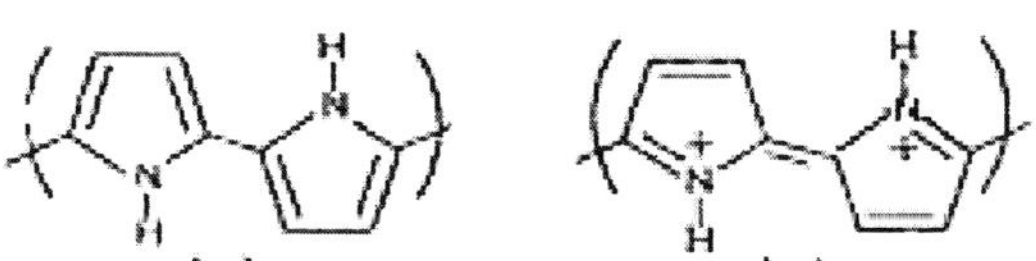

(a) (b)

As a result, the reduced form strongly absorbs at 3.4 eV (in the violet region) due to π-π^* transition and appears yellow in transmission for thin films, while the oxidised form, for which the π-π^* transition turns into intermediary interband transitions, the polymer absorbs at lower energy, 2.8-2.3 eV, and appears blue-violet.

From the optical data in figure 8, it can be distinguished three absorption areas in the visible region (2 to 3.6 eV), which may change with the degree of oxidation of PP [8]. (1) At

370 nm (≈ 3.4 eV), the strong absorption of reduced polypyrrole is associated with the interband transition (VB→CB). (2) The absorption at 2.8 eV (≈ 440 nm) is attributed to the transition valence bond to antibonding cation/dication level (VB→ACL/ADL). (3) At 2.3 eV (≈ 540 nm) is expected to take place the transition bonding to antibonding cation level (BCL→ACL). Complex calculations of theoretical concentration profiles and absorption coefficients for the species involved in the system concluded that the redox potential E_2 (PP$^{\cdot +}$/PP^{2+}) could be in the range – 0.1 V and E_1 (PP/PP$^{\cdot +}$) in the range – 0.2 V. When the polymer is fully charged probably E_2 becomes lower than E_1 and in this condition the dications can be consider as bipolarons with respect to their stability area under equilibrium.

In the case of PAN, plotting the evolution of absorbance (more suggestive the difference in absorbance between the reduced and oxidative forms of the polymer, ΔAU) at several wavelengths as a function of the potential (figure 9) it can be observed a sharp peak at 418 nm, due to oxidation of reduced PAN to cation radical (polaron) which absorbs here, and then the oxidation of this cation to dication (bipolaron).

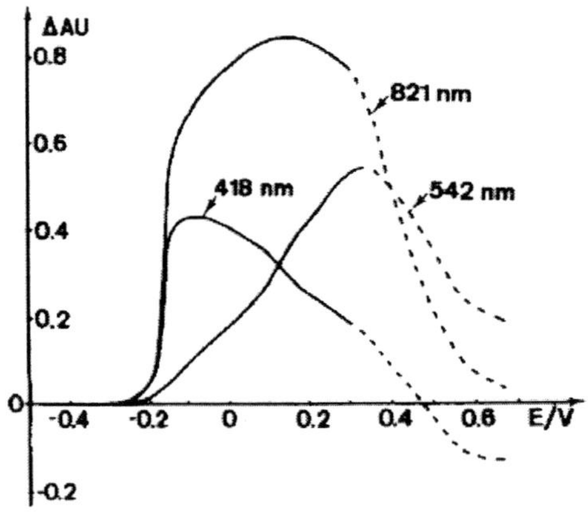

Fig. 9. Difference in absorbance vs. potential of polyaniline film in sulfuic acid

The absorbance at this wavelength may be attributed mainly to the transition of the cation radical between the bonding level of the polaron and the conduction band. After 0.3 V the dotted curve indicates that nothing can be said about the charge transfer as the film dissolves.

At 821 nm, the absorption is similar for the polaron and bipolaron states, the corresponding transition in this case being an electron promotion from the valence bond to a bonding level of the polaron-bipolaron. For the 542 nm wavelength it's difficult to make a prediction about which of the species absorb (as in this area all are active) as the slope of the curve increases almost linear until the region when the polymer dissolves.

In infrared (IR) spectra of conductive polymer films [11, 12, 13], the intensities and positions of IR bands change with the degree of delocalisation along the polymer chains, which is related to the polymer conductivity. For example, a PP film grown by voltammetry on cyclodextrin self-assembled monolayer (CD-SAM) electrode the bands at 1560 and 1480 cm^{-1} were affected by changes in the conductivity (figure 10). The ratio of the integrated absorption intensity was as low as 1.25 for very thin films (grown for 5 potentiodynamic cycles), indicating that the polymer has the typical conductivity for nanowires [12, 13]. The ratio increased with the number of cycles, being 12 for the film prepared on bare electrode (fig. 10 c).

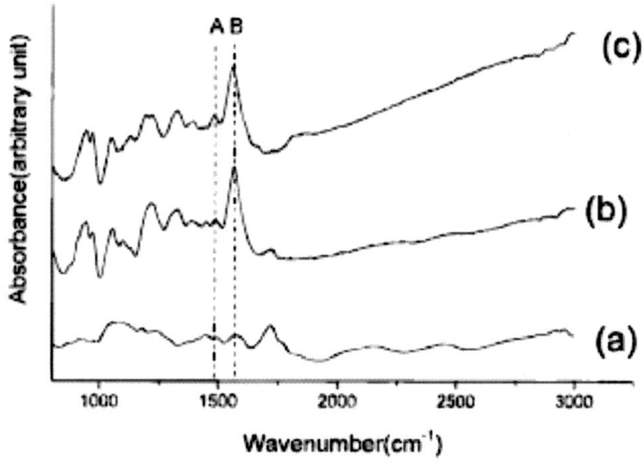

Fig. 10. IR spectra of polypyrrole films: a) at CD-SAM electrode for 5 cycles; b) at CD-SAM electrode for 12 cycles; c) at bare gold electrode for 9 cycles between – 0.5 and + 0.82 V at scan rate 50mV/s

This indicates that at the beginning, the PP growing at the CD-SAM covered electrode takes place in a single stranded form, that eventually merges with other growth centres to form a three dimensional structure with numerous cross-links.

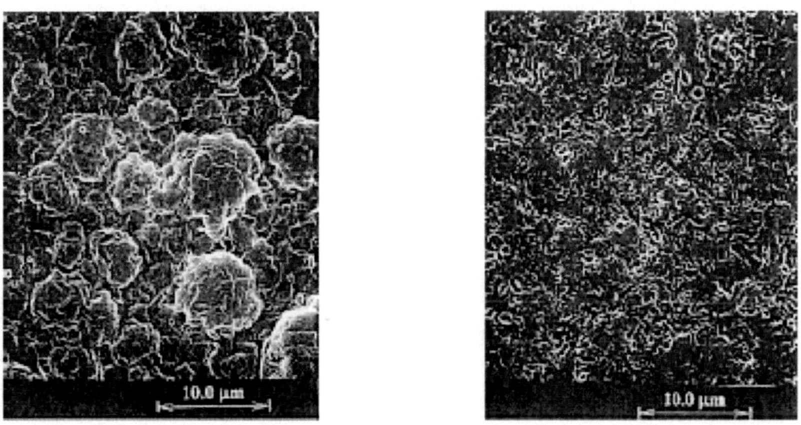

(a) (b)

Fig. 11. SEM pictures of polypyrrole films prepared on a) bare gols electrode, b) fully covered CD-SAM electrode

2.3 Morphology of the Deposited Layers

The morphology of polymer films was studied using SEM techniques or powerful electronic/optical microscopes [11, 14]. PP films prepared on bare electrode shows many valleys and mountains (see figure 11).

On CD-SAM covered electrode, PP films have relatively well-oriented structure; probably the polymer starts growing as single fibre from CD cavities and then merges into the ones at nearby sites, and in the end the fibrous structure became more aligned [11]. Investigations of temperature influence on preparation of PP films showed that polymers deposited at low temperature (10-30°C) are smooth, coherent and mechanically strong in comparison with those prepared at temperatures above 35°C [15].

Morphology of electrodeposited PAN on different types of substrates showed no influence of substrate nature [14]; the morphology of the deposits (figure 12) has the appearance of a dark blue uniform grown layer with little channels and pits due to probably the secondary reactions that compete at the anode with the main polymerisation reaction of the monomer.

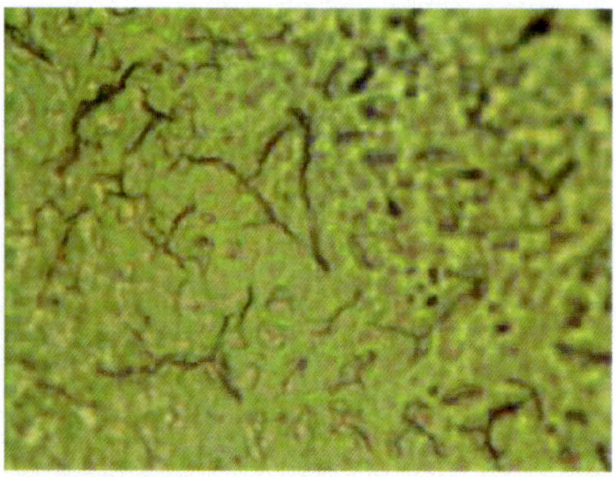

Fig. 12. Morphology of polyaniline at magnification 50 on optical microscope

3 Techniques for the Deposition of Conductive Polymers

There are various techniques for achieving conductive polymers that may be grouped in four main methods:

1) Preparation of a malleable polymer that can be easily transformed into a conjugated polymer. This means to thermally treat the initial already shaped polymer so as it becomes conductive. This method is used to obtain highly orientated thin films and fibres very anisotropic with maximum conductivity along the stretch direction.
2) Synthesis of copolymers or derivatives of a parental conjugated polymer with better properties. The method presumes a modification of the polymer structure to increase its processibility without compromising conductivity and optical properties. Similar values of conductivities for polythiophenes and PP were obtained when the hydrogen on carbon 3 on their rings was replaced with an alkyl group having at least four carbon atoms in it. Apart from the comparable conductivity with the parental polymers, the new products were soluble and able to melt. By placing carboxylic acid group or sulphonic acid group on the alkyl chain of the same polymers a water soluble version can be obtained. If the sulphonic group is used along with built-in ionisable groups, the system is able to maintain charge neutrality in its oxidised state,

so they effectively dope themselves; that is why they are called "self-doped" polymers.

3) Growing the polymer in the desired shape and form. A thin film or fibre of conducting polymer is grown by exposing to the monomer a well defined form of an insulating polymer impregnated with a catalyst. Electrochemical polymerisation is a variation of this method when the conducting polymer is deposited on an electrode during the process; it can be achieved either by potentiodynamic, or potentiostatic or galvanostatic methods.

Electropolymerisation of PP occurs at potentials above + 600 V, after the following mechanism:

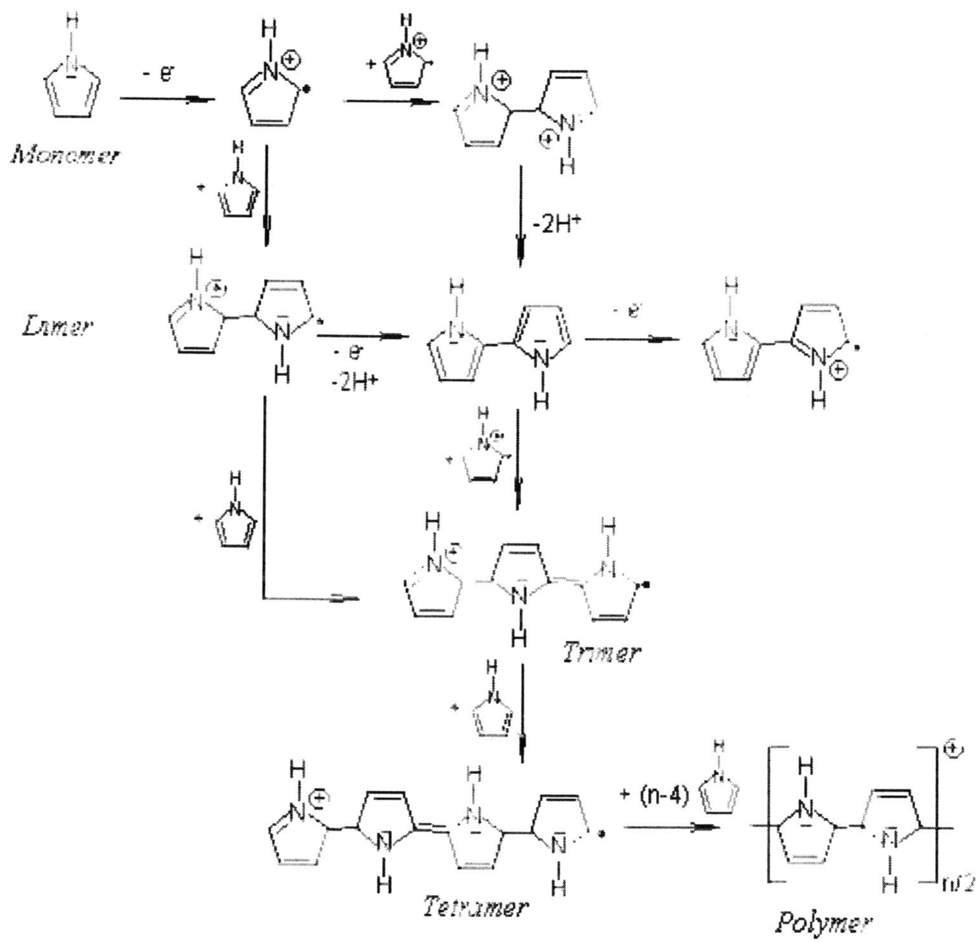

The morphology of the films depends on the nature of the electrolyte, on the crystallographic structure of the substrate (electrode), on the speed and the potential of deposition, on the presence of polyanions and surfactants, on the concentration of the initial monomer and on the pH of the solution.

For PAN the electropolymerisation mechanism in acid medium follows these steps [16]:

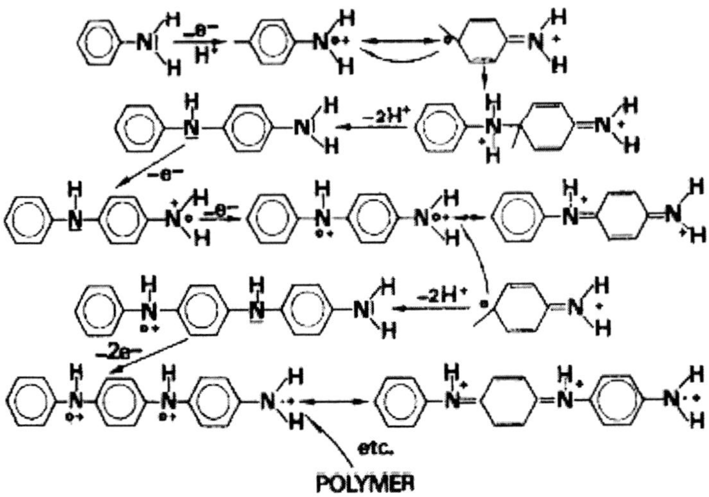

This deposition is also influenced by the nature of the electrolyte and presence of other anions and surfactants, the potential of deposition, the concentration of the monomer and the pH.

4) Manipulation of the surface active molecules into highly ordered thin films with structures and thickness controllable at molecular layer (Langmuir-Blodgett through). This technique allows control over the molecular architecture of the produced conducting films and by alternating the layers, it is possible to obtain a highly anisotropic film which conducts within the plane of the film and insulates across it. The principle consists in producing monolayers of amphiphilic molecules with hydrophilic and hydrophobic groups at the air-water surface interface of the Langmuir-Blodgett through. These are then transferred to a substrate creating a multilayer structure of molecular stacks of 2.5 nm thick.

4 Application of Conductive Polymer Films to Chemical Sensing

Conducting organic polymers have been mainly used either as material for different devices [17, 18] or as selective layers in chemical sensors. The large variety class of conductive polymers are very attractive because they may be easily and cheap fabricated in a controlled manner (especially by the electrochemical route), they operate at room temperature and they exhibit changes in conductivity at exposure to different gases. That is why a lot of their applications were directed to produce different type of sensors for chemical sensing, which will also be the main subject for discussions of this subchapter.

Unfortunately, though some substantial improvements were achieved in the stability and processibility of these polymers, another disadvantage is their lack of specificity, showing responses to a wide range of different gases and vapours. Nevertheless, this can be turned into an advantage if the materials are used to form arrays of gas-sensing elements in an "intelligent" gas sensor [19, 20]. These recent applications combine development of arrays of gas sensors with conductive polymer layers with a detection system and associated pattern recognition programs that mimics a nose, muscles or a tongue.

Another group of applications for conductive polymers are the quartz micro balances, made of gold covered single piezoelectric quartz crystal coated with a polymer using spray coating, Langmuir-Blodgett or Self- assembled monolayers techniques. The adsorption of the gas molecules creates additional mass loading in the polymer matrix that causes changes in the oscillation frequency of the crystal. Similar principle has the surface acoustic wave sensors: two separated interdigital transducers of thin metal electrodes covered with polymer on a polished piezoelectric substrate; the space between the transducers determines the wavelength. Gas absorption changes the dielectric constant and viscosity of the polymer and influences propagation of the surface waves, modifying the wavelength (frequency respectively).

Gas sensors are fabricated by chemical/ electrochemical deposition of a conductive polymer (or sometimes by Langmuir-Blodgett technique) on a narrow electrode gap (interdigitated structure) prepared on an insulating substrate. Sorption of gases into the polymer matrix changes polymer's conductivity. The introduction of an electrically neutral gas into the polymer can be seen as adding an "inert" secondary dopant which changes the polymer's properties related to changes of electronic coupling between redox sites that define a physical electron transfer pathway in the polymer [21]. If a p-type conductive polymer donates electrons to the gas its holes conductivity increases; when the same polymer acts as an electron acceptor its conductivity decreases.

As the polaron and bipolaron states are localised symmetrically within the band gap, in the middle of the band gap lays the Fermi energy level, which thermodynamically corresponds to the electrochemical potential of an electron. The energetic difference between the Fermi energy and the vacuum levels is the work function, WF. The changes in the conductivity at the substrate (electrode)/ conductive polymer contacts are determined by the difference in WF of the organic semiconductor and the metal on the electrode, where a space-charge region is created. That means the effective resistance depends on the bias voltage applied during the measurement. Although as a convenient approximation the interaction with the gas molecules can be described using forms of adsorption isotherm, its effect on the conductivity of the surface layer is much more complex and depends on the deposition method (morphology of the polymer layer) and the interface between the polymer and the insulating substrate.

Generally the WF that expresses the affinity of the conductive polymer for electrons consists of two terms: (1) the work needed to bring the electron from the bulk of the material up to the surface dipole layer (chemical potential) and (2) the energy of extraction of an electron through the surface dipole layer up to the vacuum level. From this relation, in the particular case at the interaction of organic semiconductors with gases the bulk term of the WK is predominant. This result states that by adjusting the initial value of the WF it is possible to "tune" the selectivity of the conducting polymer affinity to different gases. This tuning can be accomplished in different ways; for PAN the correlation is made with the pK of the acid dopant [22], while for PPSA, both the conductivity and the WK are controlled by electrochemical insertion of the primary dopant anion [23].

A few types of fabricated chemical sensors are presented in figure 13.

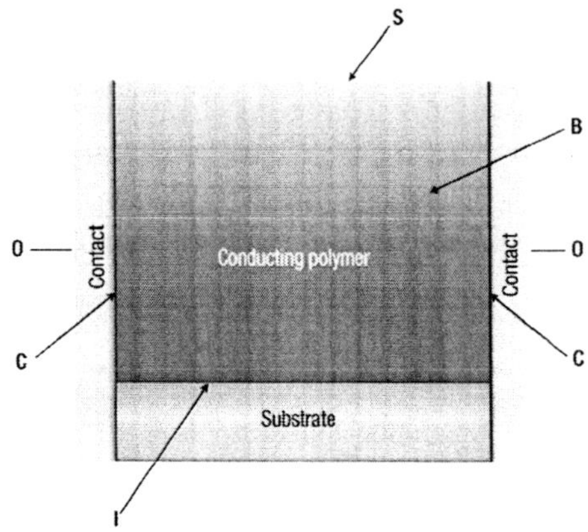

(1)
S- surface; I-interface with insulating substrate; C-interface with the contacts

(2)
Sensor array consisting of 8 CHEMFETs

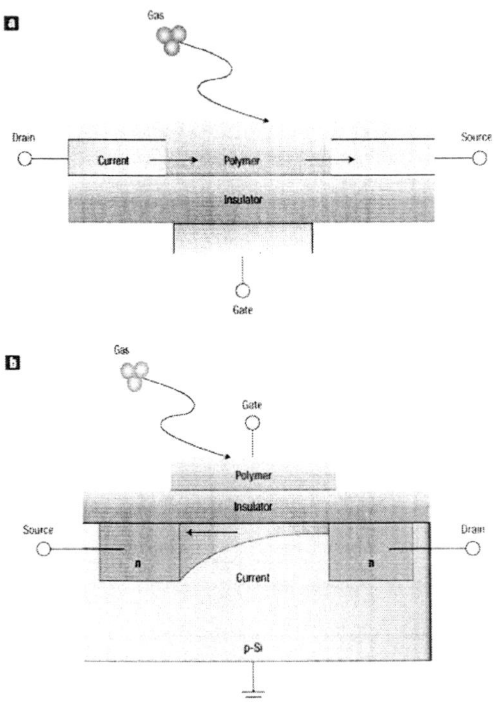

(3)
a- thin film transistor; b-insulated field-effect transistor

1) chemiresistor
2) sensor array
3) field effect transistor (FET)

Fig. 13. Different types of chemical sensors

The responses of PP to different gases and vapours are briefly presented in figure 14. Room temperature exposure to methanol and ethanol show a decrease in conductivity with stable responses and a very slow response for the ethanol [24, 25].

These are interpreted in terms of electrons in PP being more tightly bound in the presence of alcohol vapours [26]. The responses of PP to acetone, toluene and ether are smaller comparing with those obtained for alcohols. For toluene, some sensitivity was initially observed, but that disappeared over a period of tens minutes. In the case of acetone, the resistance drifted to lower conductivities, probably because of some irreversible slow interaction between the film and acetone vapours [24].

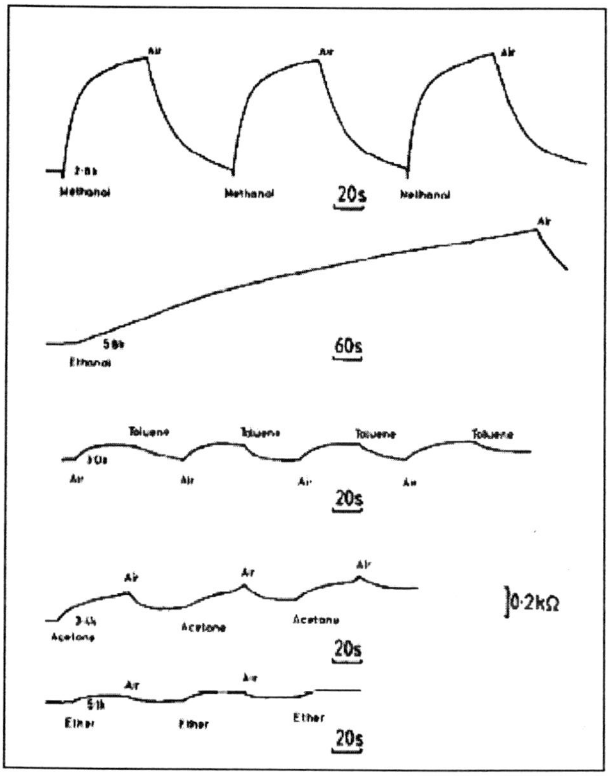

Fig. 14. Responses of polypyrrole film to saturated vapours exposures at 25°C in a flowing system. The devices were cycled repeatedly between saturated vapour and clean air

Sensors based on lead phthalocyanine (PbPc) films with conductivities 10^{-11} $(\Omega\ cm)^{-1}$ under vacuum (and 10^{-11} $(\Omega\ cm)^{-1}$ in air) were tested to NO_2 and water vapours [27]. They show different responses depending on the film thickness and structures (see as example in figure 15 the sensitivity of the film at exposure of NO_2 in air).

Lot of research effort was made in the direction of using PAN as an active polymer film to sense different types of vapours, although in most experiments it was difficult to obtain reproducible results. The average room temperature d.c. conductivities measurements for PAN deposited by various methods showed values ranging between $0.2x10^{-10}$ and $2.0x10^{-8}$ $S.cm^{-1}$ [28, 29, 14, 30]. Exposures of PAN films to vapours such are NO_x, SO_2 and NH_3 produce an increase in the conductivity of the polymer (see figure 16 [28, 29]).

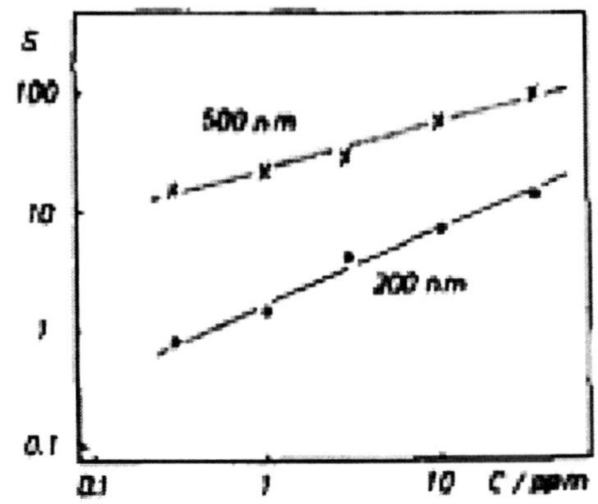

Fig. 15. Sensitivity of sensors with PbPc at NO2 exposure in air

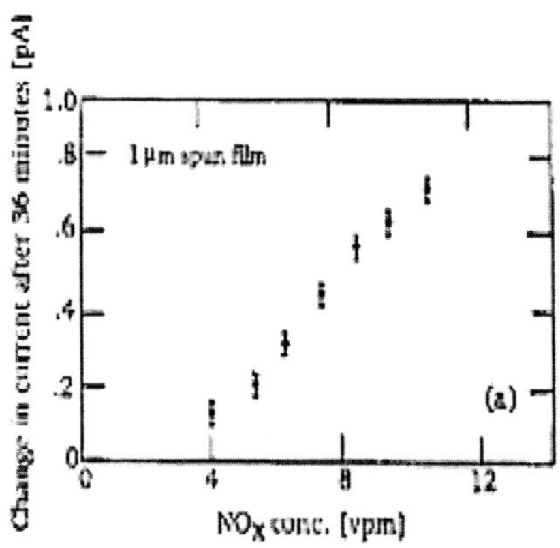

(a)

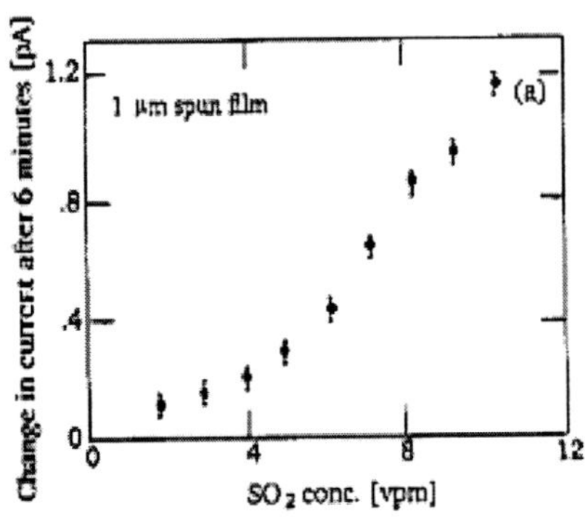

(b)

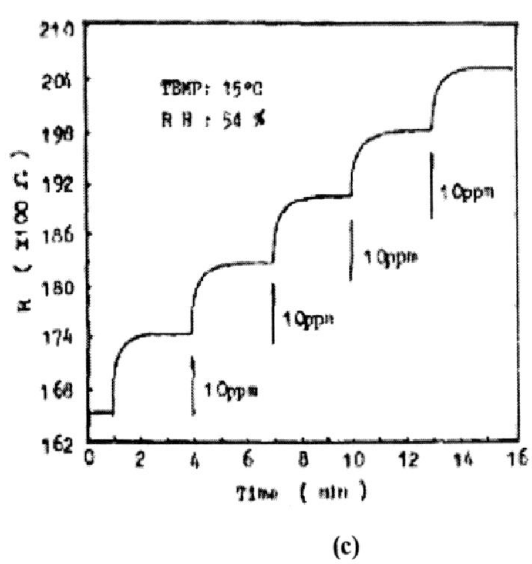

(c)

Fig. 16. The response of polyaniline to different vapours exposure a) to NO_x for a 1 μm spun film; b) to SO_2 for a 1 μm spun film; c) variation of the polymer's resistance with NH_3 concentration

For the first two analytes the responses are dependant on the film thicknesses and they exhibit a detection threshold, probably due to the conversion of phenyl rings from the polymer structure into the quinoid form on exposure to air (film oxidation) that blocks the reactive sites to be occupied by the vapours molecules. PAN has a high sensitivity to NH_3 at room temperature the range of resistance can reach 50-100 Ω/ppm, the relative change of resistance being about 5%/ 10ppm. It was demonstrated that NH_3 molecules act as donors after being adsorbed and form an electric barrier near the surface of the film and the resistance increases with the reduction of the holes number. Also exposure of PAN electrodeposited film to aqueous HCl solution increased the conductivities values with several orders of magnitude (from 1.65×10^{-9} to 3.06×10^{-2} S cm^{-1}) [14]; this was attributed to the hydrogen bonds present both in water and in HCl solution that may interact with the polymer backbone. They create a polarization of electronic charge on polymer's top surface that will result in generating additional free holes charge carriers deep in the polymer's structure responsible for the increasing in the conductivity level. The higher the number of hydrogen bonds the most important the changes in conductivity.

An interesting result was obtained for exposure of PAN to H_2S [28]. While it was expected for this reducing agent to decrease PAN's conductivity, an obvious increasing was observed. This indicates that either more than one type of reaction site is available or that a number of different reactions are possible.

CONCLUSIONS

Important steps have been made into using conductive polymers sensitivity to different analytes for fabricating various devices suitable for their detection. The main advantages for increased attention to this class of polymers are the easy and cheap techniques to deposit them, including different simple ways to achieve an improvement of their conductivity by doping. Nevertheless, with all promising results already published, there is still a lot of research to be developed in stabilizing their active structures and obtain more reproducible and reliable measurements.

REFERENCES

[1] Margolis, J. In *Conductive Polymers and Plastics;* Chapman and Hall; 1989; 2-11.
[2] Rubner, M.F. In *Molecular Electronic;* Research Studies Press; 1992; chapter 2.
[3] Cambell, I.M. In *Introduction to Synthetic Polymer;* Oxford Science Publications; 1994; 196.
[4] Haarmann, H. In *Polymers to the Year 2000 and Beyond*; John Wiley & Sons; 1993; chapter 3.
[5] Genies, E.M.; Pernaut J.M. *J. Electroanal. Chem.* **191**, 1985, 111-126.
[6] Bredas, J.L.; Themans, B.; Andre, J.M. *Phys. Rev.* **B 27**, 1983, 7827-7830.
[7] Scott, J.C.; Pfluger, P.; Krounbi, M.T.; Street, G.B. *Phys. Rev.* **B 28**, 1983, 2150-.
[8] Scott, J.C.; Bredas, J.L.; Yakushi, K.; Pfluger, P.; Street, G.B. *Synth. Met* **9**, 1984, 165-172.

[9] Genies, E.M.; Lapkowski, M. *J. Electroanal. Chem.* **220**, 1987, 67-82.
[10] Stafstrom, S. *Synth. Met.* **18** (1-3), 1987, 387-392.
[11] Lee, J-Y.; Park, S-M. *J.Electrochem. Soc.***147**(11), 2000, 4189-4195.
[12] Tian, B.; Zerbi, G. *J. Chem. Phys.* **92**, 1990, 3892-3898.
[13] Tian, B.; Zerbi, G. *J. Chem. Phys.* **92**, 1990, 3886-3891.
[14] Grigore, L.; Petty, M.C. *J. Mat. Sci.: Materials in Electronics* **14**, 2003, 389-392.
[15] Kassim, A.; Basar, Z.B.; Mahmud, H.N.M.E. *Proc. Indian Acad. Sci. (Chem. Sci.)* **114** (2), 2002, 155-162.
[16] Genies, E.M.; Tsintavis, C. *J. Electroanal. Chem.*, **195**, 1985, 109-128.
[17] Margolis, J. In *Conductive Polymers and Plastics;* Chapman and Hall; 1989; 120-121.
[18] Alcacer, L. In *Conductive Polymers Special Applications;* D. Reidel Publishing Company; 1987; pp 5.
[19] Bartlett, P.N.; Archer, P.B.M.; Ling-Chung, S.K. *Sensors and Actuator,* **19**, 1989, 125-140.
[20] Barker, P.S.; Chen, J.R.; Agbor, N.E.;.Monkman, A.P.; Mars, P.; Petty, M.C. *Sensors and Actuators* **B 17**, 1994, 143-147.
[21] Janata, J.; Josowicz, M. *Nature materials* **2**, 2003, 19-24.
[22] Hatchett, D.W.; Josowicz, M.; Janata, J. *J. Phys. Chem.* **B 103**, 1999, 10992-10998.
[23] Li, G.; Josowicz, M.; Janata, J. *Synth. Met.* **125**, 2001, 273-278.
[24] Bartlett, P.N.; Ling-Chung, S.K. *Sensors and Actuators* **20**, 1989, 287-292.
[25] Bartlett, P.N.; Archer, P.B.M.; Ling-Chung, S.K. *Sensors and Actuators* **19**, 1989, 141-150.
[26] Josowicz, M.; Janata, J. *Anal. Chem.* **58**, 1986, 514-517.
[27] Hamann, C.; Kampfrath, G.; Mueller, M. *Sensors and Actuators* **1**, 1990, 142-147.
[28] Agbor, N.E.; Petty, M.C.; Monkman, A.P. *Sensors and Actuators* **B 28**, 1995, 173-179.
[29] Jiakun, W.; Hirata, M. *Sensors and Actuators* **B 12**, 1993, 11-13.
[30] Aktar, M.; Weakliem, H.A.; Paiste, R.M.; Gaughan, K. *Synth. Met.* **26**, 1988, 203-208.

In: Polymer Research Developments
Editor: Robert K. Bregg, pp. 127-185

ISBN 1-59454-742-4

Chapter 7

Mathematical Models of Compound, Polymer Optical Fibers

J. I. Ramos*
Room I–320–D, E.T.S. Ingenieros Industriales, Universidad de Málaga
Plaza El Ejido, s/n, 29013-Málaga, SPAIN

Abstract

Asymptotic methods based on the slenderness ratio are used to obtain the leading–order equations which govern the fluid dynamics of the draw–down zone of axisymmetric, nonisothermal, Newtonian, compound liquid jets such as those employed in the manufacture of textile fibers, composite fibers and optical fibers, at low Reynolds numbers. These equations are first used to determine the steady state, linear stability, and time–dependent dynamics of isothermal compound fibers for draw ratios greater than the critical value provided by a linear stability analysis of the viscous flow regime. An analytical solution of the temperature distribution in the absence of axial conduction is then presented before undertaking time–dependent studies of the nonlinear dynamics of nonisothermal, compound fibers subject to axial velocity perturbations at either the die's exit or the take–up point. The effects of molecular orientation and crystallization on polymer optical fibers are then accounted for by adding two more equations to the leading–order equations derived asymptotically in this chapter, and some numerical results are presented. Finally, models for the air drag, latent heat effects and viscous dissipation are described.

1 Introduction

Most of the man–made fibers used in textile industries are manufactured by means of fiber spinning processes which consist of the steady extrusion of hot melts through a series of small holes in a plate (spinnerets) into ambient air; the resulting extrudates are simultaneously extended and wound up on a rapidly rotating drum (godet). Freezing takes place between the spinneret and the godet, and, usually, large extensions rates, rapid cooling, and high speeds are involved [1, 2, 3, 4, 5, 6, 7]. Fiber spinning processes are also used in the manufacture of reinforced fibers and optical fibers.

The fluid dynamics of fiber forming processes usually involves a source flow, also referred to as the forming zone, a draw–down zone, and a finishing zone. The forming zone is usually characterized by a rapidly contracting flow issuing from the bottom of a preform or pulled from an orifice at the bottom of a crucible. In drawing from preforms, we may distinguish between a very sharply contracting configuration, commonly encountered in laser heating, and a more gradual contraction that results from furnace heating. We should also include the case of an overheated preform, where the forming zone consists essentially of a liquid bath from which the fiber or filament is drawn. The different forming zones can imply fundamental differences, as far as steady state and potential instabilities are concerned.

The forming zone makes a continuous transition to the draw–down zone, where an essential part of the fiber contraction takes place. Most fiber studies have dealt with the draw–down zone of round filaments or fibers, because of its physical importance and mathematical tractability; however, the role of the forming zone on the fluid dynamics and stability of fibers cannot be ignored. The essential need for modelling this part of the flow field in melt spinning processes is obvious. Moreover, fiber deformation also occurs beyond the draw–down region in the finishing zone, but this deals with viscoelastic rather than with the fluid dynamics regime.

While textile fiber studies, in general, strive to avoid fiber rupture and gross distortion of the thread line, the drawing of optical fibers takes filament continuity for granted, and the application of optical fibers requires that the main issue in their manufacture be the sensitivity to small perturbations in the diameter. It must be noted that, when continuous drawing of textile fibers is impossible, much emphasis is placed on the prediction and control of the thread length, i.e., the filament length attainable between spontaneous ruptures.

The different requirements placed on textile and optical fibers are a consequence of the applications of these fibers. For example, emerging optical networks demand a new class of optical devices that can be reconfigured and switched, thus providing a number of important functionalities that enable what are being referred to as dynamical optical networks such as, for example, tunable optical filters, variable optical attenuators, dynamic polarization controllers, tunable compensation devices, switches, and routers [8]. Several mechanisms for manipulating light in optical networks have been demonstrated, including mechanical switching devices such as micro–electro–mechanical systems and waveguides that can be tuned and switched via principles based on piezoelectric, thermo–optic, and electro–optic effects. In many cases, all–fiber optical devices are desired since light is manipulated in the optical fiber itself.

About thirty years ago, a new optical fiber, i.e., the plastic (or polymer) optical fiber (POF), emerged, but did not find many optical applications due to its high attenuation, until about fifteen years ago when graded–index optical fibers and low–attenuation perfluorinated fibers were developed [9]. POFs are now available as high–capacity transmission media thanks to their transparency and bandwidth; they are currently replacing copper cables in short–haul communications links due to their transmission capacity, immunity to interference and small weight. In addition, POFs serve as a complement to glass fibers in short–haul communications links because they are easy to handle, flexible, and economical, although they are not used for very long distances because of their relatively high attenuation.

POFs used for optical communications are highly flexible waveguides composed of nearly transparent dielectric materials. The cross–section of these fibers is circular and, generally, consists of three layers. The inner layer is called the core, the middle one is the cladding, and the outer layer is the jacket or protective cover. Within the core, the refractive index profile may be uniform (step–index fibers) or graded (graded–index fibers), whereas the cladding is typically uniform.

Optical fibers can be classified in two main groups from the point of view of propagation: single–mode fibers which have a relative small core, and multimode fibers, whose core is large enough so that propagation may be analyzed by means of geometric ray–tracing models. Single–mode POFs exhibit a much larger attenuation than single–mode glass fibers and lack the main advantages of conventional POFs, i.e., easiness to handle and low cost connectors. As a consequence, research on these fibers has not been very active, although single–mode POFs can be easily coupled to planar waveguides and serve as components for some devices such as sensors, and their core may be doped with a dye to control absorption and nonlinear wave propagation.

Two steps are usually followed in the manufacture of POFs. First, a solid cylindrical rod, called preform, which is approximately 0.5–1 m long and several centimeters in diameter, is made; the structure of this preform determines the core and cladding refractive index profiles. The second step consists in the extrusion of the preform, which yields a POF length in the range between half a kilometer and several kilometers. Step–index POFs are manufactured by means of a melt spinning process in which fibers are obtained by either continuous or batch extrusion. In the continuous extrusion process, a monomer, an initiator and a chain transfer agent are continuously fed into the reactor and the fiber is continuously drawn from the die. This is an efficient process due to its high production rate.

In order to obtain a particular refractive index profile in the core along the radial direction, a preform with the desired profile can be employed; whenever more complex refractive index profiles are required such as graded–index ones, a technique based on drawing a preform by melting its tip with the aid of an oven is used. The preform is, in this case, a cylinder of polymer whose refractive index distribution in the radial direction can be made to coincide with that desired for the POF's core [9].

Currently, there are two main techniques to produce graded–index POFs: gel polymerization and diffusion. The gel polymerization technique requires the manufacture of a hollow cylinder made of a polymer, e.g., PMMA, which will be the POF's cladding. The hollow cylinder is filled with a mixture of the monomer from which the polymer cylinder is made, a dopant with a higher refractive index than that of the polymer, an initiator, and a chain transfer. The filled cylinder is then heated for some time and, as a consequence of the heating, the cylinder widens due to the formation of a gel phase which accelerates the polymerization reactions, i.e., the polymer is formed from the gel phase on the inner wall of the cylinder; the reaction is completed when the initially hollow cylinder becomes a solid one. The reason for the use of dopants with a higher refractive index than that of the polymer is that the light should be confined in the structure, and the mechanism for the correct distribution of the refractive index consists in the selective diffusion of the monomer and dopant materials in the gel–polymer phases created in the inner wall of the hollow cylinder. Moreover, since the molecular volume of dopants is higher than that of the monomer, the monomer diffuses faster than the dopant in the gel phase. For this reason, dopant molecules are concentrated in the central region of the cylinder, so that the refractive index is higher in this region.

Once the preform has been obtained, it is held up by a support in the fiber extrusion tower, where the lower end of the preform is heated with the aid of a furnace, so that it acquires the necessary viscosity to be extruded downward. The extrusion velocity and the speed at which the mandrel rotates are adjusted by a controlled mechanism, so that the fiber reaches the desired constant diameter.

During the extrusion process, the geometrical relationship between the core and the cladding is kept constant, even though there is a drastic reduction from the diameter of the preform to the final diameter of the extruded fiber. Finally, a coating is applied to the fiber in order to protect it against microcurvatures or microbends, improve its mechanical resistance, and ease its handling. When the coating has hardened by the effect of heat or ultraviolet radiation, the optical fiber's resistance to traction is continuously tested by forcing the coated POF to pass through a series of pulleys that exert mechanical tension on it. In this manner, the optical fiber must resist a minimum load before it is wound onto a cylindrical drum.

There have been some numerical studies on heating of single–component, polymer optical preforms [10, 11, 12]. For example, Reeve et al. [10] considered the transient heating of a polymer preform within a cylindrical furnace in their experimental and numerical studies of the manufacture of polymer optical fibers. Their numerical model based on the solution of the Navier–Stokes equations for the gases that surround the preform accounted for convective and radiative heat exchanges; the latter were dealt with by means of a net radiative heat transfer model which accounted for the furnace walls and irises. In a subsequent publication, Reeve and Mescher [11] reported experimental data on the effects of natural convection on the diameter of drawn polymer optical fibers, and showed that the air flow adjacent to the fiber may be laminar, oscillatory or chaotic, depending on the imposed thermal conditions at the furnace wall and irises. When subject to oscillatory or chaotic natural convection, Reeve and Mescher [11] found that the drawn fiber varies in diameter 2.5 to 10 times more than that measured under laminar heating conditions. Hsu and Lin [12] considered the axisymmetric diffusion of two monomers in the co–extrusion process of gradient–index plastic optical fibers.

Although there has been quite a lot research on the development of one–dimensional, mathematical models for the analysis of the drawn–down region of single–component filaments and jets under both isothermal and non–isothermal conditions at low Reynolds numbers [1, 2, 3, 4, 5, 6, 7, 13, 14], compound fibers such as those used in reinforced materials and optical fibers (which are manufactured in co–extrusion processes) have received very little attention despite the fact that the combination of two or more different materials with different properties may result in composite fibers with highly desirable properties [15, 16, 17, 18, 19, 20, 21]. For example, in the manufacture of optical fibers, the core is surrounded by a sheath of cladding material.

Park [15] used perturbations methods based on the slenderness ratio and the smallness of the Deborah number in his studies of steady, isothermal, two-phase or compound fibers consisting of a Newtonian core layer surrounded by a sheath of a non–Newtonian fluid with a Maxwell rheology. His studies result in a system of ordinary differential equations for the axial velocity component and radii of the two–phase fiber which is more manageable than the two–dimensional conservation equations of mass and linear momentum from which it was derived. Lee and Park [16] employed the one–dimensional equations developed by Schultz [14] to study the linear stability of the spinning of isothermal, bicomponent fibers characterized by a Newtonian fluid for the core and an upper–convected Maxwell fluid for the cladding, and showed that the stability of the fiber can be maintained at higher draw ratios than obtainable when the same fluid is employed for both the core and the cladding. Naboulsi and Bechtel [19] introduced a one–dimensional model of isothermal, Newtonian, bicomponent fiber filaments by integrating the three-dimensional conservation equations of mass and linear momentum over the filament cross–section, and examined the influence of density, viscosity and surface tension ratios on the steady state fluid dynamics of isothermal, compound jets at low Reynolds numbers. Ji and Yang [17] and Ji et al. [18] studied isothermal, bicomponent fibers characterized by a Newtonian fluid for the core and a Phan–Thien/Tanner fluid for the cladding.

Ramos [20, 21] developed an asymptotic model for isothermal, compound fibers consisting of Newtonian fluids, from the time–dependent axisymmetric form of the conservation equations, obtained some analytical solutions under steady state conditions, performed a linear stability analysis of the steady solution and carried out time–dependent numerical simulations below and above the critical draw ratio predicted by linear stability theory in order to determine the dynamic response of isothermal compound fibers to time–dependent axial velocity variations at the die's exit and/or take–up point.

The linear stability analysis for incompressible, isothermal compound fibers or jets performed by Ramos [21] for what he refers to as the viscous regime, provides the same stability condition on the draw ratio as analogous analyses for planar film casting [22], single–component fibers [23, 24, 25, 26], single–component annular jets [27] and bicomponent or compound, annular liquid jets [28], when there is no surface tension and pressure differences across the annular jet.

Near the critical draw ratio, bifurcation theory for isothermal, single–component, round fibers yields a stable supercritical Hopf bifurcation, the frequency of which decreases as the winder speed is increased near the critical one [29]. For draw ratios higher than the critical one, Schultz et al. [29] found that the oscillation amplitude compares favorably with experimental data for isothermal fibers; however, they found quantitative differences in the disturbance frequency. This discrepancy can be attributed to the rheology of the fibers used in the experiments, for it was observed that, with viscoelastic fluids, the frequency decreases with the amplitude.

As stated above, previous studies of the drawn–down region of bicomponent or compound fibers at low Reynolds numbers [15, 16, 17, 18, 19, 20, 21] have only considered isothermal, steady state flows or determined the linear stability of these flows; in addition, except for that of Ramos [21], none of these studies has considered the possible steady state solutions of isothermal, steady, bicomponent fibers, and determined the effects of the downstream boundary conditions, forcing and fluid dynamics parameters on the nonlinear dynamics of isothermal, compound liquid fibers. Moreover, there does not seem to be any studies concerning the cooling as well as the nonlinear response of nonisothermal, bicomponent or compound fibers to perturbations in the fiber's geometry, axial velocity and temperature, perturbations in the operating conditions, etc., analogous to those performed by, for example, Geyling [2], for single–component, round fibers. It must be noted that compound fibers reduce to single–component ones whenever the core and cladding are made of exactly the same material, and, therefore, any model for compound fibers should reduce to a model for single–component, round fibers when the materials of the core and cladding are identical.

The objective of this chapter is to develop by means of asymptotic methods based on the slenderness ratio a set of mathematical models for compound fibers consisting of Newtonian fluids for both the core and the cladding, i.e, for the inner and outer jets, respectively. To achieve this objective, we have used a step–by–step procedure which corresponds to the chronological order (dating back to 1998) followed by the author on this subject. We, thus, first start in Section 2 with the axisymmetric form of the mass, linear momentum and energy equations for Newtonian fluids together with the kinematic, dynamic and thermal boundary conditions at the symmetry axis, core–cladding interface and compound fiber's outer radius, before undertaking the asymptotic analysis of these equation for slender fibers at low Reynolds numbers in Section 3. This section does not consider molecular orientation and crystallization.

In Section 4, we study the steady–state dynamics of isothermal compound fibers by, first, providing some analytical solutions to the leading–order equations resulting from the asymptotic analysis. We then study analytically the linear stability of the viscous flow regime, and present both steady state results for some values of the nondimensional parameters and time–dependent results for isothermal, compound fibers subject to excitations in the axial velocity component at either the die's exit and/or the take–up point. The characterization of the time–dependent results is made in terms of time histories, power spectra and phase diagrams for the core radius at the take–up point. Nonisothermal compound fibers are considered in Section 5 where, first, an analytical solution for the temperature along the fiber is provided, and then some sample results for both steady and time–dependent compound fibers are presented; the latter are to be compared with the time–dependent results presented in Section 4 for isothermal, compound fibers in order to assess the enhancement of stability produced by cooling.

In Sections 2–5, no reference is made to molecular orientation and crystallization and, therefore, these sections are applicable to amorphous polymers. The effects of molecular orientation and crystallization are dealt with in Section 6 which first describes the physics of these phenomena before presenting some available models for them. In the same section, the effects of molecular orientation and crystallization are included in the leading–order equations presented in Section 3 as additive and multiplicative terms, respectively, in the dynamic viscosity laws for the core and cladding. The resulting model accounts for the molecular orientation and crystallization of both the cladding and core and is of great interest for the study of compound, polymer/plastic optical fibers when their rheology is of the Newtonian type. Section 6 also presents some numerical results that illustrate the effects of the molecular orientation and crystallization on nonisothermal compound fibers, Finally, in Section 7, some conclusions and extensions of the one–dimensional equations derived in this chapter are presented. The extensions include models for the air drag and latent heat effects, and accounting for viscous dissipation.

It must be stated at the outset that the models presented here correspond to the draw–down zone of melt spinning processes of compound fibers, which include single–component, round fibers; no models for the forming and finishing zones are presented, although some references dealing with these zones are provided. Moreover, the models correspond to Newtonian fluids.

Throughout the chapter, emphasis has been placed on the assumptions (and limitations) made on the derivation of the equations, and enough details have been provided especially on the asymptotic analysis presented in Section 3; however, the amount of details differs greatly from section to section, and the list of references includes what the author considers are the most important papers on the subject which, as stated before, includes some references to works on single–component, round fibers. Finally, it should be mentioned that, as far as the author knows, most of the results presented here are new and have not been previously published.

2 Formulation

Consider an axisymmetric, compound liquid jet such as the one shown schematically in Figure 1, consisting of two immiscible, incompressible (constant density) Newtonian fluids. The inner (round) and outer (annular) jets correspond to $0 \leq r \leq R_1(t,x)$ and $R_1(t,x) \leq r \leq R_2(t,x)$, respectively, where r, x and t denote the radial and axial coordinates, and time, respectively. The fluid dynamics of the compound jet are governed by the following conservation equations of mass, linear momentum and energy

$$\frac{\partial u_i}{\partial x} + \frac{1}{r}\frac{\partial (v_i r)}{\partial r} = 0, \tag{1}$$

$$\rho_i\left(\frac{\partial u_i}{\partial t} + u_i\frac{\partial u_i}{\partial x} + v_i\frac{\partial u_i}{\partial r}\right) = -\frac{\partial p_i}{\partial x} + \frac{\partial}{\partial x}\left(2\mu_i\frac{\partial u_i}{\partial x}\right) + \frac{1}{r}\frac{\partial}{\partial r}\left(r\mu_i\left(\frac{\partial u_i}{\partial r} + \frac{\partial v_i}{\partial x}\right)\right) + \rho_i g, \tag{2}$$

$$\rho_i\left(\frac{\partial v_i}{\partial t}+u_i\frac{\partial v_i}{\partial x}+v_i\frac{\partial v_i}{\partial r}\right)=-\frac{\partial p_i}{\partial r}+\frac{1}{r}\frac{\partial}{\partial r}\left(2\mu_i r\frac{\partial v_i}{\partial r}\right)+\frac{\partial}{\partial x}\left(\mu_i\left(\frac{\partial u_i}{\partial r}+\frac{\partial v_i}{\partial x}\right)\right)-2\mu_i\frac{v_i}{r^2}, \quad (3)$$

$$\rho_i C_i\left(\frac{\partial T_i}{\partial t}+u_i\frac{\partial T_i}{\partial x}+v_i\frac{\partial T_i}{\partial r}\right)=K_i\frac{\partial^2 T_i}{\partial x^2}+\frac{K_i}{r}\frac{\partial}{\partial r}\left(r\frac{\partial T_i}{\partial r}\right)+\mu_i\phi_i, \quad (4)$$

where u and v are the axial and radial velocity components, respectively, ρ, C, K and μ are the density, specific heat, thermal viscosity and dynamic viscosity, respectively, p is the pressure, T is the temperature, g is the gravitational acceleration, the subscripts $i = 1,2$ denote the fluids of the inner (round) and outer (annular) jets, respectively, and the dissipation rate (except for μ) is given by

$$\phi_i=2\left(\left(\frac{\partial v_i}{\partial r}\right)^2+\left(\frac{\partial u_i}{\partial x}\right)^2+\left(\frac{v_i}{r}\right)^2+\frac{1}{2}\left(\frac{\partial u_i}{\partial r}+\frac{\partial v_i}{\partial x}\right)^2\right). \quad (5)$$

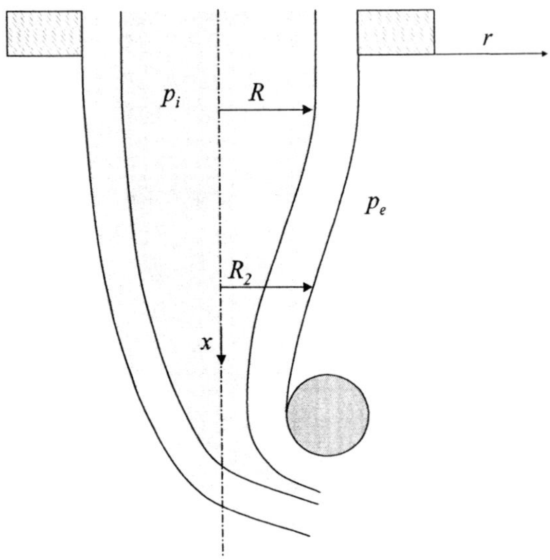

Figure 1: Schematic of a compound liquid jet. (In the text, the inner jet radius is referred to as R_1.)

In the above equations, it has been assumed that the density, specific heat and thermal conductivity are constant. The dynamic viscosity is a function of the temperature and may be given by Arrhenius–type expressions such as, for example,

$$\mu_i=\psi_i\exp\left(-\frac{F_i}{\mathbf{R}T}\right), \quad (6)$$

where ψ_i is a constant pre–exponential factor, F is the activation energy and $\mathbf{R}$ is the universal gas constant. Other expressions for the dynamic viscosity may be written as

$$\mu_i=\psi_i\exp\left(-\frac{E_i}{\mathbf{R}(T-T_i^m)}\right), \quad (7)$$

or

$$\mu_i = \psi_i \exp\left(-G_i(T - T_i^m)\right), \tag{8}$$

where T^m denotes the melting temperature and E and G are constants. Equations (7) and (8) indicate that the viscosity increases in an exponential manner as T decreases below T^m.

Equations (1)–(4) are subjected to kinematic, dynamic and thermal boundary conditions at the jet's interfaces, $R_1(t,x)$ and $R_2(t,x)$, where R_1 corresponds to the interface between the inner (core) and outer (cladding) jets, and symmetry conditions. The kinematic conditions establish that the compound jet's interfaces are material surfaces where the shear stress is continuous, and the jump in normal stresses across the interface is balanced by surface tension. The symmetry and kinematic, dynamic and thermal boundary conditions at the jet's interfaces for compound liquid jets may be written as

$$v_1(0,x,t) = 0, \qquad \frac{\partial u_1}{\partial r}(0,x,t) = 0, \qquad \frac{\partial p_1}{\partial r}(0,x,t) = 0, \tag{9}$$

$$v_2(R_2,x,t) = \frac{\partial R_2}{\partial t} + u_2(R_2,x,t)\frac{\partial R_2}{\partial x}, \tag{10}$$

$$v_i(R_1,x,t) = \frac{\partial R_1}{\partial t} + u_i(R_1,x,t)\frac{\partial R_1}{\partial x}, \qquad i = 1,2, \tag{11}$$

$$2\mu_2\left(\frac{\partial v_2}{\partial r} - \frac{\partial u_2}{\partial x}\right)\frac{\partial R_2}{\partial x} + \mu_2\left(\frac{\partial u_2}{\partial r} + \frac{\partial v_2}{\partial x}\right)\left(1 - \left(\frac{\partial R_2}{\partial x}\right)^2\right) = 0, \qquad r = R_2, \tag{12}$$

$$\begin{aligned} 2\mu_2\left(\frac{\partial v_2}{\partial r} - \frac{\partial u_2}{\partial x}\right)\frac{\partial R_1}{\partial x} + \mu_2\left(\frac{\partial u_2}{\partial r} + \frac{\partial v_2}{\partial x}\right)\left(1 - \left(\frac{\partial R_1}{\partial x}\right)^2\right) = \\ 2\mu_1\left(\frac{\partial v_1}{\partial r} - \frac{\partial u_1}{\partial x}\right)\frac{\partial R_1}{\partial x} + \mu_1\left(\frac{\partial u_1}{\partial r} + \frac{\partial v_1}{\partial x}\right)\left(1 - \left(\frac{\partial R_1}{\partial x}\right)^2\right), \qquad r = R_1, \end{aligned} \tag{13}$$

$$\begin{aligned} &2\mu_2\frac{\partial u_2}{\partial x}\left(\frac{\partial R_2}{\partial x}\right)^2 + 2\mu_2\frac{\partial v_2}{\partial r} - 2\mu_2\left(\frac{\partial u_2}{\partial r} + \frac{\partial v_2}{\partial x}\right)\frac{\partial R_2}{\partial x} \\ &+ (p_e - p_2)\left(1 + \left(\frac{\partial R_2}{\partial x}\right)^2\right) = -\sigma_2 J_2, \qquad r = R_2, \end{aligned} \tag{14}$$

$$\begin{aligned} &2\mu_2\frac{\partial u_2}{\partial x}\left(\frac{\partial R_1}{\partial x}\right)^2 + 2\mu_2\frac{\partial v_2}{\partial r} - 2\mu_2\left(\frac{\partial u_2}{\partial r} + \frac{\partial v_2}{\partial x}\right)\frac{\partial R_1}{\partial x} + (p_1 - p_2)\left(1 + \left(\frac{\partial R_1}{\partial x}\right)^2\right) \\ &-2\mu_1\frac{\partial u_1}{\partial x}\left(\frac{\partial R_1}{\partial x}\right)^2 - 2\mu_1\frac{\partial v_1}{\partial r} + 2\mu_1\left(\frac{\partial u_1}{\partial r} + \frac{\partial v_1}{\partial x}\right)\frac{\partial R_1}{\partial x} = \sigma_1 J_1, \qquad r = R_1, \end{aligned} \tag{15}$$

$$T_1(R_1,x,t) = T_2(R_1,x,t), \qquad -K_1\nabla T_1\cdot\mathbf{n}_1 = -K_2\nabla T_2\cdot\mathbf{n}_1, \qquad r = R_1, \tag{16}$$

$$\frac{\partial T_1}{\partial r}(0,x,t) = 0, \qquad -K_2 \nabla T_2 \cdot \mathbf{n}_2 = \mathbf{q}_2, \qquad r = R_2, \tag{17}$$

where

$$J_i = \frac{\left(1 + \left(\frac{\partial R_i}{\partial x}\right)^2\right)^{\frac{1}{2}}}{R_i} - \frac{\frac{\partial^2 R_i}{\partial x^2}}{\left(1 + \left(\frac{\partial R_i}{\partial x}\right)^2\right)^{\frac{1}{2}}}, \tag{18}$$

σ_2 denotes the surface tension at the outer jet's outer surface, σ_1 is the surface tension at the interface between the inner and outer jets, $\mathbf{n}_1$ and $\mathbf{n}_2$ denote the outward unit normals at R_1 and R_2, respectively, $\mathbf{q}_2$ is the heat flux at the outer surface of the outer jet, $\nabla T = \left(\frac{\partial T}{\partial x}, \frac{\partial T}{\partial r}\right)$, and p_e is the pressure of the gases surrounding the compound liquid jet. These gases are assumed to be dynamically passive, i.e., p_e is only a function of time, since, in general, they have smaller density and dynamic viscosity than those of liquids. This implies that the gases surrounding the liquid may not introduce strong velocity variations along each cross section of the jet, although they may affect its dynamics. In addition, p_e can be set to zero.

The heat flux $\mathbf{q}_2$ includes the effects of convection and radiation, i.e., $\mathbf{q}_2 = h_c(T(R_2,x,t) - T_\infty) + \sigma\varepsilon^*(T(R_2,x,t)^4 - T_\infty^4)$, where h_c is the convective film heat transfer coefficient, T_∞ is the temperature of the gases that surround the compound fiber, σ is the Stefan–Boltzmann constant and ε^* is the emissivity of the compound fiber's outer surface. The heat flux can also be written as $\mathbf{q}_2 = h(T(R_2,x,t) - T_\infty)$, where $h = h_c + \sigma\varepsilon^*(T(R_2,x,t)^2 + T_\infty^2)(T(R_2,x,t) + T_\infty)$, which depends, amongst other parameters, on the fiber's temperature at R_2 and the velocity of the fiber. For example, Kase and Matsuo [30] determined experimentally that, for single–component, round jets, $h_c = C\frac{K_\infty}{R(x)}\left(\frac{2\rho_\infty u(x)R(x)}{\mu_\infty}\right)^m$ where K_∞, ρ_∞, and μ_∞ are the thermal conductivity, density and dynamic viscosity, respectively, of the air surrounding the round fiber at room temperature, $u(x)$ and $R(x)$ are the local axial velocity component and radius of the jet, respectively, and C and m are determined from experimental data.

Equation (11) implies that $(\mathbf{v}_2 - \mathbf{v}_1) \cdot \mathbf{n}_1 = 0$, i.e., the normal velocity at the outer–inner jet interface is continuous, where $\mathbf{v} = (u,v)$ denotes the velocity vector. If the fluids are viscous, then $(\mathbf{v}_2 - \mathbf{v}_1) \cdot \mathbf{t}_1 = 0$ at $r = R_1(t,x)$ where $\mathbf{t}_1 = (\cos\theta_1, \sin\theta_1)$, $\tan\theta_1 = \frac{\partial R_1}{\partial x}$, is the unit vector tangent to $r = R_1(t,x)$ and the tangential velocity components at this interface are continuous, i.e.,

$$u_1(R_1,x,t) = u_2(R_1,x,t). \tag{19}$$

In addition to the above boundary conditions in the radial direction, conditions in the axial direction must also be provided. If the compound jet emerges from a nozzle or die, there is a stress singularity at the nozzle–jet's interface due to the relaxation of the velocity profile from no-slip conditions at the nozzle walls to the free-surface flow away from the nozzle. This relaxation may result in a jet contraction or swelling which implies that the radial velocity component is of importance near the nozzle. Moreover, the stress singularity at the nozzle exit and the jet contraction or swelling near the nozzle may result in a relatively important radial pressure gradient near the nozzle; therefore, an accurate analysis of the flow near the nozzle requires a full solution of the Navier–Stokes equations within the nozzle and in the free surface flow, and requires the use of numerical methods.

In this chapter, a long wavelength approximation is used to reduce Eqs. (1)–(19) to a more manageable (and easier to solve) one–dimensional set of equations.

3 Asymptotic Analysis of Nonisothermal Compound Liquid Jets at Low Reynolds Numbers

For slender jets at low Reynolds number, i.e., $\varepsilon = \frac{R_0}{\lambda} << 1$, it is convenient to nondimensionalize r, x, t, u, v, ρ, C, K and p with respect to R_0, λ, λ/u_0, u_0, v_0, ρ_0, C_0, K_0 and $\mu u_0/\lambda$, respectively, where R and λ denote a characteristic radius and a characteristic wave length in the axial direction, e.g., the die's exit radius and the distance from the die's exit to the take–up point, respectively, u_0 is a characteristic (constant) axial velocity component, $v_0 = R_0 u_0/\lambda$, and ρ_0 and μ_0 are a reference density and viscosity, respectively. The temperature may be nondimensionalized as $(T - T_r)/\Delta T$ where $\Delta T = (T_0 - T_\infty)$, and $T_r = T_0$ and T_∞ denote a reference temperature at the die's exit and the temperature of the surroundings, respectively. Alternatively, the temperature can be nondimensionalized with respect to $\Delta T = T_0$ with $T_r = 0$. Since, the density of the fluids has been assumed to be constant, ρ_0 can be taken equal to, for example, ρ_2, although we have preferred to introduce a reference density for nondimensionalization purposes throughout the chapter; the same comment applies to the reference values employed for the nondimensionalization of the specific heats and thermal conductivities.

Using this nondimensionalization, Eqs. (1)–(19) become

$$\frac{\partial u_i}{\partial x} + \frac{1}{r}\frac{\partial (v_i r)}{\partial r} = 0, \tag{20}$$

$$\varepsilon \rho_i Re L(u_i) = -\varepsilon^2 \frac{\partial p_i}{\partial x} + \frac{\partial}{\partial x}\left(2\mu_i \varepsilon^2 \frac{\partial u_i}{\partial x}\right) + \frac{1}{r}\frac{\partial}{\partial r}\left(r\mu_i\left(\frac{\partial u_i}{\partial r} + \varepsilon^2 \frac{\partial v_i}{\partial x}\right)\right) + \rho_i \frac{Re}{Fr}, \tag{21}$$

$$\varepsilon \rho_i Re L(v_i) = -\frac{\partial p_i}{\partial r} + \frac{1}{r}\frac{\partial}{\partial r}\left(2\mu_i r \frac{\partial v_i}{\partial r}\right) + \frac{\partial}{\partial x}\left(\mu_i\left(\frac{\partial u_i}{\partial r} + \varepsilon^2 \frac{\partial v_i}{\partial x}\right)\right) - 2\mu_i \frac{v_i}{r^2}, \tag{22}$$

$$\varepsilon \rho_i C_i Pe L(T_i) = \varepsilon^2 K_i \frac{\partial^2 T_i}{\partial x^2} + \frac{K_i}{r}\frac{\partial}{\partial r}\left(r\frac{\partial T_i}{\partial r}\right) + \mu_i \Psi_i, \tag{23}$$

$$v_1(0,x,t) = 0, \qquad \frac{\partial u_1}{\partial r}(0,x,t) = 0, \qquad \frac{\partial p_1}{\partial r}(0,x,t) = 0, \tag{24}$$

$$v_2(R_2,x,t) = \frac{\partial R_2}{\partial t} + u_2(R_2,x,t)\frac{\partial R_2}{\partial x}, \tag{25}$$

$$v_i(R_1,x,t) = \frac{\partial R_1}{\partial t} + u_i(R_1,x,t)\frac{\partial R_1}{\partial x}, \tag{26}$$

$$2\varepsilon^2\left(\frac{\partial v_2}{\partial r} - \frac{\partial u_2}{\partial x}\right)\frac{\partial R_2}{\partial x} + \left(\frac{\partial u_2}{\partial r} + \varepsilon^2\frac{\partial v_2}{\partial x}\right)\left(1 - \varepsilon^2\left(\frac{\partial R_2}{\partial x}\right)^2\right) = 0, \qquad r = R_2, \tag{27}$$

$$\frac{\mu_2}{\mu_1}\left(2\varepsilon^2\left(\frac{\partial v_2}{\partial r}-\frac{\partial u_2}{\partial x}\right)\frac{\partial R_1}{\partial x}+\left(\frac{\partial u_2}{\partial r}+\varepsilon^2\frac{\partial v_2}{\partial x}\right)\left(1-\varepsilon^2\left(\frac{\partial R_1}{\partial x}\right)^2\right)\right)=$$

$$2\varepsilon^2\left(\frac{\partial v_1}{\partial r}-\frac{\partial u_1}{\partial x}\right)\frac{\partial R_1}{\partial x}+\left(\frac{\partial u_1}{\partial r}+\varepsilon^2\frac{\partial v_1}{\partial x}\right)\left(1-\varepsilon^2\left(\frac{\partial R_1}{\partial x}\right)^2\right), \qquad r=R_1, \tag{28}$$

$$\mu_2\left(2\varepsilon^2\frac{\partial u_2}{\partial x}\left(\frac{\partial R_2}{\partial x}\right)^2+2\frac{\partial v_2}{\partial r}-2\left(\frac{\partial u_2}{\partial r}+\varepsilon^2\frac{\partial v_2}{\partial x}\right)\frac{\partial R_2}{\partial x}\right)$$

$$+(p_e-p_2)\left(1+\varepsilon^2\left(\frac{\partial R_2}{\partial x}\right)^2\right)=-\frac{1}{\varepsilon Ca}J_2, \qquad r=R_2, \tag{29}$$

$$\mu_2\left(2\varepsilon^2\frac{\partial u_2}{\partial x}\left(\frac{\partial R_1}{\partial x}\right)^2+2\frac{\partial v_2}{\partial r}-2\left(\frac{\partial u_2}{\partial r}+\varepsilon^2\frac{\partial v_2}{\partial x}\right)\frac{\partial R_1}{\partial x}\right)$$

$$-\mu_1\left(2\varepsilon^2\frac{\partial u_1}{\partial x}\left(\frac{\partial R_1}{\partial x}\right)^2+2\frac{\partial v_1}{\partial r}-2\left(\frac{\partial u_1}{\partial r}+\varepsilon^2\frac{\partial v_1}{\partial x}\right)\frac{\partial R_1}{\partial x}\right)=$$

$$-(p_1-p_2)\left(1+\varepsilon^2\left(\frac{\partial R_1}{\partial x}\right)^2\right)+\frac{\sigma_1}{\sigma_2}\frac{1}{\varepsilon Ca}J_1, \qquad r=R_1, \tag{30}$$

$$T_1(R_1,x,t)=T_2(R_1,x,t), \tag{31}$$

$$K_1\left(\frac{\partial T_1}{\partial r}-\varepsilon^2\frac{\partial T_1}{\partial x}\frac{\partial R_1}{\partial x}\right)=K_2\left(\frac{\partial T_2}{\partial r}-\varepsilon^2\frac{\partial T_2}{\partial x}\frac{\partial R_1}{\partial x}\right), \qquad r=R_1, \tag{32}$$

$$\frac{\partial T_1}{\partial r}(0,x,t)=0, \tag{33}$$

$$K_2\left(\frac{\partial T_2}{\partial r}-\varepsilon^2\frac{\partial T_2}{\partial x}\frac{\partial R_2}{\partial x}\right)=Bi(T(R_2,x,t)-T_\infty), \qquad r=R_2, \tag{34}$$

where

$$\psi_i=2Br\left(\varepsilon^2\left(\frac{\partial v_i}{\partial r}\right)^2+\varepsilon^2\left(\frac{\partial u_i}{\partial x}\right)^2+\varepsilon^2\left(\frac{v_i}{r}\right)^2+\frac{1}{2}\left(\frac{\partial u_i}{\partial r}+\varepsilon^2\frac{\partial v_i}{\partial x}\right)^2\right). \tag{35}$$

$$J_i=\frac{\left(1+\varepsilon^2\left(\frac{\partial R_i}{\partial x}\right)^2\right)^{\frac{1}{2}}}{R_i}-\varepsilon^2\frac{\frac{\partial^2 R_i}{\partial x^2}}{(1+\varepsilon^2(\frac{\partial R_i}{\partial x})^2)^{\frac{1}{2}}}, \tag{36}$$

$$L(u)=\frac{\partial u}{\partial t}+u\frac{\partial u}{\partial x}+v\frac{\partial u}{\partial r}, \tag{37}$$

and

$$Re = \frac{\rho_0 u_0 R_0}{\mu_0}, \qquad Fr = \frac{u_0^2}{gR_0}, \qquad Ca = \frac{\mu_0 u_0}{\sigma_2}, \tag{38}$$

$$Pe = RePr, \qquad Pr = \frac{\mu_0 C_0}{K_0}, \qquad Br = \frac{\mu_0 u_0^2}{K_0 \Delta T}, \qquad Bi = \frac{hR_0}{K_0}, \tag{39}$$

denote the Reynolds, Froude, capillary, thermal Péclet, Prandtl, Brinkman and Biot numbers, respectively, and the same symbols have been used for dimensional and dimensionless quantities for the sake of brevity. Note that the surface tensions have been assumed to be constant, i.e., independent of the temperature, and h contains the effects of convection and radiation as discussed previously.

The nondimensional viscosity law corresponding to Eq. (8) can be written as

$$\mu_i = \psi_i \exp(G_i(1 - T_i)), \tag{40}$$

when the temperature is nondimensionalized with respect to T_0, so that μ_i increases as T_i decreases. Analogous expressions can also be derived for other viscosity laws, i.e., Eqs. (6) or (7).

Equations (20)–(37) depend on ε, Re, Fr, Ca, p_e, Pr, Br, Bi, $\frac{\sigma_1}{\sigma_2}$, $\frac{C_1}{C_2}$, $\frac{K_1}{K_2}$, $\frac{\rho_1}{\rho_2}$, $\frac{\psi_1}{\psi_2}$, G_1 and G_2. In addition, the solution to these equations also depends on the boundary conditions at the die's exit and take–up point. This large set of parameters does not allow us to obtain simpler equations except for slender geometries and small values of the Reynolds, Biot and Brinkman numbers. We, therefore, consider typical operating conditions in the manufacture of compound fibers which are usually characterized by small values of the Reynolds, Biot and Brinkman numbers.

For small Reynolds numbers, $Re = \varepsilon\bar{R}$ with $\bar{R} = O(1)$, and, depending on the magnitude of the Froude, Biot, Brinkman and capillary numbers, several flow regimes can be identified. Here, we consider $Fr = \bar{F}/\varepsilon$, $Ca = \bar{C}/\varepsilon$, $Bi = \varepsilon^2\bar{B}$, $Br = \varepsilon^4\bar{b}$ where $\bar{F} = O(1)$, $\bar{C} = O(1)$, $\bar{B} = O(1)$ and $\bar{b} = O(1)$ which correspond to small gravitational effects, small surface tension, small heat convection and small heat generation due to viscous dissipation. Substitution of these values and expansion of the dependent variables R_1, R_2, u_i, v_i, p_i and T_i, where $i = 1,2$ as

$$\phi = \phi_0 + \varepsilon^2\phi_2 + O(\varepsilon^4), \tag{41}$$

where ϕ denotes dependent variables, into the governing equations and boundary conditions, together with the expansion of the boundary conditions at $R_1(t,x)$ and $R_2(t,x)$ about $R_{10}(t,x)$ and $R_{20}(t,x)$, respectively, yield asymptotic expansions which at leading order, i.e., at $O(\varepsilon^0)$, correspond to

$$\frac{\partial u_{i0}}{\partial x} + \frac{1}{r}\frac{\partial(v_{i0}r)}{\partial r} = 0, \qquad i = 1,2, \tag{42}$$

$$\frac{1}{r}\frac{\partial}{\partial r}\left(r\mu_{i0}\frac{\partial u_{i0}}{\partial r}\right) = 0, \qquad i = 1,2, \tag{43}$$

$$-\frac{\partial p_{i0}}{\partial r}+\frac{1}{r}\frac{\partial}{\partial r}\left(r\mu_{i0}\frac{\partial v_{i0}}{\partial r}\right)+\frac{\partial}{\partial x}\left(r\mu_{i0}\frac{\partial u_{i0}}{\partial r}\right)-2\mu_{i0}\frac{v_{i0}}{r^2}=0, \qquad i=1,2, \tag{44}$$

$$\frac{1}{r}\frac{\partial}{\partial r}\left(r\frac{\partial T_{i0}}{\partial r}\right)=0, \qquad i=1,2, \tag{45}$$

$$v_{10}(0,x,t)=0, \qquad \frac{\partial u_{10}}{\partial r}(0,x,t)=0, \qquad \frac{\partial p_{10}}{\partial r}(0,x,t)=0, \tag{46}$$

$$\frac{\partial u_{20}}{\partial r}(R_{20},x,t)=0, \qquad \frac{\mu_{20}}{\mu_{10}}\frac{\partial u_{20}}{\partial r}(R_{10},x,t)=\frac{\partial u_{10}}{\partial r}(R_{10},x,t), \tag{47}$$

$$2\mu_{20}\frac{\partial v_{20}}{\partial r}(R_{20},x,t)-2\mu_{20}\frac{\partial u_{20}}{\partial r}(R_{20},x,t)\frac{\partial R_{20}}{\partial x}+p_e-p_{20}(R_{20},x,t)=-\frac{1}{\bar{C}R_{20}}, \tag{48}$$

$$2\mu_{20}\frac{\partial v_{20}}{\partial r}(R_{10},x,t)-2\mu_{20}\frac{\partial u_{20}}{\partial r}(R_{10},x,t)\frac{\partial R_{10}}{\partial x}+p_{10}(R_{10},x,t)-p_{20}(R_{10},x,t)$$
$$-2\mu_{10}\frac{\partial v_{10}}{\partial r}(R_{10},x,t)+2\mu_{10}\frac{\partial u_{10}}{\partial r}(R_{10},x,t)\frac{\partial R_{10}}{\partial x}=\frac{\sigma_1}{\sigma_2}\frac{1}{\bar{C}R_{10}}, \tag{49}$$

$$\frac{\partial T_{10}}{\partial r}(0,x,t)=0, \qquad \frac{K_2}{K_1}\frac{\partial T_{20}}{\partial r}(R_{10},x,t)=\frac{\partial T_{10}}{\partial r}(R_{10},x,t), \qquad \frac{\partial T_{20}}{\partial r}(R_{20},x,t)=0, \tag{50}$$

for Eqs. (20)–(24) and (27)–(34). The solutions to these equations can be written as

$$u_{i0}=B_i(t,x), \qquad p_{i0}=D_i(t,x), \qquad T_{i0}=E_i(t,x), \qquad i=1,2, \tag{51}$$

$$v_{i0}=\frac{C_i}{r}-\frac{\partial B_i}{\partial x}\frac{r}{2}, \qquad i=1,2, \tag{52}$$

which imply that, at leading–order the axial velocity component $B_1=B_2\equiv B$ (cf. Eq. (19)), $E_1=E_2$ (cf. Eq. (50)), and the pressure and temperature are only functions of the axial coordinate and time. $C_1(t,x)$ and $C_2(t,x)$ can be determined from the kinematic conditions provided by Eqs. (25) and (26) at leading–order and $v_{10}(0,x,t)=0$ (cf. Eq. (46)) as

$$C_1=C_2=\frac{\partial}{\partial t}\left(\frac{R_{10}^2}{2}\right)+\frac{\partial}{\partial x}\left(B\frac{R_{10}^2}{2}\right)=\frac{\partial}{\partial t}\left(\frac{R_{20}^2}{2}\right)+\frac{\partial}{\partial x}\left(B\frac{R_{20}^2}{2}\right)=0, \tag{53}$$

which, in turn, imply that

$$\frac{\partial A_1}{\partial t}+\frac{\partial}{\partial x}(BA_1)=0, \qquad \frac{\partial A_2}{\partial t}+\frac{\partial}{\partial x}(BA_2)=0, \tag{54}$$

where $A_1=\frac{R_{10}^2}{2}$ and $A_2=\frac{R_{20}^2-R_{10}^2}{2}$. Equation (54) corresponds to the mass conservation at leading–order.

It must be noted that, since the dynamic viscosity is a function of the temperature, i.e., $\mu_i=\mu(T_{i0}+\varepsilon^2T_{i2}+O(\varepsilon^4))$, and a Taylor's series expansion about T_{i0} shows that $\mu(T_i)=\mu_{i0}+\varepsilon^2\mu_{i2}+O(\varepsilon^4)$, where $\mu_{i0}=\mu_i(T_{i0})$ and $\mu_{i2}=\frac{d\mu_i}{dT}(T_{i0})T_{i2}$.

From Eqs. (48), (49) and (52), one can easily obtain the following equations

$$p_{20} = D_2 = p_e + \frac{1}{\bar{C}}\frac{1}{R_{20}} - \mu_{20}\frac{\partial B}{\partial x}, \tag{55}$$

$$p_{10} = D_1 = p_e + \frac{1}{\bar{C}}\left(\frac{1}{R_{20}} + \frac{\sigma_1}{\sigma_2}\frac{1}{R_{10}}\right) - \mu_{10}\frac{\partial B}{\partial x}. \tag{56}$$

Equations (51), (52) and (54)–(56) contain four unknowns, i.e., $B = B_1 = B_2$, R_{10}, R_{20}, and $E_1 = E_2 \equiv E$; therefore, these equations are not a closed system. In order to close this system of equations, it is necessary to go to higher orders in the asymptotic expansion.

At $O(\varepsilon^2)$, the axial momentum equations yield

$$\frac{\partial}{\partial r}\left(r\mu_{i0}\frac{\partial u_{i2}}{\partial r}\right) = r(M_i + G_i), \qquad i = 1,2, \tag{57}$$

where

$$M_i = \bar{R}\rho_i\frac{DB}{Dt} - \frac{\bar{R}}{\bar{F}}\rho_i + \frac{\partial D_i}{\partial x} - \frac{\partial}{\partial x}\left(2\mu_{i0}\frac{\partial B}{\partial x}\right), \qquad G_i = -\frac{1}{2r}\frac{\partial^2 B}{\partial x^2}\frac{\partial}{\partial r}(r^2\mu_{i0}), \tag{58}$$

The shear stress conditions at the interfaces (cf. Eqs. (27) and (28)), together with v_{i0} (cf. Eq. (52)) and the no–slip condition provided by Eq. (19) at second–order yield, after lengthy algebra,

$$\begin{aligned}(\rho_1 A_1 + \rho_2 A_2)\frac{DB}{Dt} &= (\rho_1 A_1 + \rho_2 A_2)\frac{\bar{R}}{\bar{F}} + \frac{\partial}{\partial x}\left((3\mu_{10}A_1 + 3\mu_{20}A_2)\frac{\partial B}{\partial x}\right)\\ &+ A_1\frac{1}{\bar{C}}\left(\frac{1}{R_{20}^2}\frac{\partial R_{20}}{\partial x} + \frac{\sigma_1}{\sigma_2}\frac{1}{R_{10}^2}\frac{\partial R_{10}}{\partial x}\right) + A_2\frac{1}{\bar{C}R_{20}^2}\frac{\partial R_{20}}{\partial x},\end{aligned} \tag{59}$$

where $\frac{DB}{Dt} = \frac{\partial B}{\partial t} + B\frac{\partial B}{\partial x}$.

The energy Eq. (23) at second–order yields

$$T_{i2} = \left(\rho_i C_i \bar{P}\frac{DT_{i0}}{Dt} - \frac{\partial^2 T_{i0}}{\partial x^2}\right)\frac{r^2}{4} + H_i \ln r + R_i, \qquad i = 1,2, \tag{60}$$

where $\bar{P} = \bar{R}Pr$, and H_i and R_i are functions of x and t. Equation (60) subject to the symmetry condition at $r = 0$ for $i = 1$ and the thermal boundary conditions at the interfaces (at second–order) yields

$$(\rho_1 C_1 A_1 + \rho_2 C_2 A_2)\bar{P}\frac{DE}{Dt} = \frac{\partial}{\partial x}\left((A_1 K_1 + A_2 K_2)\frac{\partial E}{\partial x}\right) - \frac{\bar{B}}{K_2}R_{20}(E - T_\infty), \tag{61}$$

which clearly shows that the compound jet behaves as a single–component jet with a thermal inertia and conductivity which are proportional to the cross–sectional areas and the thermal capacities and conductivities of the materials that compose the jet.

Equations (54), (59) and (61) reduce to those derived by Schultz and Davis [14, 24] and Gupta and Schultz [31] for steady, single–component jets under isothermal and nonisothermal conditions, respectively.

Equations (54), (59) and (61) constitute a system of four one–dimensional equations for R_{10}, R_{20}, B and E. This system is asymptotically valid for small Reynolds, Biot and Brinkman numbers and large values of the Froude and capillary numbers, is much simpler than Eqs. (1)–(19), and includes the effects of the temperature through Eq. (61) and the dependence of the viscosity on temperature in the linear momentum equations (cf. Eq. (40)). However, this system of one–dimensional equations implies that the leading–order temperatures and axial velocity components of the inner and outer jets are the same. Moreover, no latent heat and no air drag effects have been considered, and viscous dissipation has been assumed to be small.

It must be noted that Eqs. (54), (59) and (61) are also valid for $Br = O(\varepsilon^2)$ because of Eq. (35) at leading–order, and their validity requires that either the differences in axial velocity or the differences in temperature across the compound fiber's cross–section be sufficiently small [14, 24, 31], or the corrections to both the inner and outer surface's radius be small. Either of these criteria can be employed to verify the validity of the one–dimensional Eqs. (54), (59) and (61) derived in this section, by using, if necessary, the results of numerical simulations.

3.1 Axial and Interfacial Stresses

The (dimensional) axial stresses on the fiber are equal to $2\mu_i \frac{\partial u_i}{\partial x} - p_i$, which can be integrated in the radial direction to determine the (local) axial tensile forces on the compound jet. These axial forces can be nondimensionalized with respect to $\varepsilon\pi\mu_0 u_0 R_0$ and their leading–order values can be written in dimensionless form as

$$F_x^{(i)} = A_1 \left(2\mu_{i0} \frac{\partial B_i}{\partial x} - D_i \right). \tag{62}$$

Equation (62) can be written, using the results previously obtained in this section, as (cf. Eqs. (55) and (56))

$$F_x^{(1)} = A_1 \left(3\mu_{10} \frac{\partial B_1}{\partial x} - p_e - \frac{1}{\bar{C}} \left(\frac{1}{R_{20}} + \frac{\sigma_1}{\sigma_2} \frac{1}{R_{10}} \right) \right), \tag{63}$$

$$F_x^{(2)} = A_2 \left(3\mu_{20} \frac{\partial B_2}{\partial x} - p_e - \frac{1}{\bar{C} R_{20}} \right), \tag{64}$$

The stresses exerted at the interface $R_1(t,x)$ can be decomposed into normal and tangential components here referred to as interfacial normal and shear tractions, and written as

$$\sigma_n = \frac{1}{1+\tan^2\theta} \left((\sigma_{xx} - p)\tan^2\theta - 2\sigma_{xr}\tan\theta + \sigma_{rr} - p \right), \tag{65}$$

$$\sigma_t = \frac{1}{1+\tan^2\theta} \left((\sigma_{rr} - \sigma_{xx})\tan\theta + \sigma_{xr}(1-\tan^2\theta) \right), \tag{66}$$

where $\tan\theta = \frac{\partial R}{\partial x}$, and σ_{ij} denotes the components of the viscous stress tensor.

Upon introducing the dimensionless defined previously and nondimensionalizing σ_n and σ_t with respect to $\frac{\mu_0 u_0}{\lambda}$, it is easy to show that the interfacial stresses at $r = R_{10}(t,x)$ are

$$\sigma_n^{(1)} = -p_e - \frac{1}{\bar{C}}\left(\frac{1}{R_{20}} + \frac{\sigma_1}{\sigma_2}\frac{1}{R_{10}}\right) + O(\varepsilon^2), \tag{67}$$

$$\sigma_n^{(2)} = -p_e - \frac{1}{\bar{C}}\frac{1}{R_{20}} + O(\varepsilon^2), \tag{68}$$

$$\sigma_t^{(1)} = -\mu_{10}\left(3\frac{\partial R_{10}}{\partial x}\frac{\partial B}{\partial x} + \frac{R_{10}}{2}\frac{\partial^2 B}{\partial x^2}\right) + O(\varepsilon), \tag{69}$$

$$\sigma_t^{(2)} = -\mu_{20}\left(3\frac{\partial R_{10}}{\partial x}\frac{\partial B}{\partial x} + \frac{R_{10}}{2}\frac{\partial^2 B}{\partial x^2}\right) + O(\varepsilon), \tag{70}$$

so that

$$\sigma_n^{(1)} - \sigma_n^{(2)} = -\frac{1}{\bar{C}}\frac{\sigma_1}{\sigma_2}\frac{1}{R_{10}} + O(\varepsilon^2), \qquad \sigma_t^{(1)} = \sigma_t^{(2)}. \tag{71}$$

4 Isothermal, Compound Fibers

In this section, we first determine some analytical solutions for steady, isothermal, compound jets or fibers, then we analyze the linear stability of the viscous flow regime, and, finally, we present some results for both steady and unsteady, isothermal, compound fibers.

4.1 Analytical Solutions for Steady, Isothermal, Compound Fibers

For steady, isothermal, compound fibers, Eq. (54) has the following solutions

$$A_1 B = Q_1, \qquad A_2 B = Q_2, \tag{72}$$

where $Q_i, i = 1,2$, are constants, $Q_1 + Q_2 = Q_T$ and $B = B_1 = B_2$.

Viscous regime. This regime is characterized by $\bar{R} = 0$, finite values of $\bar{F}$, and no surface tension, i.e., Eq. (59) becomes

$$\frac{d}{dx}\left(3\left(A_2 + \frac{\mu_1}{\mu_2}A_1\right)\frac{dB}{dx}\right) = 0, \tag{73}$$

and has the following solution

$$B(x) = \beta\exp(\alpha x), \tag{74}$$

where $\frac{1}{\alpha} = 3\left(Q_2 + \frac{\mu_1}{\mu_2}Q_1\right)$ and β is an integration constant. Note that $\beta = 1$ because $B(0) = 1$ and $\alpha = \ln B(1)$. The values of R_{10} and R_{20} can be easily determined from Eq. (72) as

$$R_{10}(x) = \sqrt{\frac{2Q_1}{\beta}}\exp\left(-\frac{1}{2}\alpha x\right), \qquad R_{20}(x) = \sqrt{\frac{2Q_T}{\beta}}\exp\left(-\frac{1}{2}\alpha x\right). \tag{75}$$

Viscous–capillary regime. This regime is characterized by Eq. (72), $\bar{R}=0$ and $\frac{\bar{R}}{F}=0$, i.e., Eq. (59) becomes

$$\frac{d}{dx}\left(3\left(A_2+\frac{\mu_1}{\mu_2}A_1\right)\frac{dB}{dx}\right)+\frac{1}{2\bar{C}}\left(\frac{dR_{20}}{dx}+\frac{\sigma_1}{\sigma_2}\frac{dR_{10}}{dx}\right)=0, \tag{76}$$

and has the following solution

$$B_1=\left(\delta\exp\left(\frac{\alpha}{2P}x\right)+\frac{Q}{\alpha}\right)^2, \tag{77}$$

where α and δ are integration constants, $P=3\left(Q_2+\frac{\mu_1}{\mu_2}Q_1\right)$, $Q=\frac{1}{\sqrt{2\bar{C}}}\left(\sqrt{Q_T}+\sqrt{Q_1}\frac{\sigma_1}{\sigma_2}\right)$. Note that δ and α depend on the axial velocity at the take–up point, i.e., they depend on $B(1)$. The values of R_{10} and R_{20} can be easily determined from Eq. (72), but they are not shown here.

Viscous–gravitationless regime. This regime corresponds to Eq. (72), $Fr=\infty$ and $\bar{R}\neq 0$, i.e., Eq. (59) becomes

$$\bar{R}\left(A_2+\frac{\rho_1}{\rho_2}A_1\right)B\frac{dB}{dx}=\frac{d}{dx}\left(3\left(A_2+\frac{\mu_1}{\mu_2}A_1\right)\frac{dB}{dx}\right), \tag{78}$$

which has the following solution

$$B(x)=\frac{\alpha\beta}{P}\frac{\exp(\alpha x)}{1-\beta\exp(\alpha x)}, \tag{79}$$

where α and $\beta=\frac{P}{\alpha+P}$ are integration constants which can be easily determined from the conditions $B(0)=1$ and $B(1)$, and

$$P=\bar{R}\frac{Q_2+\frac{\rho_1}{\rho_2}Q_1}{3\left(Q_2+\frac{\mu_1}{\mu_2}Q_1\right)}, \tag{80}$$

Viscous–gravitationless–capillary regime. This regime corresponds to Eq. (72), $\bar{R}\neq 0$ and $\frac{\bar{R}}{F}=0$, i.e., Eq. (59) becomes

$$Re\left(A_2+\frac{\rho_1}{\rho_2}A_1\right)B\frac{dB}{dx}=\frac{d}{dx}\left(3\left(A_2+\frac{\mu_1}{\mu_2}A_1\frac{dB}{dx}\right)\right)+\frac{1}{2\bar{C}}\left(\frac{dR_{20}}{dx}+\frac{\sigma_1}{\sigma_2}\frac{dR_0}{dx}\right), \tag{81}$$

which can be easily integrated upon making the change $B=z^2$ and has different solutions depending on the roots of $Pz^3+\alpha z-Q=0$ where P and Q have been defined previously, and α is an integration constant.

Let us define $q=\frac{\alpha}{3P}$ and $r=\frac{Q}{2P}$. Then, the solution to Eq. (81) is

$$\prod_{i=1}^{3}(z-z_i)^{\Pi_{j=1,j\neq i}^{3}(z_i-z_j)^{-1}}=\beta\exp\left(\frac{P}{2}x\right), \tag{82}$$

if $q^3 + r^2 < 0$ and z_i are the different (real) roots (z^*) of

$$Pz^3 + \alpha z - Q = 0, \tag{83}$$

$$-\frac{1}{2}\frac{1}{(z-z^*)^2} = \beta + \frac{P}{2}x, \tag{84}$$

which corresponds to $q^3 + r^2 = 0$ and three identical (real) roots of Eq. (83),

$$\left(\frac{z-z_1}{z-z_2}\right)^{\frac{1}{(z_2-z_1)^2}} \exp\left(\frac{(z_2-z_1)}{(z-z_1)}\right) = \beta \exp\left(\frac{P}{2}(z_2-z_1)^2 x\right), \tag{85}$$

which corresponds to $q^3 + r^2 = 0$ and two identical (real) roots z_1 of Eq. (83), and

$$\beta + \frac{P}{2}x = -\frac{1}{(z_1-z_r)^2 + z_i^2}$$
$$\left(\frac{z_1-z_r}{z_i}\arctan\frac{z-z_r}{z_i} + \ln\frac{\left((z-z_r)^2+z_i^2\right)^{\frac{1}{2}}}{z-z_1}\right), \tag{86}$$

which corresponds to $q^3 + r^2 > 0$, a real root (z_1) and two complex conjugate ones ($z_r \pm \mathrm{i} z_i$), $\mathrm{i}^2 = -1$, of Eq. (83), and β and α are integration constants which can be determined from $B(0) = 1$ and depend on the value of the take–up velocity, i.e., $B(1)$.

4.2 Linear Stability of the Viscous Flow Regime

The linear stability of the analytical solutions presented in the previous section requires, in general, the use of numerical techniques due to the spatial dependence of the steady state values B, R_{10} and R_{20} on x. However, for the viscous regime analyzed in the previous section, it is possible to perform a linear stability analysis analytically, because this regime is governed by Eqs. (72), (74) and (cf. Eq. (73))

$$\frac{\partial}{\partial x}\left(3\left(A_2 + \frac{\mu_1}{\mu_2}A_1\right)\frac{\partial B_1}{\partial x}\right) = 0. \tag{87}$$

Let us define the steady state solutions obtained for the viscous regime in the previous section with the subscript *ss*, so that these steady solutions may be written as

$$B_{ss}(x) = \exp(\alpha x), \qquad A_{1_{ss}}(x) = Q_1 \exp(-\alpha x), \qquad A_{2_{ss}}(x) = Q_2 \exp(-\alpha x), \tag{88}$$

and perturb this steady solution as

$$B(t,x) = B_{ss}(1+b), \qquad A_1(t,x) = A_{1_{ss}}(1+a_1), \qquad A_2(t,x) = A_{2_{ss}}(1+a_2), \tag{89}$$

where b, a_1 and a_2 are functions of t and x.

Substitution of Eqs. (88) and (89) into Eqs. (54) and (59) and neglecting non-linear terms in perturbations yield the following linear partial differential equations

$$A_{1_{ss}}\frac{\partial a_1}{\partial t}+Q_1\left(\frac{\partial a_1}{\partial x}+\frac{\partial b}{\partial x}\right)=0, \tag{90}$$

$$A_{2_{ss}}\frac{\partial a_2}{\partial t}+Q_2\left(\frac{\partial a_2}{\partial x}+\frac{\partial b}{\partial x}\right)=0, \tag{91}$$

$$\alpha Q_2\frac{\partial a_2}{\partial x}+\frac{\mu_1}{\mu_2}\alpha Q_1\frac{\partial a_1}{\partial x}+\alpha\left(Q_2+\frac{\mu_1}{\mu_2}Q_1\right)\frac{\partial b}{\partial x}+\left(Q_2+\frac{\mu_1}{\mu_2}Q_1\right)\frac{\partial^2 b}{\partial x^2}=0, \tag{92}$$

which, upon using

$$a_1=\hat{a}_1(x)\exp(\sigma t),\qquad a_2=\hat{a}_2(x)\exp(\sigma t),\qquad b=\hat{b}(x)\exp(\sigma t), \tag{93}$$

where σ is a complex number, integration of the perturbed axial momentum equation and use of $\hat{a}_1(0)=\hat{a}_2(0)=\hat{b}(0)=\hat{b}(1)=0$, can be written as

$$\sigma\exp(-\alpha x)\hat{a}_1+\frac{d\hat{a}_1}{dx}+\frac{d\hat{b}}{dx}=0, \tag{94}$$

$$\sigma\exp(-\alpha x)\hat{a}_2+\frac{d\hat{a}_2}{dx}+\frac{d\hat{b}}{dx}=0, \tag{95}$$

$$\alpha Q_2\hat{a}_2+\frac{\mu_1}{\mu_2}\alpha Q_1\hat{a}_1+\alpha\left(Q_2+\frac{\mu_1}{\mu_2}Q_1\right)\hat{b}+\left(Q_2+\frac{\mu_1}{\mu_2}Q_1\right)\frac{d\hat{b}}{dx}=\delta, \tag{96}$$

where δ is an integration constant.

Equations (94)–(96) imply that

$$\frac{d\hat{b}}{dx}(1)=-\sigma\exp(-\alpha)\hat{a}_1(1)-\frac{d\hat{a}_1}{dx}(1)=-\sigma\exp(-\alpha)\hat{a}_2(1)-\frac{d\hat{a}_2}{dx}(1), \tag{97}$$

$$\delta=\alpha Q_2\hat{a}_2(1)+\frac{\mu_1}{\mu_2}\alpha Q_1\hat{a}_1(1)+\left(Q_2+\frac{\mu_1}{\mu_2}Q_1\right)\frac{d\hat{b}}{dx}(1), \tag{98}$$

$$\frac{d\hat{b}}{dx}(0)=-\frac{d\hat{a}_1}{dx}(0)=-\frac{d\hat{a}_2}{dx}(0),\qquad \delta=\left(Q_2+\frac{\mu_1}{\mu_2}Q_1\right)\frac{d\hat{b}}{dx}(0), \tag{99}$$

which together with Eqs. (94)–(96) yield

$$\hat{a}_1=\hat{a}_2,\qquad (\alpha-\sigma\exp(-\alpha))\hat{a}_1(1)=\frac{d\hat{a}_1}{dx}(1)-\frac{d\hat{a}_1}{dx}(0), \tag{100}$$

$$\frac{d^2P}{dx^2}+\sigma\exp(-\alpha x)\frac{dP}{dx}=0, \tag{101}$$

where $P=Q_2\hat{a}_2+Q_1\frac{\mu_1}{\mu_2}\hat{a}_1$.

Integration of Eq. (101) and use of the boundary conditions given by Eqs. (97)–(99) yield the following eigenvalue problem

$$\int_{\frac{\sigma}{\alpha e^{\alpha}}}^{\frac{\sigma}{\alpha}} \frac{e^{z}}{z} dz = \frac{e^{\frac{\sigma}{\alpha e^{\alpha}}} - e^{\frac{\sigma}{\alpha}}}{1 - e^{\frac{\sigma}{\alpha e^{\alpha}}}}, \tag{102}$$

which coincides exactly with that derived by Davis and Schultz [24] for round jets; it also coincides with that derived by Ramos for isothermal, compound jets [21], isothermal, planar film casting processes [22], and isothermal, hollow, compound fibers [27].

The eigenvalue whose real part is zero corresponds to $\alpha_c = 3.00650, B(1) = 20.21$ and $\sigma_i = 14.011$, where σ_i denotes the imaginary part of σ. Equation (102) clearly shows that the eigenvalue σ only depends on α, i.e., the axial velocity at the take-up point, and the results obtained by Schultz and Davis [24] for round jets at low Reynolds numbers apply to compound jets. This is not surprising, for, in the viscous regime, gravitational and inertia effects (which depend on the density ratio) are absent from this regime,.

Stability analyses similar to the one performed in this section for the viscous regime can also be performed for the other flow regimes for which analytical solutions were obtained in the previous section under steady conditions. The stability of these regimes as well as any other steady solution requires the use of numerical techniques based upon the discretization of the equations for the perturbed quantities.

4.3 Sample Results for Steady, Isothermal, Compound Fibers

As shown in previous sections, the nonlinear dynamics of steady, isothermal, Newtonian, compound liquid jets at low Reynolds numbers depends on $\bar{R}$, $\frac{\bar{R}}{F}$, $\bar{C}$, p_e, $\frac{\rho_1}{\rho_2}$, $\frac{\mu_1}{\mu_2}$, $\frac{\sigma_1}{\sigma_2}$, Q_1, Q_2, and $B_{ss}(1)$. In this section, we present some sample results that have been obtained by solving numerically the steady state equations corresponding to steady, isothermal, compound jets. The numerical technique for solving these equations employs central differences for the diffusion terms and upwind differences for the convection ones, and the number of grid points in the x–direction was varied in order to obtain grid–independent results. For most of the results presented in this section, the number of grid points in the axial direction was, at least 2001, which correspond to a grid spacing $\Delta x = 5 \times 10^{-4}$.

Some sample results illustrating the steady state compound jet's geometry, axial velocity component, axial traction on the core and the ratio of the axial traction in the core to that of the cladding are presented in Figures 2 and 3. Figure 2 corresponds to Cases 1, 17 and 18 of Table 1 that list the parameters used in the calculations and $B(1) = 200$, and indicates that the axial velocity increases rapidly near the downstream or take–up point where it exhibits a boundary layer structure; the thickness of this boundary layer increases as the Reynolds number is decreased. Figure 2 also shows that the axial traction on the inner round jet or core also increases quite rapidly near the take–up point, except at low Reynolds numbers for which it increases rather smoothly from the upstream to the downstream boundaries. The compound jet's geometry illustrated in Figure 2 clearly shows the jet's contraction near the nozzle exit at low Reynolds numbers; the contraction at higher Reynolds numbers is large

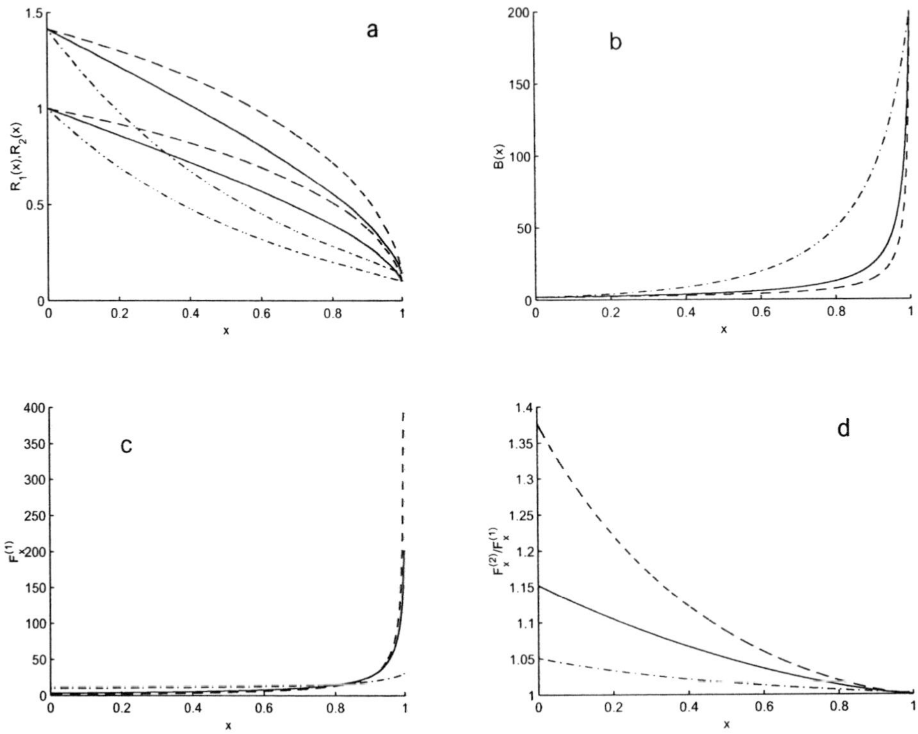

Figure 2: Compound jet's geometry (a), axial velocity component (b), axial traction on the inner jet (c) and ratio of axial traction on the outer jet to that on the inner one (d) for Cases 1, 17 and 18 of Table 1. (Solid lines: Case 1; dashed lines: Case 17; dashed–dotted lines: Case 18).

at the take–up point where the axial velocity component is largest. Figure 2 also indicates that the ratio of the axial traction on the outer jet to that on the inner one (hereon, referred to as the axial traction ratio) is largest at the upstream boundary and increases as the Reynolds number is increased; this ratio decreases downstream and tends towards a value equal to unity.

For Cases 1, 2 and 3 of Table 1 which correspond to different values of $\frac{\bar{R}}{\bar{F}}$, very few differences in the leading–order axial velocity component and axial traction of the inner jet were observed; however, the compound jet's contraction increased whereas the axial traction ratio decreased as $\frac{\bar{R}}{\bar{F}}$ was decreased, thus indicating that an increase in the gravitational acceleration increases the jet contraction. The ratio of traction forces at the nozzle exit or die was found to be about 1.16, 1.15 and 1.12 for $\frac{\bar{R}}{\bar{F}} = 0.1$, 1 and 10, respectively, and decreased towards unity at the downstream boundary.

The compound jet's geometry, leading–order axial velocity and axial traction force on the inner jet were not found to be very sensitive to the capillary number for $\bar{C} = 1$, 10 and 0.5 (Cases 1, 4 and 5 of Table 1), and the effects of the capillary number were found to be

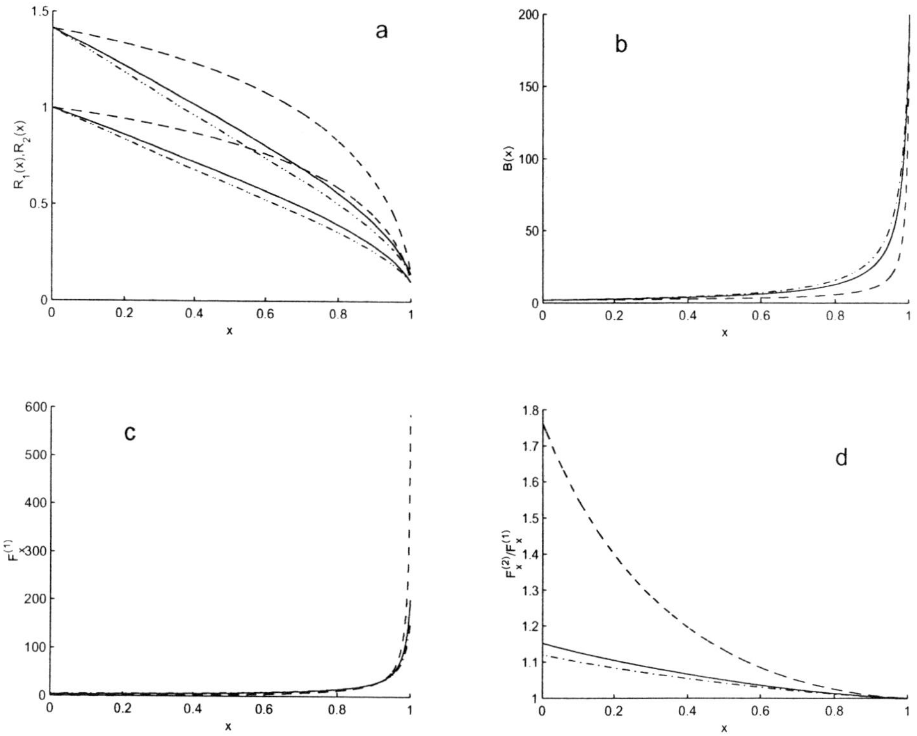

Figure 3: Compound jet's geometry (a), axial velocity component (b), axial traction on the inner jet (c) and ratio of axial traction on the outer jet to that on the inner one (d) for Cases 1, 7 and 8 of Table 1. (Solid lines: Case 1; dashed lines: Case 7; dashed–dotted lines: Case 8).

negligible for $\bar{C} = 100$, i.e., Case 6 of Table 1; however, the ratio of axial traction forces at the upstream boundary was found to be equal to about 1.45, 1.15 and 1.02 for $\bar{C} = 0.5$, 1 and 10, respectively.

The contraction of the steady state compound jet's radii, the axial traction on the inner jet and the gradient of the axial velocity component at the downstream boundary were found to increase as $\frac{\rho_1}{\rho_2}$ was increased as indicated in Figure 3 which corresponds to Cases 1, 7 and 8 of Table 1; both the axial traction force on the inner jet and its gradient increase at the downstream boundary as $\frac{\rho_1}{\rho_2}$ is increased. The ratio of axial traction forces at the upstream boundary was found to be equal to about 1.75, 1.15 and 1.12 for $\frac{\rho_1}{\rho_2} = 5$, 1 and 0.5, respectively.

The contraction of the steady state compound jet's radii, the axial traction on the inner jet and the gradient of the axial velocity component at the downstream boundary were found to increase as $\frac{\mu_1}{\mu_2}$ was decreased as shown in Figure 4 which corresponds to Cases 1, 9 and 10 of Table 1; the axial traction force on the inner jet increases whereas its gradient at the downstream boundary decreases as $\frac{\mu_1}{\mu_2}$ is increased. The ratio of axial traction forces at the

Table 1: Values of the parameters used in the steady–state, isothermal, compound fibers.

Case	$\bar{R}$	$\frac{\bar{R}}{F}$	$\bar{C}$	$\frac{\rho_1}{\rho_2}$	$\frac{\mu_1}{\mu_2}$	$\frac{\sigma_1}{\sigma_2}$	Q_1	Q_2
1	1	1	1	1	1	1	1	1
2	1	10	1	1	1	1	1	1
3	1	0.1	1	1	1	1	1	1
4	1	1	0.5	1	1	1	1	1
5	1	1	10	1	1	1	1	1
6	1	1	100	1	1	1	1	1
7	1	1	1	5	1	1	1	1
8	1	1	1	0.5	1	1	1	1
9	1	1	1	1	10	1	1	1
10	1	1	1	1	0.5	1	1	1
11	1	1	1	1	1	0.1	1	1
12	1	1	1	1	1	0.5	1	1
13	1	1	1	1	1	1	2	1
14	1	1	1	1	1	1	0.1	1
15	1	1	1	1	1	1	1	2
16	1	1	1	1	1	1	1	0.1
17	2	1	1	1	1	1	1	1
18	0.1	1	1	1	1	1	1	1

upstream boundary was found to be equal to about 3.75, 1.15 and 0.20 for $\frac{\mu_1}{\mu_2} = 0.5$, 1 and 10, respectively.

The compound jet's geometry, leading–order axial velocity and axial traction force on the inner jet were not found to be very sensitive to $\frac{\sigma_1}{\sigma_2}$ and p_e; however, the ratio of axial traction forces at the upstream boundary was found to be equal to about 1.15, 1.07 and 1.01 for $\frac{\sigma_1}{\sigma_2} = 1$, 0.5 and 0.1, respectively, and equal to about 1.18 and 1.15 for $p_e = 1$ and 0, respectively.

For $R_{10}(0) = 1$, it has been observed that an increase in Q_1 results in an increase in $B(0)$, a decrease in $R_{20}(0)$, and a large contraction of the compound jet near the take–up point. Both $B(1)$ and the axial traction force on the inner jet at the take–up point increase as Q_1 is increased, while the ratio of axial traction forces at the nozzle exit is about 21 for $Q_1 = 0.1$. An increase in Q_2 with $B(0) = 1$ and $R_{10}(0) = 1$ results in an increase in $R_{20}(x)$; however, the axial velocity component and the axial traction force are not very sensitive to Q_2. The ratio of axial traction forces at the upstream boundary was found to be equal to about 2.23 and 0.20 for $Q_2 = 2$ and 0.1, respectively.

4.4 Sample Results for Unsteady, Isothermal, Compound Fibers

Under transient conditions, the nonlinear dynamics of isothermal, Newtonian, compound fibers depends on $\bar{R}$, $\frac{\bar{R}}{F}$, $\bar{C}$, p_e, $\frac{\rho_1}{\rho_2}$, $\frac{\mu_1}{\mu_2}$, $\frac{\sigma_1}{\sigma_2}$, Q_1, Q_2, and $B_{ss}(1)$, and the location, amplitude and frequency of the forcing. In this section, axial velocity perturbations at either the nozzle exit or the take–up point have been employed, i.e.,

$$B(t,x_j) = B_{ss}(x_j)(1 + a_j \sin S_j t), \tag{103}$$

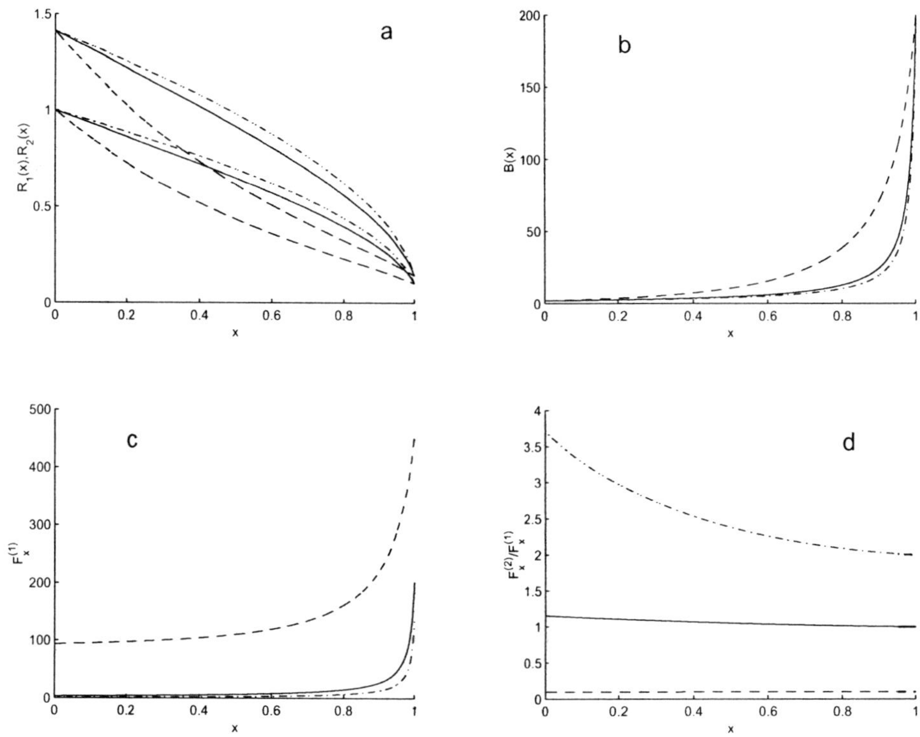

Figure 4: Compound jet's geometry (a), axial velocity component (b), axial traction on the inner jet (c) and ratio of axial traction on the outer jet to that on the inner one (d) for Cases 1, 9 and 10 of Table 1. (Solid lines: Case 1; dashed lines: Case 9; dashed–dotted lines: Case 10).

where a_j and S_j denote the (nondimensional) amplitude and frequency of the imposed forcing at x_j, and $(j,x_j) = (i,0)$ and $(e,1)$ denote the nozzle exit and the take–up point, respectively.

Some sample results which illustrate the nonlinear dynamics of isothermal, compound liquid jets when subject to axial velocity perturbations at the nozzle exit, i.e., Eq. (103), are presented in Figures 5–10 and Tables 2–4. Unless otherwise stated, these time–dependent results were obtained numerically by discretizing the time derivatives by means of an implicit, first–order accurate (in time) backward finite difference formula; the advective derivatives were discretized by means of first–order accurate upwind formulae, whereas the diffusion terms were discretized by means of second–order accurate finite difference expressions. The computations were performed on double precision arithmetic, at least 2001 grid points, and a time–step equal to at most 10^{-4}; the time step was selected in a case–by–case manner in order to accurately capture fast transient phenomena. When strange or chaotic behavior or holes in the phase diagrams were observed, the computations were repeated and performed on quadruple precision with a time step equal to 10^{-8}. The results presented in

Figures 5–8 and Tables 2–4 correspond to $B_{ss}(0) = 1$, $R_{10}(t,0) = \hat{R}$ and $B_{ss}(1) = 2\frac{Q_1}{\hat{R}^2}$. The parameters that appear Tables 2–4 correspond to the basic set employed as a reference in the calculations, and only the parameter under the corresponding heading was varied while maintaining the other parameters equal to those of the basic reference values.

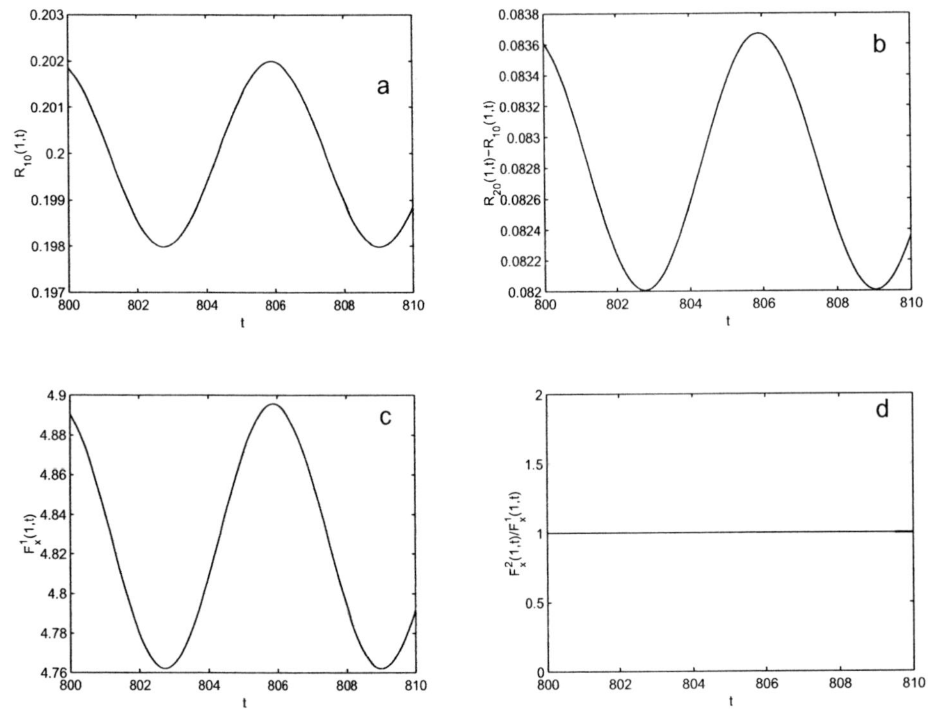

Figure 5: $R_{10}(1,t)$ (a), $R_{20}(1,t) - R_{10}(1,t)$ (b), $F_x^1(1,t)$ (c) and $\frac{F_x^2}{F_x^1}$ (d) on the compound jet at the take–up point as functions of time for sinusoidal axial velocity fluctuations at the die's exit. ($\bar{R} = 10^{-4}$, $\frac{\bar{R}}{\bar{F}} = 0$, $\bar{C} = 10^{39}$, $\frac{\rho_1}{\rho_2} = 1$, $\frac{\mu_1}{\mu_2} = 1$, $\frac{\sigma_1}{\sigma_2} = 1$, $p_e = 0$, $Q_1 = 0.50$, $Q_2 = 1 - Q_1$, $B_{ss}(0) = 1$, $\hat{R} = 0.20$, $B_{ss}(1) = 2\frac{Q_1}{\hat{R}^2}$, $a_i = 0.05$, $a_e = 0$, $S_i = 1$, $S_e = 1$).

Figure 5 indicates that the compound jet's radii and axial traction force on the inner jet at the take–up point are periodic functions of time which have the same frequency as that of the imposed axial velocity perturbation at the nozzle exit or die, i.e., at $x = 0$, for $B_{ss}(1) = 25$. The amplitude of the compound jet's radius at the interface between the inner and outer jets is about 0.002 which is much smaller than the mean value of the inner jet's radius. This periodic behavior is also observed in both the power spectrum and phase diagram of $R_{10}(t,1)$ exhibited in Figure 6, where $U_{10} = \frac{\partial R_{10}}{\partial t}$. The power spectrum is characterized by a single peak at a frequency equal to that of the imposed velocity perturbations, i.e., $\frac{1}{2\pi}$, and the phase diagram is a circumference. The results presented in Figure 5 also indicate that the axial velocity component and the axial traction on the inner jet at the take–up point

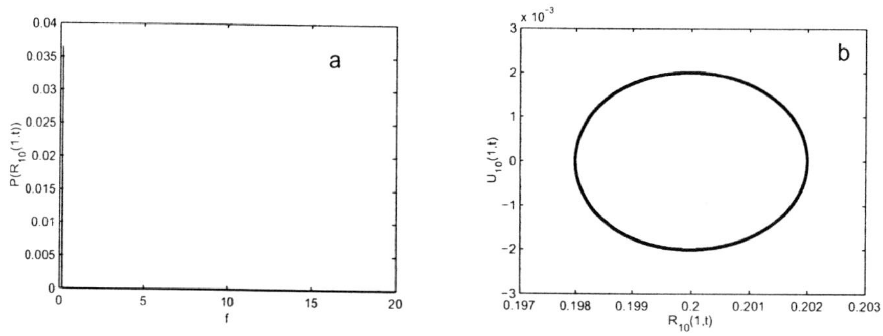

Figure 6: Power spectrum (left) and phase diagram (right) of $R_{10}(1,t)$ for sinusoidal axial velocity fluctuations at the die's exit. ($\bar{R} = 10^{-4}$, $\frac{\bar{R}}{\bar{F}} = 0$, $\bar{C} = 10^{39}$, $\frac{\rho_1}{\rho_2} = 1$, $\frac{\mu_1}{\mu_2} = 1$, $\frac{\sigma_1}{\sigma_2} = 1$, $p_e = 0$, $Q_1 = 0.50$, $Q_2 = 1 - Q_1$, $B_{ss}(0) = 1$, $\hat{R} = 0.20$, $B_{ss}(1) = 2\frac{Q_1}{\hat{R}^2}$, $a_i = 0.05$, $a_e = 0$, $S_i = 1, S_e = 1$).

are periodic functions of time, whereas the ratio of axial traction forces is constant, i.e., the axial traction on the outer jet is also a periodic function of time which has the same frequency as that of the imposed axial velocity component perturbations.

For $B_{ss}(1) \approx 44.5$, the results presented in Figure 7 indicate that the compound jet's radii are spiky, and their maximum values seem to be modulated with a frequency of the order of that of the imposed velocity perturbations; an analogous comment applies to the axial traction force on the inner jet at the take–up point. Figure 7 also indicates that the ratio of axial traction forces is constant, i.e., the axial traction on the outer jet exhibits similar trends to those of the axial traction on the inner one.

The power spectrum corresponding to Figure 7 is characterized by several peaks, and the frequency associated with these peaks is not related to that of the applied velocity perturbations as indicated in Figure 8. Moreover, the phase diagram presented in Figure 8 has a duck's beak shape, presents corrugations on its periphery, is thick, and contains some holes, i.e., regions which are never visited; in addition, the largest and smallest values of $R_{10}(t,1)$ are larger and smaller, respectively, than those presented in Figure 6. The values of $\frac{\partial R_{10}}{\partial t}(t,1)$ are larger than those presented in Figure 6, and the thickness of the phase diagram is related to the modulation of the spikes observed in the time history of $R_{10}(t,1)$ shown in Figure 7.

For $B_{ss}(1) = 100$, the time histories presented in Figure 9 seem to be analogous to those of Figure 7, except that the largest and smallest values of $R_{10}(t,1)$ increase and decrease, respectively, as $B_{ss}(1)$ is increased; the axial traction force also increases as $B_{ss}(1)$ is increased in accord with the large gradient of the axial velocity at the take–up point. Moreover, the amplitude of the modulation of the spikes in the jet's radii and axial traction force on the inner jet at the take–up point for $B_{ss}(1) = 100$ are smaller than those for $B_{ss}(1) \approx 44.5$. This is also observed in the phase diagram illustrated in Figure 10 which is much thinner than that of Figure 8, and contains many holes. This diagram also shows that there is a high density of points near $\frac{\partial R_{10}}{\partial t}(t,1) = 0$ in agreement with the time histories

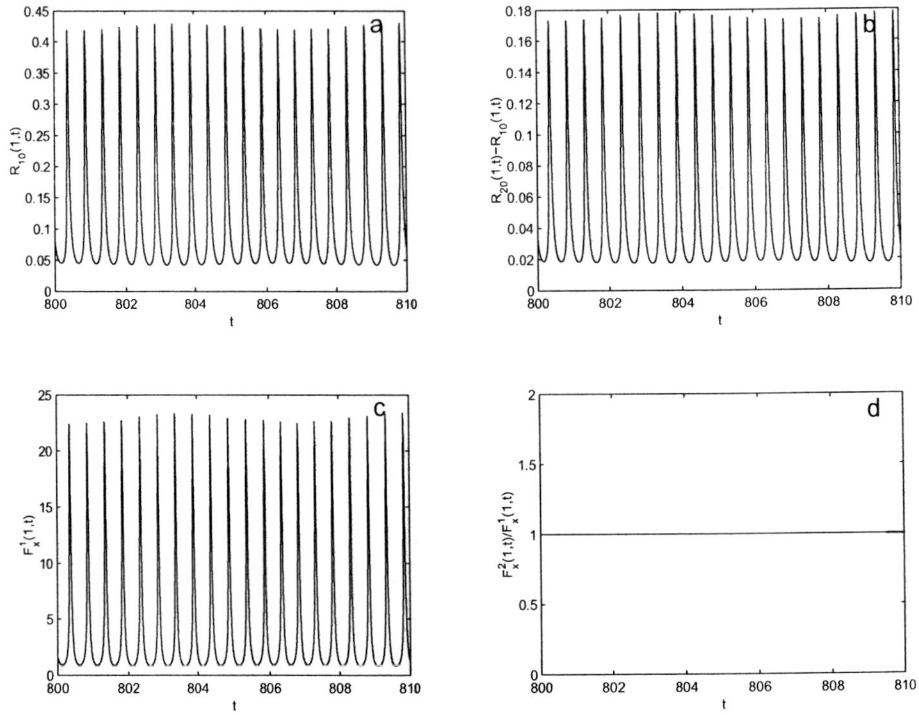

Figure 7: $R_{10}(1,t)$ (a), $R_{20}(1,t) - R_{10}(1,t)$ (b), $F_x^1(1,t)$ (c) and $\frac{F_x^2}{F_x^1}$ (d) on the compound jet at the take–up point as functions of time for sinusoidal axial velocity fluctuations at the die's exit. ($\bar{R} = 10^{-4}$, $\frac{\bar{R}}{\bar{F}} = 0$, $\bar{C} = 10^{39}$, $\frac{\rho_1}{\rho_2} = 1$, $\frac{\mu_1}{\mu_2} = 1$, $\frac{\sigma_1}{\sigma_2} = 1$, $p_e = 0$, $Q_1 = 0.50$, $Q_2 = 1 - Q_1$, $B_{ss}(0) = 1$, $\hat{R} = 0.15$, $B_{ss}(1) = 2\frac{Q_1}{\hat{R}^2}$, $a_i = 0.05$, $a_e = 0$, $S_i = 1$, $S_e = 1$).

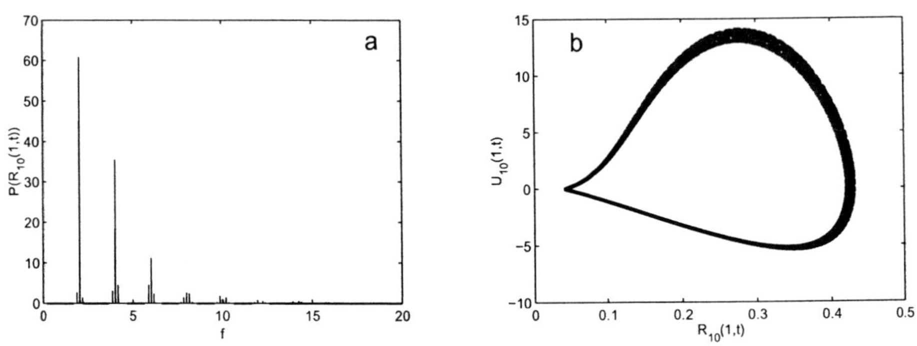

Figure 8: Power spectrum (left) and phase diagram (right) of $R_{10}(1,t)$ for sinusoidal axial velocity fluctuations at the die's exit. ($\bar{R} = 10^{-4}$, $\frac{\bar{R}}{\bar{F}} = 0$, $\bar{C} = 10^{39}$, $\frac{\rho_1}{\rho_2} = 1$, $\frac{\mu_1}{\mu_2} = 1$, $\frac{\sigma_1}{\sigma_2} = 1$, $p_e = 0$, $Q_1 = 0.50$, $Q_2 = 1 - Q_1$, $B_{ss}(0) = 1$, $\hat{R} = 0.15$, $B_{ss}(1) = 2\frac{Q_1}{\hat{R}^2}$, $a_i = 0.05$, $a_e = 0$, $S_i = 1$, $S_e = 1$).

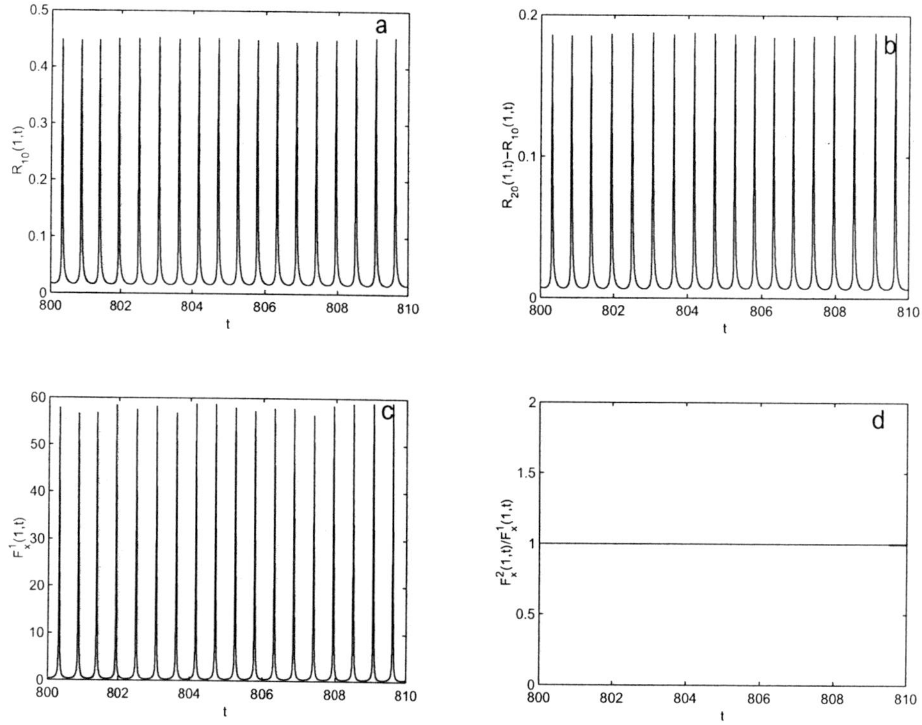

Figure 9: $R_{10}(1,t)$ (a), $R_{20}(1,t) - R_{10}(1,t)$ (b), $F_x^1(1,t)$ (c) and $\frac{F_x^2}{F_x^1}$ (d) on the compound jet at the take–up point as functions of time for sinusoidal axial velocity fluctuations at the die's exit. ($\bar{R} = 10^{-4}$, $\frac{\bar{R}}{\bar{F}} = 0$, $\bar{C} = 10^{39}$, $\frac{\rho_1}{\rho_2} = 1$, $\frac{\mu_1}{\mu_2} = 1$, $\frac{\sigma_1}{\sigma_2} = 1$, $p_e = 0$, $Q_1 = 0.50$, $Q_2 = 1 - Q_1$, $B_{ss}(0) = 1$, $\hat{R} = 0.10$, $B_{ss}(1) = 2\frac{Q_1}{\hat{R}^2}$, $a_i = 0.05$, $a_e = 0$, $S_i = 1$, $S_e = 1$).

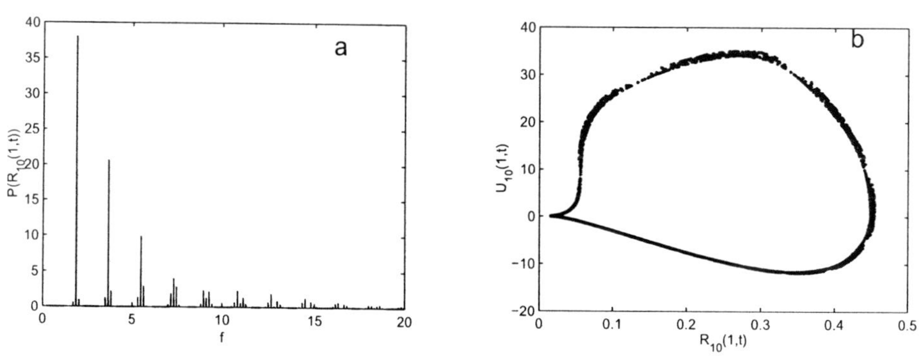

Figure 10: Power spectrum (left) and phase diagram (right) of $R_{10}(1,t)$ for sinusoidal axial velocity fluctuations at the die's exit. ($\bar{R} = 10^{-4}$, $\frac{\bar{R}}{\bar{F}} = 0$, $\bar{C} = 10^{39}$, $\frac{\rho_1}{\rho_2} = 1$, $\frac{\mu_1}{\mu_2} = 1$, $\frac{\sigma_1}{\sigma_2} = 1$, $p_e = 0$, $Q_1 = 0.50$, $Q_2 = 1 - Q_1$, $B_{ss}(0) = 1$, $\hat{R} = 0.10$, $B_{ss}(1) = 2\frac{Q_1}{\hat{R}^2}$, $a_i = 0.05$, $a_e = 0$, $S_i = 1$, $S_e = 1$).

presented in Figure 9.

The power spectrum presented in Figure 10 is also broader and the frequencies corresponding to its peaks are smaller than those of Figure 8. The broadening of the power spectrum and the appearance of holes in the phase diagram as $B_{ss}(1)$ is increased beyond the critical value provided by the linear stability analysis for the viscous flow regime reported previously (cf. Section 4), are indicative of the presence of chaos. However, the transition from the periodic motion presented in Figures 5 and 6 to the chaotic ones of Figures 9 and 10 as $B_{ss}(1)$ is increased does not seem to follow a standard scenario such as that of Ruelle–Takens, at least, for sufficiently large values of $B_{ss}(1)$; in fact, it has been observed that the transition to chaos for large values of $B_{ss}(1)$ is rather abrupt or explosive, whereas, for smaller values of $B_{ss}(1)$, it has been observed that $R_{10}(t,1)$ evolves from a fixed point to a periodic motion and then to a quasiperiodic motion and chaos as $B_{ss}(1)$ is increased. The results presented in Figures 8 and 10 also indicate that the largest value of $\frac{\partial R_{10}}{\partial t}(t,1)$ increases as $B_{ss}(1)$ is increased.

The phase diagrams of Figures 8 and 10 have a duck's beak shape, and exhibit many holes. It must be pointed out that the phase diagrams illustrated in Figures 6, 8 and 10 were obtained for $1000 \leq t \leq 1200$ and that, for a time step equal to 10^{-8}, they contain 10^{10} points. The broadening of both the power spectrum and the phase diagrams as well as the increase in the number of holes in the phase diagram as $B_{ss}(1)$ increases indicate the presence of strange attractors and chaotic behavior.

It is worth remarking that the frequencies of the quasiperiodic and chaotic motions observed in previous figures as well as in other calculations not presented here do not seem to be related to that of the imposed velocity perturbations. Moreover, phase diagrams and power spectra similar to those presented in Figures 6, 8 and 10 have been observed for sinusoidal axial velocity perturbations of amplitudes equal to 0.01 and 1 at the nozzle exit or die. However, for sinusoidal axial velocity perturbations at the downstream boundary and $B_{ss}(1) = 100$, the phase diagrams had more empty regions and the power spectra were broader than those of Figure 10, even for $a_e = 0.01$. When axial velocity perturbations were imposed at both the upstream and the downstream boundaries, the power spectra were narrower than those for axial velocity perturbations imposed at only the downstream boundary and slightly broader than those of Figure 10. Moreover, the phase diagrams when sinusoidal axial velocity perturbations were imposed at both the upstream and the downstream boundaries were more filled than those for sinusoidal axial velocity perturbations imposed at only the downstream boundary, but presented more empty regions or holes than that of Figure 10. When sinusoidal axial velocity perturbations were imposed at either the downstream or at both the upstream and the downstream boundaries, it was observed that some kind of tongues emerged from the main beak structure, thus resulting in phase diagrams with multiple leaves.

Table 2 shows that the largest value of $R_{10}(t,1)$ increases, whereas the smallest value of $R_{10}(t,1)$ decreases as $B_{ss}(1)$ is increased for sinusoidal excitations; a similar behavior is observed in the axial traction force at the take–up or downstream point. Table 2 also shows that $R_{10}(t,1)$ is not very sensitive to the amplitude and frequency of the imposed velocity

Table 2: Effects of the steady–state take–up downstream boundary conditions on the maximum and minimum values of $R_{10}(t,1)$ and axial traction forces on the inner and outer jets at the take–up point, and maximum spectral power (P) and frequency (f) associated with the maximum power of $R_{10}(t,1)$, for $\bar{R} = 10^{-4}$, $\frac{\bar{R}}{F} = 0$, $\bar{C} = 10^{39}$, $\frac{\rho_1}{\rho_2} = 1$, $\frac{\mu_1}{\mu_2} = 1$, $\frac{\sigma_1}{\sigma_2} = 1$, $p_e = 0$, $Q_1 = 0.50$, $Q_2 = 1 - Q_1$, $B_{ss}(0) = 1$, $\hat{R} = 0.10$, $B_{ss}(1) = 2\frac{Q_1}{\hat{R}^2}$.

$\hat{R}$	$R_{max}(t,1)$	$R_{min}(t,1)$	$F^{(1)}_{max}$	$F^{(1)}_{min}$	$F^{(2)}_{max}$	$F^{(2)}_{min}$	P	f
$0.10^{a,1,4}$	0.4527	0.0155	58.8597	0.3386	58.8597	0.3386	38.0239	1.8299
$0.15^{a,1,5}$	0.4310	0.0419	23.2809	0.8251	23.2809	0.8251	60.7773	2.0099
$0.20^{a,1,6}$	0.2019	0.1980	4.8904	4.7621	4.8904	4.7621	0.0365	0.16
$0.10^{b,2,4}$	0.4520	0.0156	59.4557	0.3478	59.4557	0.3478	39.2607	1.8299
$0.15^{b,2,5}$	0.4287	0.0423	23.5574	0.8522	23.5574	0.8522	61.3542	2.0099
$0.20^{b,2,6}$	0.2020	0.1982	4.8539	4.7987	4.8539	4.7987	0.0365	0.16
$0.10^{c,3,4}$	0.4532	0.0158	58.5327	0.3437	58.5327	0.3437	38.8697	1.8299
$0.15^{c,3,5}$	0.4286	0.0430	23.2061	0.8470	23.2061	0.8470	58.1195	2.0099
$0.20^{c,3,6}$	0.2000	0.2000	4.9155	4.7326	4.7326	4.8539	6.4670e-22	0.16

[a]The upstream axial velocity is sinusoidally excited with an amplitude and frequency equal to 0.01 and 1, respectively; [1]identical results for $a_i = 0.05$ and 0.10; [b]the downstream axial velocity is sinusoidally excited with an amplitude and frequency equal to 0.01 and 1, respectively; [2]identical results for $a_e = 0.05$ and 0.10; [c]the upstream and downstream axial velocities are sinusoidally excited with an amplitude and frequency equal to 0.01 and 1, respectively; [3]identical results for $a_i = a_e = 0.05$ and 0.10; [4]phase diagram with holes; [5]filled phase diagram; [6]periodic.

perturbations, but it is sensitive to the location of these perturbations. In addition, Table 2 shows that the frequency of the maximum power first increases as $B_{ss}(1)$ is increased, and then decreases; a periodic $R_{10}(t,1)$ results if the axial velocity is perturbed at both the upstream and downstream boundaries with the same amplitude and frequency if $B_{ss}(1) =$ 25, even though this value exceeds that determined in the linear stability analysis for the viscous flow regime performed previously in this chapter (cf. Section 4), and indicates that nonlinear effects cause a supercritical Hopf bifurcation for values of $B_{ss}(1)$ larger than but close to the critical draw ratio determined from the linear stability analysis of the viscous flow regime. There is, however, no contradiction between the results of Table 2 and those of the stability analysis because the latter corresponds to $\bar{R} = 0$, whereas the former accounts for non–zero Reynolds numbers.

Table 2 also shows that the difference between the largest and the smallest traction forces on the inner and outer jets at the take–up point increases as $B_{ss}(1)$ is increased. Table 2 also indicates that the difference between the largest and smallest values of $R_{10}(t,1)$ is larger for axial velocity perturbations imposed at the upstream than at the downstream boundary; however, the frequency corresponding to the maximum power of the spectrum of $R_{10}(t,1)$ is independent of the location of the imposed velocity perturbations.

Table 3 shows that $R_{10}(t,1)$ is a periodic function of time for $\hat{R} = 0.30$ when axial velocity perturbations are applied at either the upstream or the downstream boundaries; the frequency of this periodic motion coincides with that of the imposed sinusoidal axial veloc-

Table 3: Effects of the location of sinusoidal axial velocity perturbations on the maximum and minimum values of $R_{10}(t,1)$ and axial traction forces on the inner and outer jets at the take–up point, and maximum spectral power (P) and frequency (f) associated with the maximum power of $R_{10}(t,1)$, for $\bar{R}=10^{-4}$, $\frac{\bar{R}}{\bar{F}}=0$, $\bar{C}=10^{30}$, $\frac{\rho_1}{\rho_2}=1$, $\frac{\mu_1}{\mu_2}=1$, $\frac{\sigma_1}{\sigma_2}=1$, $p_e=0$, $Q_1=Q_2=0.50$, $B_{ss}(0)=1$, $\hat{R}=0.30$, $B_{ss}(1)=2\frac{Q_1}{\hat{R}^2}$.

(a_i,a_e,S_i,S_e)	$R_{max}(t,1)$	$R_{min}(t,1)$	$F^{(1)}_{max}$	$F^{(1)}_{min}$	$F^{(2)}_{max}$	$F^{(2)}_{min}$	P	f
$(0.01,0,1,1)^{a,1,4}$	0.3028	0.2970	3.6516	3.5695	3.6516	3.5695	0.0822	0.16
$(0,0.01,1,1)^{b,2,4}$	0.3031	0.2972	3.6370	3.5816	3.6370	3.5816	0.0822	0.16
$(0.01,0.01,1,1)^{c,3,4}$	0.3000	0.3000	3.6768	3.5400	3.6768	3.5400	8.2114e-22	0.16

[a]The upstream axial velocity is sinusoidally excited with an amplitude and frequency equal to 0.01 and 1, respectively; [1]identical results for $a_i=0.05$ and 0.10; [b]the downstream axial velocity is sinusoidally excited with an amplitude and frequency equal to 0.01 and 1, respectively; [2]identical results for $a_e=0.05$ and 0.10; [c]the upstream and downstream axial velocities are sinusoidally excited with an amplitude and frequency equal to 0.01 and 1, respectively; [3]identical results for $a_i=a_e=0.05$ and 0.10; [4]periodic.

ity fluctuations. Table 3 also shows that, when the axial velocity perturbations are imposed at both the upstream and downstream boundaries, $R_{10}(t,1)$ remains constant, i.e., there is a cancellation between the perturbations introduced by the upstream and downstream forcings. It must be noted that the value of $\hat{R}=0.30$ corresponds to about $B_{ss}(1)=11$ which is below the critical value provided by the linear stability analysis of the viscous flow regime.

The effects of the nondimensional fluid dynamics parameters on $R_{10}(t,1)$ and the axial traction force at the take–up point are illustrated in Table 4. This table shows that, for $\bar{R}=1$, the compound jet behaves in a periodic manner with a frequency equal to that of the imposed sinusoidal axial velocity fluctuations, and there is very little difference between the largest and smallest values of $R_{10}(t,1)$. The largest and smallest values of $R_{10}(t,1)$ increase and decrease, respectively, as $\bar{R}$ is decreased on account of the large gradients of the axial velocity at the downstream boundary; a similar comment applies to the axial traction force at the take–up point. For the Reynolds number of the basic set of parameters of Table 4, the effects of the Froude number are small; however, the largest value of $R_{10}(t,1)$ decreases as $\frac{\bar{R}}{\bar{F}}$ is increased on account of the increase in the gravitational pull.

Table 4 also shows that the effects of the capillary number are small, although a decrease in $\bar{C}$ results in an increase in the largest value of $R_{10}(t,1)$. The effects of $\frac{\rho_1}{\rho_2}$, $\frac{\mu_1}{\mu_2}$ and $\frac{\sigma_1}{\sigma_2}$ on the nonlinear dynamics of isothermal, compound liquid jets have been found to be small for the values of the parameters employed in Table 4. As should be expected, the largest value of $R_{10}(t,1)$ increases as p_e is decreased.

For $\hat{R}=0.15$, $\bar{R}=10^{-4}$, $\frac{\bar{R}}{\bar{F}}=0$, $\bar{C}=10^{30}$, $\frac{\rho_1}{\rho_2}=\frac{\mu_1}{\mu_2}=\frac{\sigma_1}{\sigma_2}=1$, $p_e=0$, $Q_1=0.5$, $Q_2=1-Q_1$, $B_{ss}(0)=1$ and $B_{ss}(1)=2\frac{Q_1}{\hat{R}^2}$, the phase diagram of $R_{10}(t,1)$ was found to be rather similar to but wider than that of Figure 10 for $a_i=0.05$ and 0.10, $a_e=0$, and $S_i=1$. However, for the same values of the parameters and $\bar{R}=10^{-1}$, it was rather smooth, i.e, it

Table 4: Effects of the fluid dynamics parameters on the maximum and minimum values of $R_{10}(t,1)$ and axial traction forces on the inner and outer jets at the take–up point, and maximum spectral power (*P*) and frequency (*f*) associated with the maximum power of $R_{10}(t,1)$, for sinusoidal upstream axial velocity of amplitude and frequency equal to 0.01 and 1, respectively.

Parameter	$R_{max}(t,1)$	$R_{min}(t,1)$	$F_{max}^{(1)}$	$F_{min}^{(1)}$	$F_{max}^{(2)}$	$F_{min}^{(2)}$	P	f
$\bar{R} = 0.1$[1]	0.2737	0.0307	59.2482	1.5227	59.2482	1.5227	38.6010	2.1449
$\bar{R} = 1$[2]	0.1009	0.0990	53.6308	51.6645	53.6308	51.6645	0.0092	0.16
$\frac{\bar{R}}{\bar{F}} = 1$[3]	0.4426	0.0161	56.8320	0.3583	56.8320	0.3583	38.5327	1.8849
$\frac{\bar{R}}{\bar{F}} = 0.01$[3]	0.4537	0.0154	59.1257	0.3354	59.1257	0.3354	31.7847	1.8299
$\bar{C} = (10^5)$[3]	0.4527	0.0155	58.8154	0.3386	58.8154	0.3386	37.7074	1.8299
$\bar{C} = (10^2)$[3]	0.4543	0.0154	59.0518	0.3359	59.0539	0.3360	32.6179	1.8249
$\frac{\rho_1}{\rho_2} = 10$[3]	0.4521	0.0156	58.9173	0.3418	58.9173	0.3418	28.9642	1.8299
$\frac{\rho_1}{\rho_2} = 0.1$[3]	0.4535	0.0155	58.6993	0.3371	58.6993	0.3371	38.7148	1.8299
$\frac{\mu_1}{\mu_2} = 10$[3]	0.4529	0.0155	590.8526	3.3816	59.0853	0.3382	38.7382	1.8299
$\frac{\mu_1}{\mu_2} = 0.1$[3]	0.4535	0.0154	5.9194	0.0336	59.1943	0.3364	37.7181	1.8299
$\frac{\sigma_1}{\sigma_2} = 10$[3]	0.4527	0.0155	58.8597	0.3386	58.8597	0.3386	38.0239	1.8299
$Q_1 = 0.75$[3]	0.3775	0.0193	70.0343	0.7411	23.3448	0.2470	41.0396	3.0798
$Q_1 = 0.25$[3]	0.5650	0.0121	38.4789	0.1076	115.4367	0.3227	31.2553	0.7950

[1] filled phase diagram; [2] periodic; [3] phase diagram with holes. The basic set of parameters is $\bar{R} = 10^{-4}$, $\frac{\bar{R}}{\bar{F}} = 0$, $\bar{C} = 10^{39}$, $\frac{\rho_1}{\rho_2} = 1$, $\frac{\mu_1}{\mu_2} = 1$, $\frac{\sigma_1}{\sigma_2} = 1$, $p_e = 0$, $Q_1 = 0.50$, $Q_2 = 1 - Q_1$, $B_{ss}(0) = 1$, $\hat{R} = 0.10$, $B_{ss}(1) = 2\frac{Q_1}{R^2}$.

did not have the duck's beak shape of Figure 10, and it did not present any hole; for $\bar{R} = 1$, the phase diagram was a circumference, i.e., the dynamics was periodic.

For $\hat{R} = 0.30$, $\bar{R} = 10^{-4}$, $\frac{\bar{R}}{\bar{F}} = 0$, $\bar{C} = 10^{30}$, $\frac{\rho_1}{\rho_2} = \frac{\mu_1}{\mu_2} = \frac{\sigma_1}{\sigma_2} = 1$, $p_e = 0$, $Q_1 = 0.5$, $Q_2 = 1 - Q_1$, $B_{ss}(0) = 1$ and $B_{ss}(1) = 2\frac{Q_1}{R^2}$, $R_{10}(t,1)$ is a periodic function of time which has the same frequency as that of the applied sinusoidal axial velocity perturbation for $a_i = 0.01, 0.05$ and 0.10, $a_e = 0$, $S_i = 1$, and $S_e = 1$. The same comment applies for $\hat{R} = 0.30$, $\bar{R} = 10^{-4}$, $\frac{\bar{R}}{\bar{F}} = 0$, $\bar{C} = 10^{30}$, $\frac{\rho_1}{\rho_2} = \frac{\mu_1}{\mu_2} = \frac{\sigma_1}{\sigma_2} = 1$, $p_e = 0$, $Q_1 = 0.5$, $Q_2 = 1 - Q_1$, $B_{ss}(0) = 1$, $B_{ss}(1) = 2\frac{Q_1}{R^2}$, $a_i = 0$, $a_e = 0.01$, 0.05 and 0.10, $S_i = 1$, and $S_e = 1$. However, for $\hat{R} = 0.30$, $\bar{R} = 10^{-4}$, $\frac{\bar{R}}{\bar{F}} = 0$, $\bar{C} = 10^{30}$, $\frac{\rho_1}{\rho_2} = \frac{\mu_1}{\mu_2} = \frac{\sigma_1}{\sigma_2} = 1$, $p_e = 0$, $Q_1 = 0.5$, $Q_2 = 1 - Q_1$, $B_{ss}(0) = 1$, $B_{ss}(1) = 2\frac{Q_1}{R^2}$, $R_{10}(t,1)$ exhibits a fixed point for $a_i = a_e = 0.01, 0.05$ and 0.10, and $S_e = S_i = 1$, whereas the axial traction force at the take–up point is a periodic function with a frequency equal to that of the imposed velocity perturbations.

The results presented in this section and others not shown here indicate that $R_{10}(t,1)$ evolves from a fixed point, to a periodic orbit and then to quasiperiodic and chaotic motions as $B_{ss}(1)$ is increased; however, the transition, if one exists, from quasiperiodic to chaotic responses is rather abrupt, and the phase diagram in the chaotic region may present holes, i.e., regions which are not visited, and corrugations for large values of $B_{ss}(1)$. It has also been observed that the power spectrum of $R_{10}(t,1)$ broadens and presents many holes as $B_{ss}(1)$ is increased. The width of the spectrum and the topology of the phase diagrams have also been found to depend on the fluid dynamics parameters that characterize isothermal, compound liquid jets at low Reynolds numbers, and the location of the imposed axial velocity perturbations. The amplitude and frequency of these perturbations were found to play a very small role for sufficiently large values of $B_{ss}(1)$, thus indicating that the nonlinear dynamics of compound liquid jets are not very much affected by the amplitude and frequency of the imposed velocity perturbations for very large values of $B_{ss}(1)$.

The nonlinear, time–dependent, isothermal studies presented above do not exhibit the routes to chaos discussed by Yarin et al. [25] for isothermal, round jets at low Reynolds numbers and Gospodinov and Yarin [26] for isothermal, annular jets. Yarin et al. [25] claimed that the cross–section of Newtonian isothermal round fibers may vary aperiodically for draw ratios equal to about 30 and larger when periodic variations of the input cross–section are imposed, and that the route to chaos may be smooth, via period doubling, or explosive, via abrupt disappearance of quasiperiodic solutions. On the other hand, Gospodinov and Yarin [26] claimed that, even for draw ratios equal to 400, the mid–radius of annular jets exhibits a multimodal quasiperiodic character. The results for isothermal, compound jets presented here indicate, however, that the critical draw ratio for the viscous regime can be exceeded without the appearance of chaos if the Reynolds number is sufficiently large but still smaller than unity, and that the route to chaos is via quasiperiodic motions for very small amplitudes of the forcing, but, for amplitudes equal to or higher than about 0.01 and draw ratios sufficiently large, the transition to chaos, if one exists, is abrupt. Moreover, the phase diagrams employed by Yarin et al. [25] are not really phase diagrams because they consider $B(t,x_j)$ versus $a(t,x_j)$ where a is the fiber radius; on the other hand, the phase diagrams presented in this chapter refer to the time history of $R_{10}(t,1)$.

5 Nonisothermal, Compound Fibers

In this section, we first show an analytical solution for the temperature of steady, nonisothermal, compound fiber, and then some results corresponding to both steady and unsteady, nonisothermal compound fibers.

5.1 Analytical Solution for Steady, Nonisothermal, Compound Fibers

Equation (61) represents a two–point boundary value problem which is to be solved subject to $E(0)=T(0)=1$ and a boundary condition at the take–up point. A plausible downstream boundary condition for the temperature is

$$\frac{d}{dx}\left((K_1A_1+K_2A_2)\frac{dE}{dx}\right)(1)=0, \tag{104}$$

which corresponds to no heat flux at the downstream boundary. Unfortunately, this boundary condition does not allow us to obtain closed–form analytical solutions to Eq. (61). However, if axial conduction is neglected and $\frac{\bar{B}}{P}=St=O(1)$ where St is the Stanton number,

Eq. (61) becomes under steady state conditions a first–order ordinary differential equation whose solution may be written as

$$E(x) = T_\infty + C\exp\left(-\frac{St}{\rho_1 C_1 Q_1 + \rho_2 C_2 Q_2}\int_0^x R_{20}(x)dx\right), \quad (105)$$

where the constant C can be easily determined from the condition at the die's exit, i.e., $E(0) = 1$, as $C = (1 - T_\infty)$.

For the viscous flow regime, considered Section 4.1, the leading–order temperature of the compound fiber may be written as

$$E(x) = T_\infty + C\exp\left(-\frac{St}{\rho_1 C_1 Q_1 + \rho_2 C_2 Q_2}\sqrt{\frac{2Q_T}{\beta}}\frac{2}{\alpha}\left(1 - \exp\left(-\frac{\alpha x}{2}\right)\right)\right). \quad (106)$$

For $\alpha\lambda << 1$, a Taylor's series expansion of the inner exponential in Eq. (106) shows that the temperature decreases in an exponential fashion along the fiber axis and the rate of decrease increases as $\bar{B}$ is increased. Moreover, in the absence of swelling, the integral in Eq. (105) has the following upper and lower bounds $R_{20}L$ and $\frac{Q_T}{B(1)}L$, respectively, which also indicate that the temperature decreases in an exponential fashion along the fiber axis and its rate of decrease increases as $\bar{B}$ is increased.

Analytical solutions can also be obtained whenever axial conduction is neglected, for the steady flow regimes considered previously in Section 4. These solutions are not reported here.

5.2 Results for Steady, Nonisothermal, Compound Fibers

In addition to the nondimensional parameters that govern the dynamics of steady, isothermal, compound fibers, one must add $\bar{P}$ or Pr, $\bar{B}$, $\frac{C_1}{C_2}$, $\frac{K_1}{K_2}$, T_∞, $\frac{\psi_1}{\psi_2}$, G_1 and G_2 in order to determine the cooling of steady, nonisothermal, compound fibers, where G_1 and G_2 are the nondimensional activation energies and ψ_1 and ψ_2 are the pre–exponential factors of the viscosity laws for the inner and outer jets, respectively (cf. Eq. (40)). In this section, we present some sample results in order to illustrate the distribution of temperature along the compound fiber for several parameters. It must be emphasized that, as stated previously, the solidification of the compound fiber is accounted for through the increase in the dynamic viscosity, since latent heat effects have been neglected. It should also be emphasized that, since the asymptotic analysis performed in Section 3 indicates that the temperatures of the inner and outer jets are identical at leading–order, we assume that the temperature of the two materials at the die's exit is identical, although the two materials may solidify at different axial locations if the activation energies of their viscosity laws are quite different from each other.

The results presented in this section correspond to $\bar{R} = 1$, $\frac{\bar{R}}{\bar{F}} = 1$, $\bar{C} = 10^{39}$, $\frac{\sigma_1}{\sigma_2} = \frac{\rho_1}{\rho_2} = \frac{\psi_1}{\psi_2} = 1$, $Q_1 = Q_2 = 0.5$, $T_\infty = 0$ and the values of the parameters shown in Table 5, where $\bar{P}_1 = \frac{\rho_1 C_1 u_0 R_0}{\varepsilon K_1}$ and $\bar{P}_2 = \frac{\rho_2 C_2 u_0 R_0}{\varepsilon K_2}$ denote the thermal Péclet numbers for the inner and outer jets, respectively, and were obtained with a similar numerical technique to the one employed to study steady, isothermal, compound fibers.

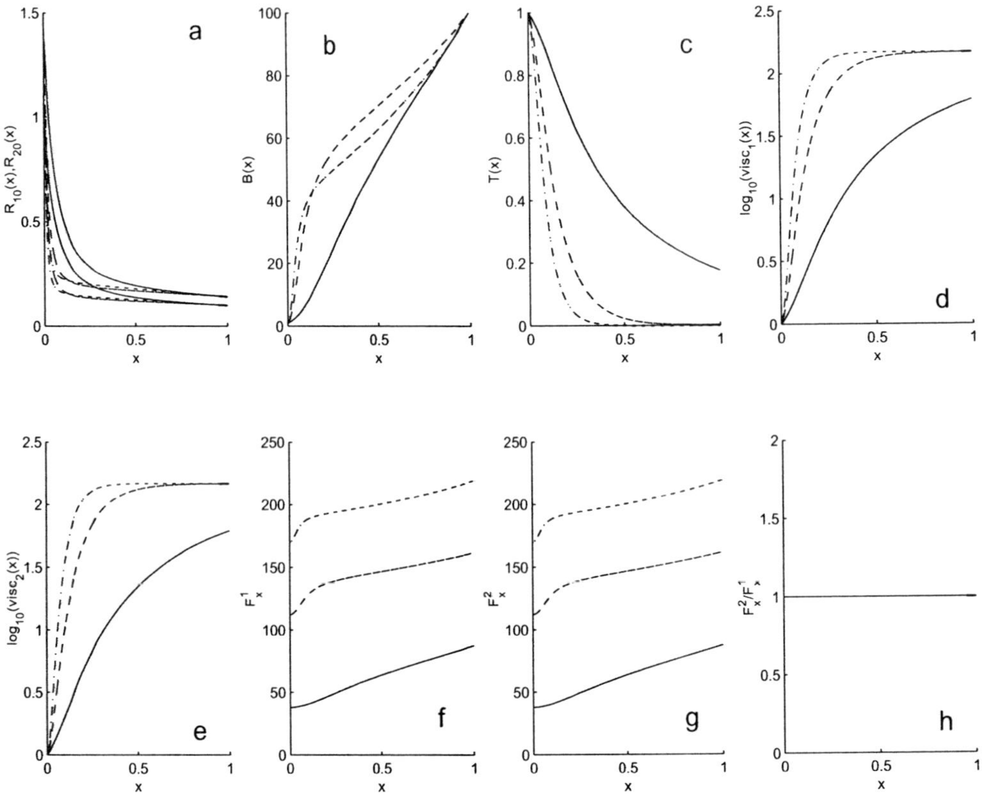

Figure 11: Compound's jet geometry (a), axial velocity component (b), temperature (c), dynamic viscosity of the inner (d) and outer (e) jets, axial traction on the inner (f) and outer jets (g) and axial traction ratio (h) for Cases 2, 3 and 4 of Table 5. (Solid line: Case 2; dashed line: Case 3; dashed–dotted line: Case 4).

Figure 11 shows the compound fiber's geometry, axial velocity component, temperature, dynamic viscosities of the inner and outer jets, traction forces on the inner and outer jets, and the traction ratio for Cases 2, 3 and 4 of Table 5. This figure indicates that the jet contraction occurs closer to the die's exit as the Biot number is increased; the viscosities of the materials also increase as the Biot number increases but their rate of increase decreases as $\bar{B}$ is increased.

Figure 11 shows that the axial velocity component does not reach a constant value for Cases 2, 3 and 4, although the convexity of $B(x)$ depends very much on the Biot number. The axial forces increase as the Biot number is increased, but their ratio is constant.

Figure 12 indicates that the jet contraction moves towards the die's exit as the activation energy G_1 of the viscosity law is increased. This is also illustrated in the axial velocity profile which reaches a constant value for $x \geq 0.2$, and in the viscosity of the inner jet which

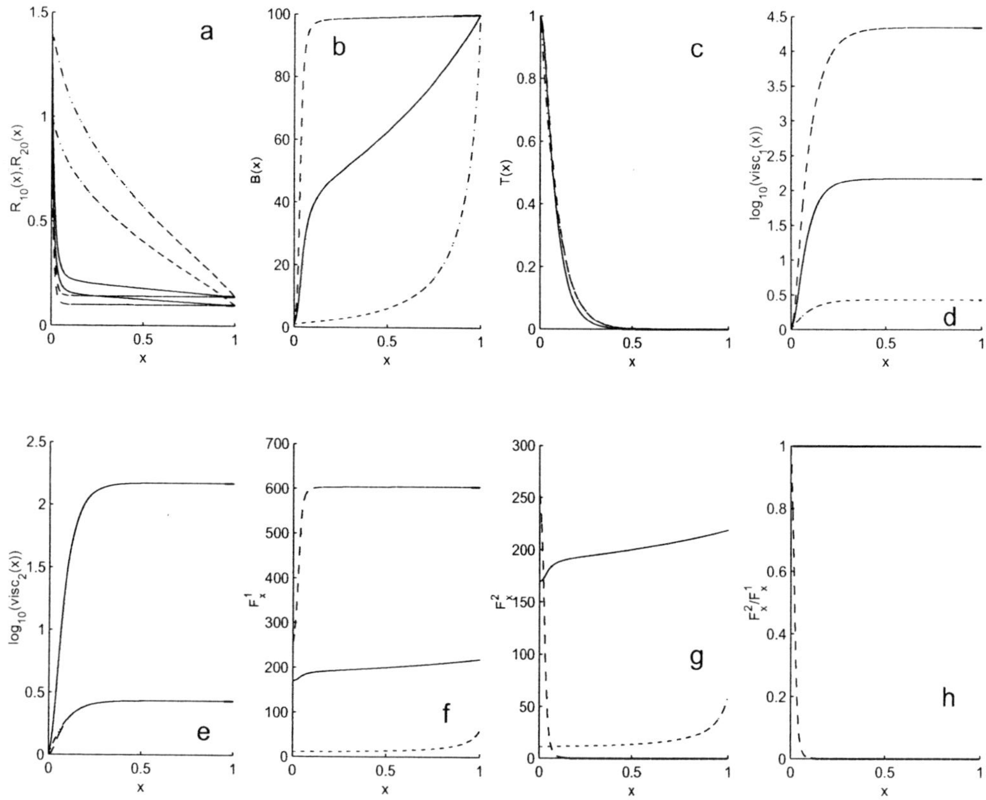

Figure 12: Compound's jet geometry (a), axial velocity component (b), temperature (c), dynamic viscosity of the inner (d) and outer (e) jets, axial traction on the inner (f) and outer jets (g) and axial traction ratio (h) for Cases 4, 5 and 6 of Table 5. (Solid line: Case 4; dashed line: Case 5; dashed–dotted line: Case 6).

reaches a value close to 10^4. As stated previously, latent heat effects were not considered and the fact that the compound jet reaches a constant geometry is due to the increase of the dynamic viscosity as the compound fiber is cooled.

Figure 12 also shows that the axial forces on the inner and outer materials depend very much on the activation energy of the viscosity law, and the axial traction ratio is not a constant for $G_1 \neq G_2$; in fact, Figure 12 shows that the axial traction force may even tend to zero, thus indicating that, for Cases 4, 5 and 6 of Table 5, most of the traction is taken by the inner jet when $G_1 > G_2$. Similar results to those of Figure 12 were found for Cases 4, 7 and 8, except that the greatest axial traction corresponded to the outer jet when $G_2 > G_1$.

The results illustrated in Figure 13 correspond to Cases 4, 9 and 10 of Table 5 and exhibit the effects of the thermal Péclet number on the compound jet. As this number is increased, the thermal inertia increases, while the heat losses decrease for the constant Biot number used in this section. As a consequence, the dynamic viscosity increases as the Péclet number is decreased, but there are no differences in the axial forces of the inner and outer

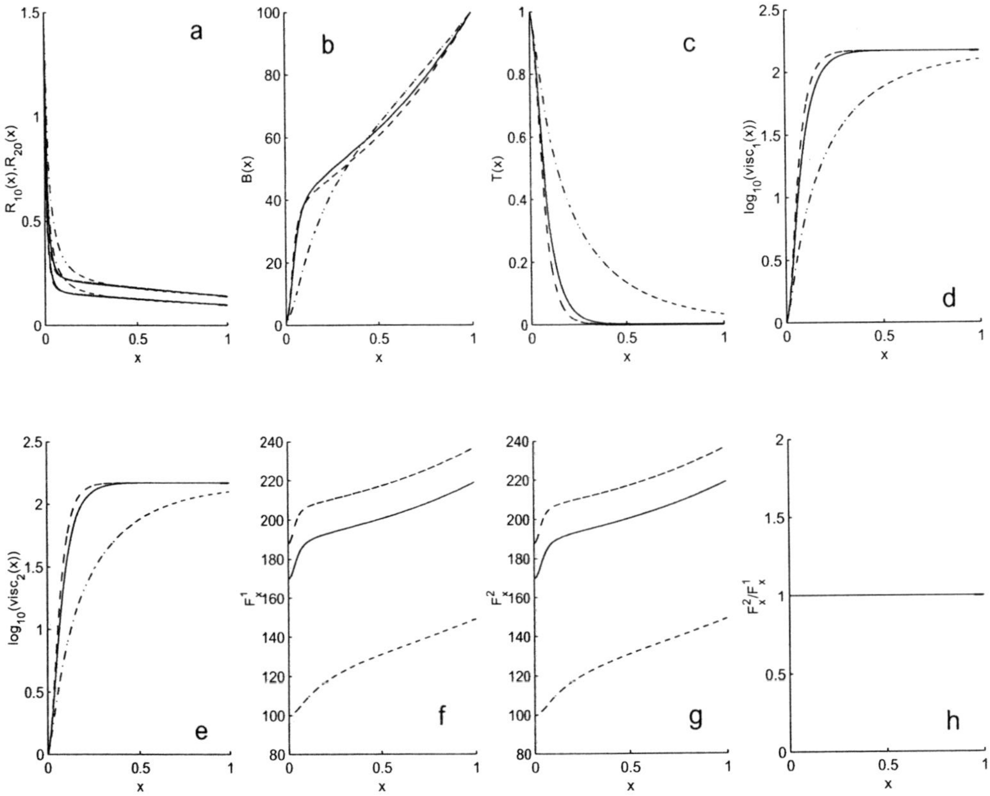

Figure 13: Compound's jet geometry (a), axial velocity component (b), temperature (c), dynamic viscosity of the inner (d) and outer (e) jets, axial traction on the inner (f) and outer jets (g) and axial traction ratio (h) for Cases 4, 9 and 10 of Table 5. (Solid line: Case 4; dashed line: Case 9; dashed–dotted line: Case 10).

jets; however, the Péclet number does affect the convexity of the axial velocity profile, and the fiber's contraction approaches the die's exit as the Péclet number is decreased. Similar results to the ones shown in Figure 13 were found for Cases 4, 11 and 12 of Table 5.

5.3 Results for Unsteady, Nonisothermal, Compound Fibers

Figure 14 presents some sample results for time–dependent, nonisothermal compound jets subject to sinusoidal axial velocity perturbations at either the die's exit and/or the take–up point (cf. Eq. (103)). These results correspond to $\bar{R} = 10^{-4}$, $\frac{\bar{R}}{\bar{F}} = 1$, $\bar{C} = 10^{39}$, $\frac{\sigma_1}{\sigma_2} = \frac{\rho_1}{\rho_2} = \frac{\psi_1}{\psi_2} = \frac{C_1}{C_2} = \frac{K_1}{K_2} = 1$, $Q_1 = Q_2 = 0.5$, $p_e = 0$, $T_\infty = 0$, $B_{ss}(1) = 400$, $G_1 = G_2 = 5$, $\bar{P}_1 = \bar{P}_2 = 1$, $T_\infty = 0$, and $\bar{B} = 10$, where $\bar{P}_1 = \frac{\rho_1 C_1 u_0 R_0}{\varepsilon K_1}$ and $\bar{P}_2 = \frac{\rho_2 C_2 u_0 R_0}{\varepsilon K_2}$ denote the thermal Péclet numbers of the inner and outer jets. The amplitude and frequency of the imposed axial velocity fluctuations were $a_i = a_e = 0.1$ and $S_i = S_e = 1$, respectively (cf. Eq. (103)).

Figure 14 shows that imposed sinusoidal axial velocity fluctuations at either the die's

Table 5: Values of the parameters used for steady–state, nonisothermal, compound fibers.

Case	G_1	G_2	$\bar{P}_1$	$\bar{P}_2$	$\bar{B}$
1	5	5	1	1	5
2	5	5	1	1	10
3	5	5	1	1	50
4	5	5	1	1	100
5	10	5	1	1	100
6	1	5	1	1	100
7	5	1	1	1	100
8	5	10	1	1	100
9	5	5	0.1	1	100
10	5	5	10	1	100
11	5	5	1	0.1	100
12	5	5	1	10	100

exit or the take–up point do not have strong effects on the compound fiber's geometry and temperature at the take–up point. The fiber's radii, axial velocity component, and temperature at the take–up point oscillate in a periodic fashion with a frequency that coincides with that of the imposed fluctuations. The axial velocity component and temperature oscillations at the take–up point are in phase, and this phase coincides with those of the axial forces at the take–up point and differs ninety degrees from those of the dynamic viscosities of the inner and outer jets. Figure 14 also shows that the amplitude of the axial velocity oscillations at the take–up point is 4 percent. Similar results to those illustrated in Figure 14 have also been observed for $1 \geq \bar{R} \geq 10^{-4}$, $0.01 \leq \frac{\bar{R}}{\bar{F}} \leq 1$, $1 \leq \bar{C} \leq 10^{39}$, $0.1 \leq \frac{\rho_1}{\rho_2} \leq 1$, $0.1 \leq \frac{\psi_1}{\psi_2} \leq 1$, $0.1 \leq \frac{\sigma_1}{\sigma_2} \leq 1$, $0.1 \leq \frac{C_1}{C_2} \leq 1$, $0.1 \leq \frac{K_1}{K_2} \leq 1$, $0.1 \leq Q_1 \leq 1$, $0.1 \leq Q_2 \leq 1$, $10 \leq B_{ss}(1) \leq 2000$, $1 \leq G_1 \leq 10$, $1 \leq G_2 \leq 10$, $1 \leq \bar{P}_1 \leq 100$, and $1 \leq \bar{P}_2 \leq 100$, although the amplitude of the oscillations was found to depend slightly on these parameters and, to a larger extent, on the activation energies of the dynamic viscosity law.

Figure 15 shows the power spectrum and phase diagram of the compound fiber's inner radius at the take–up point for the same parameters as those of Figure 14, and indicates that this radius is a periodic function of time which has the same frequency as that of the imposed, sinusoidal, axial velocity component at the die's exit. A comparison between Figure 15 and Figure 6 indicates that the amplitude and frequency of the oscillations of $R_{10}(t,1)$ of nonisothermal, compound jets is much smaller than those of isothermal ones, and much smaller than those corresponding to the chaotic behavior exhibited in Figures 8 and 9.

A comparison between Figure 14 and Figures 5, 7 and 9 and between Figure 15 and Figures 6, 8 and 10 indicates that cooling enhances the stability of the compound fiber largely due to the high increase of the viscosity with the (decrease of) temperature. Note that Figures 7 and 9 correspond to $B_{ss}(1) = 44$ and 100, respectively, and exhibit chaotic behavior, whereas Figure 15 corresponds to $B_{ss}(1) = 400$ and exhibits periodic oscillations. This implies that the cooling of the fiber allows to operate at higher draw ratios than those determined from a linear stability analysis for the viscous flow regime of isothermal, compound

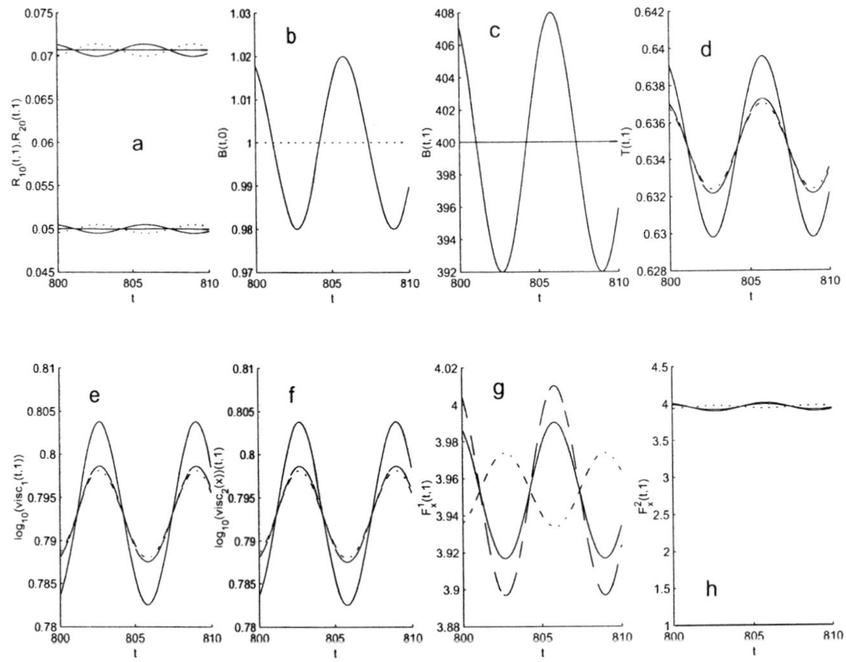

Figure 14: Compound's jet geometry at the take–up point (a), axial velocity component at the die's exit (b) and at the take–up point (c), temperature (d), dynamic viscosity of the inner (e) and outer (f) jets, and axial traction on the inner (g) and outer jets (h) for nonsisothermal, compound fibers subject to imposed sinusoidal axial velocity fluctuations at the die's exit and/or take–up point . (Solid line: fluctuations at both the die's exit and take–up point, $a_i = a_e = 0.1$, $S_i = S_e = 1$; dashed line: fluctuations at the die's exit only, $a_i = 0.1$, $S_i = 1$, $a_e = 0$; dashed–dotted line: fluctuations at the take–up point only, $a_e = 0.1$, $S_e = 1$, $a_i = 0$).

fibers.

The results presented in Figures 14 and 15 are consistent with those derived by Pearson et al. [32] who showed that freezing the material while still in the threadline clearly enhances the stability of single–component fibers operating in the viscous flow regime. Therefore, most industrially important melt spinning processes are inherently stable because of in–line freezing of the fiber. This is not to say that the oscillations in the fiber's geometry at the take–up point exhibited in Figure 15 are not important, because indeed they are important when polymer optical fibers are used for long–distance communications.

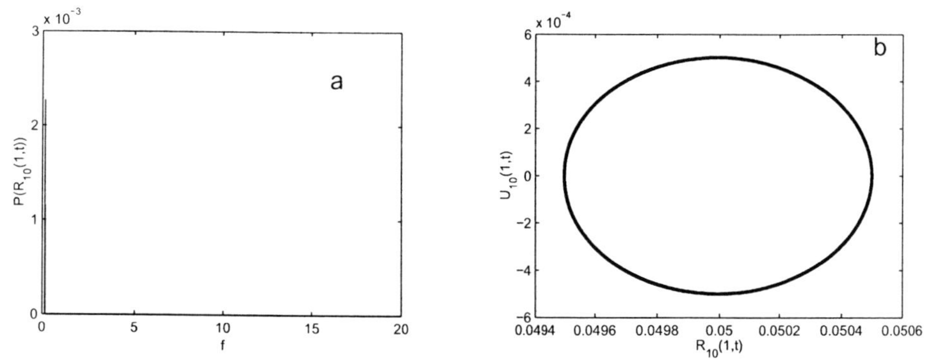

Figure 15: Power spectrum (left) and phase diagram (right) of $R_{10}(1,t)$ for sinusoidal axial velocity fluctuations at the die's exit only. ($a_i = 0.1$, $a_e = 0$, $S_i = 1$).

6 Polymer Orientation and Crystallization

The asymptotic one–dimensional model presented in Section 3 accounts for the conservation of mass, linear momentum and energy of compound fibers consisting of Newtonian fluids, and can be easily modified to include other rheologies [33, 34, 35] such as, for example, power–law viscosity law, viscoelasticity, etc. Moreover, the model does not account for the polymer orientation and flow–induced crystallization, and it is, therefore, valid for amorphous materials and polymers. In this section, we first present some comments on polymer orientation and crystallization, we then review some one–dimensional models for polymer orientation and crystallization in melt spinning of round fibers or filaments, and, finally, we present an orientation and crystallization model for melt spinning of compound fibers and some steady–state results.

6.1 Orientation and Crystallization

Most plastic products are manufactured by heating the polymer to above its melting temperature and then cooling it in a mold as in injection molding or deforming the melt while simultaneously cooling it to get the desired shape as in film blowing and melt spinning. The properties of the final product depend on the processing conditions to which the polymer is subject during its manufacture. Furthermore, depending on the molecular structure and processing conditions, the final product can be in either an amorphous or semi–crystalline state. Polymers that are unable to crystallize on cooling below their glass transition temperature, form amorphous solids, and, if these solids are formed by deforming the polymer while cooling it through the glass transition temperature, they can exhibit strong anisotropy. As the deformed amorphous polymer melt cools below its glass transition temperature, its molecules lose their mobility and become frozen in this oriented configuration.

Polymers having a regular or ordered molecular structure form a semi–crystalline solid when kept at a temperature below the melting one for a sufficiently long time. Under quiescent conditions, the crystallization process can be very slow especially at temperatures just below the equilibrium melting temperature, and the resulting solid has a spherulitic morphology.

The rate of crystallization depends on the molecular orientation in the melt; when subject to deformations that align the polymer molecules, the rate of crystallization increases dramatically, and, when the temperature drops below the glass transition temperature, there is a cessation of molecular motion and the crystallization rate decreases and may stop. As the crystallinity increases, it retards the crystallization process and decreases the mobility of the polymer molecules in the amorphous phase.

Some polymers such as, for example, polyethylene, find applications at temperatures between their melting and glass transition temperatures. At these temperatures, the solid consists essentially of rigid crystals and a flexible amorphous phase and it is therefore solid and tough.

The properties of the final product depend on the final morphology of the amorphous and semi–crystalline phases; the morphology, in turn, depends on the thermal and deformation histories undergone by the material during processing. Moreover, many products such as fibers and films are subjected to large inelastic deformations after manufacture, and, during the course of these deformations, further crystallization takes place, and the morphology and properties of the product evolve with the deformation.

Liquid–crystalline solutions may phase separate even under isothermal conditions due to the electrostatic and steric interactions that result from the rigidity of the polymer chain backbones even at low polymer concentrations, and form anisotropic phases with a degree of molecular orientation even in the quiescent state as evidenced by their ability to exhibit birefringence under static conditions. During processing, e.g., isothermal spinning, these molecules are aligned by the flow, thus yielding products with highly anisotropic properties [36]. For example, ultrastrength textile fibers such as Vectran and Kevlar achieve their distinguished commercial properties (tensile modulus) as a result of the interaction between the anisotropic molecular–scale structure of the melt or solution, the macroscopic hydrodynamics of spinning flows and nonsisothermal effects including radial heat transfer through the round fiber free surface, temperature dependence of material properties, crystallization, and phase change.

The phase change from the amorphous to the semi–crystalline stage is referred to as a *phase transition*. Early work on phase transitions considered that heat conduction was dominant and assumed that the temperature was the basic variable. Mean–phase field models were subsequently developed; these models introduce both the temperature and an order parameter, modify the energy equation to account for the order parameter, and introduce an additional equation the origin of which can be traced back to the Landau—Ginzburg theory of phase transitions. However, in most situations of practical interest, other mechanisms than conduction, e.g., convection in the liquid and deformation of the solid, take place and must be considered by introducing the kinematics of both the liquid and solid. Moreover, these mean–phase field models are unable to predict the mechanical properties of the newly formed solid, and, in many materials, after crystallization is initiated, the material goes through an intermediate set of states where it is a mixture of a liquid and a solid, i.e., a mush, before being completely transformed into a solid. Although there have been some studies on mushy zones, these studies have not addressed the issues of changes in symmetry associated with the phase change and the large deformations of the solid.

In the liquid phase, there are density fluctuations caused by thermal agitation [37, 38]. These fluctuations may eventually create small *clusters* or aggregates of polymer molecules having the same properties as the crystalline phase; the small crystals are continuously being created and destroyed by fluctuations because the formation of a crystal involves the creation of an interface between the liquid and the crystal and its consequent energy cost, i.e., the creation of crystals is a competitive process whereby there is a decrease of energy due to the fact that the chemical potential of the crystal phase is lower than that of the liquid and an increase of energy and an energy cost associated with the creation of interfaces. Moreover, surface effects are dominant in clusters of small size and, as a consequence, their growth is not energetically favorable, and the small crystals tend to dissolve. There exists, however, a critical size beyond which volume effects dominate over surface ones, and the growth of the cluster is favored by a global reduction of energy. The size that determines the stability of the clusters is called *critical size*, and the process of formation of crystals of size larger than or equal to the critical size is called *nucleation*. There are two basic types of nucleation: homogeneous nucleation which occurs in the bulk of a pure substance and heterogeneous nucleation which takes place in the presence of impurities, pre–existing crystals, boundaries, etc.

Nucleation is the first step in the crystallization process, for it determines the appearance of the first crystal nuclei which are the germ of the second stage of crystallization, i.e., growth. In this second stage, nuclei larger than the critical size tend to grow either through the addition of monomers or by acting as sites of heterogeneous nucleation, e.g., nucleation on the surface of a growing crystal.

Crystals grow freely until they begin to completely fill the whole space. Once they hit each other, the growth stops at the contact surface. This phenomenon, referred to as impingement, determines the final morphology of the processed polymer. In the final stage of crystallization, the impingement impedes the subsequent growth of the crystals and, therefore, amorphous matter remains trapped among clusters. Furthermore, the clusters themselves are not fully crystalline because they may contain some amorphous inclusions. This trapped material can eventually join the main crystal structure and, therefore, increase the degree of crystallinity in a process referred to as secondary crystallization.

As stated previously, the first step in the kinetics of phase transitions is nucleation. This is an activated process where an energy barrier has to be overcome in order to form nuclei of a critical size, beyond which the new phase grows spontaneously. Models for the crystallization of polymers have been mainly based on the Avrami–Kolmogorov equation [39, 40, 41, 42] that is based on the theory of filling the space through the nucleation and growth of one phase into another. This equation was initially developed for isothermal, quiescent crystallization, although it has been extended to account for non–isothermal conditions [43, 44, 45, 46]. Experimental results for polyethylene and polypropylene show that there is a temperature drop before crystallization is initiated, after which there is plateau where vigorous crystallization takes place at an almost constant temperature, followed by a temperature drop to the ambient value after the cessation of crystallization.

When phase transitions occur in a flowing polymer melt, crystallization takes place under strain and non–isothermal conditions, and the morphology of the processed polymer depends on the temperature and deformation history. In fact, it has been observed that crystallization may be enhanced by shear, pressure and strain, and that shear may increase the nucleation density, whereas the crystallization rate decreases as the cooling rate is increased. Usually, a highly oriented crystalline morphology is obtained under isothermal and non–isothermal deformations, in contrast with the spherulite morphology observed under quiescent conditions. In spun fibers, for example, lamellae are found to be perpendicular to the fiber axis and enhanced crystallization is caused by chain extension arising from entanglements between molecules. Orientation in the melt usually accelerates the phase transition process, but the molecular orientation may be enhanced or relaxed depending on the deformation history and the relaxation time of the melt.

Flow–induced crystallization in melts has been modelled in a number of different manners. A frequently used approach is based on the Avrami–Kolmogorov equation which is modified to account for enhanced crystallization due to the flow. In this approach the effect of the flow is accounted for by the inclusion of an orientation factor which depends on the flow [6]. Another approach is based on an extension of the work of Flory [47] on the stress–induced crystallization of rubber. Such an extension assumes that the temporary network junctions play the same role as the chemical cross–links in the theory of rubber crystallization and is based on the decrease in entropy of the stretched molecules and the tendency of the polymer to crystallize. A third approach is based on the formulation of constitutive equations that involve the concept of multiple natural configurations and obtain evolution equations for the natural configuration and mass fraction of the crystalline material by maximizing a prescribed rate of dissipation [48]. This approach combines continuum mechanics and thermodynamics, assumes that effects are additive, and results in different models that depend on the forms used for the internal energy, entropy and dissipation rate.

6.2 Review of Previous Melt Spinning Models which Include Orientation and/or Crystallization

Models that only account for the orientation–hydrodynamics interactions in fiber spinning processes and, therefore, neglect energy effects and phase changes have been developed by Forest and co–workers [49, 50] for liquid crystalline, round polymer fibers. These authors considered the evolution of the microstructure in isothermal flows by starting from three–dimensional Doi–Edwards averaged kinetic equations before undertaking the study of the spinline crystallization under isothermal [51] and nonisothermal flow conditions [52]. It is commonly accepted that crystallization kinetics is triggered by a combination of orientation, stress and temperature [6].

The one–dimensional models of fiber spinning of liquid–crystalline polymers of Forest et al. [52] include the mass, linear momentum and energy conservation equations, and two equations for the degree of orientation, s, defined as the ensemble average of the alignment of the molecular direction to the axial direction, and the degree of crystallinity, θ, which can be expressed as

$$\frac{\partial s}{\partial t}+B\frac{\partial s}{\partial x}=(1-s)(2s+1)\frac{\partial B}{\partial x}-\frac{\sigma}{\lambda}U(s), \tag{107}$$

$$\frac{\partial \theta}{\partial t}+B\frac{\partial \theta}{\partial x}=k_A(s)(\theta_\infty-\theta), \tag{108}$$

where σ is an anisotropic drag parameter ($0<\sigma\leq 1$, $\sigma=1$ for isotropic models, and $\sigma\approx 0.5$ for rigid–rod molecular models), λ is the molecular relaxation time of the liquid–crystalline polymer, $k_A(s)=k_A(0)\exp(a_2s^2)$ is the linearized growth rate, $k_A(0)$ is the amorphous growth rate [6], a_2 is a constant, θ_∞ is the ultimate degree of crystallinity, $U(s)=s\left(1-\frac{N}{3}(1-s)(2s+1)\right)$ is a bulk free energy which is related to the intermolecular potential, N is the dimensionless density of the liquid–crystalline polymer, the viscosity that appears in the one–dimensional linear momentum equation is replaced by $\mu_{eff}=\mu+\frac{2}{3}\alpha\lambda s^2$, and α is a parameter that relates the kinetic energy to the inertial energy of the liquid–crystalline polymer.

It is well known that non–isothermal effects due to the rapid cooling by the surrounding quench air and the low thermal conductivity of fibers can result in non–uniform fiber microstructure and (mechanical, electrical, chemical, optical) properties. As a result, skin–core differentiation or molecular orientation and structure can be observed in many fibers. Although considerable efforts have been made to understand the dynamics of the melting spinning process and the relationship between spinning conditions and fiber structure, most previous studies on melting spinning processes are based primarily on one–dimensional models which include a single–phase two–component submodel to describe stress–induced crystallization along the spinning line [53, 54, 55, 56]. However, the results of these models do not provide any information about radial variations and on the implications of these variations on the fiber microstructure. Moreover, if one–dimensional models of the draw–down region of fiber spinning processes overestimate the heat transfer at the fiber's outer surface, they predict inaccurate temperature fields which, in turn, influence the average axial velocity, tensile stress, crystallinity, and free–surface shape due to the nonlinear couplings between the mass, linear momentum and energy equations (cf. Eqs. (40), (54), (59) and (61)). Through the dependence of the crystallization kinetics on temperature and molecular orientation, a strong crystallinity profile is developed, and the radial variation of the molecular orientation and crystallinity has a strong influence on the fiber microstructure and properties, i.e., physical, mechanical, thermal, chemical, electrical and optical properties, such as tensile modulus and optical birefringence.

The one–dimensional model of fiber spinning of round fibers developed by Doufas et al. [53, 54, 55, 56] includes the combined effects of flow–induced crystallization, viscoelasticity, filament cooling, air drag, inertia, surface tension, latent heat and gravity. It also includes both an amorphous phase (simulated as a modified Giesekus fluid) and a semi–crystalline phase (approximated as rigid rods that grow and orient in the flow field) and are coupled through the stress and linear momentum balance and the feedback of the crystallinity to the system of relaxation times, predicts neck–like deformation and associated strain softening, and shows that crystallization occurs mostly after the freeze point.

The microstructure model prior to the onset of crystallization used by Doufas et al. [53, 54, 55, 56] assumes that the polymer melt exiting the spinneret is coarse–grained as a concentration of n non–linear dumbbell molecules per unit volume. In a mean field approximation, each of the polymer melts is assumed to contain N_0 flexible statistical links of length l. The microstructure, i.e., orientation and stretching, of the melt is characterized by the conformational tensor $\mathbf{c} = <\mathbf{RR}>$ that represents the second moment of the end–to–end vector $\mathbf{R}$ of the polymer chains [34], and the bracket denotes averaging with respect to the distribution function of the melt phase. Moreover, the conformational tensor is assumed to be determined from a differential equation and allows to determine the extra stress tensor.

Doufas et al. [53, 54, 55, 56] also assumed that crystallization begins when the axial temperature in the filament reaches the equilibrium melting temperature, although experimental data suggest that crystallization in high–speed melt spinning occurs below that temperature. After the onset of crystallization, it is assumed that crystallization corresponds to the transfer of N statistical links from the amorphous (melt) phase to the semi–crystalline phase at an arbitrary position along the spin line under both undercooling and stress; this implies that stretching and crystallization may occur simultaneously, and the degree of transformation from the amorphous to the semi–crystalline phase is defined as $z = \frac{N}{N_0}$. In general, the semi–crystalline phase consists of amorphous and crystalline regions, and the absolute degree of crystallinity is defined as the ratio of the mass of the pure crystalline material to the total mass of the system, i.e., $\theta = z\theta_\infty$, where θ_∞ is the degree of crystallinity within the semi–crystalline phase, i.e., the ratio of the mass of pure crystalline material to the mass of the semi–crystalline phase.

In order to account for the effects of crystallization on the rheology of melt spinning processes, one must account for the rheological behavior of the amorphous phase (the melt), the semi–crystalline phase, and the rate of crystallization. From the rheological point of view, it may be assumed that the two phases (amorphous and semi–crystalline) act in parallel and, therefore, the strain rate tensor is the same for both phases, while the total stress is simply the superposition of the individual stresses.

A simple model to account for the crystallization (if any) of spun filaments under non-isothermal conditions consists in approximating the *thermal* crystallization rate by a quasi–static Avrami–Kolmogorov equation, e.g.,

$$\frac{d}{dt}[-\ln(1-\theta)]^{\frac{1}{n}} = K_{st}(T,\Delta p), \tag{109}$$

where n denotes the Avrami exponent (usually assumed to be an integer $1 \leq n \leq 4$), and K_{st} is the steady state crystallization rate which depends on the temperature and spinning stress, Δp, which may be approximated by

$$K_{st}(T,\Delta p) = K_{st}(T,0)\exp[C_3(\Delta p)^2], \tag{110}$$

where C_3 is a constant and $K_{st}(T,0)$ is the crystallization rate of an isotropic crystallization [63]. This *thermal* crystallization rate can be added to the *athermal* one and to flow– and shear–induced crystallization rates in order to determine the whole rate.

Another approach to the analysis of the microstructure formed during fiber spinning is to solve the one–dimensional mass and linear momentum equations, i.e., Eqs. (54) and (59), respectively, together with the two–dimensional energy equation, i.e., Eq. (23) and a two–dimensional equation for the degree of crystallization, e.g., [57, 58, 59]. This approach has to be carried out numerically, but it is not clear that the use of hybrid one– and two–dimensional formulations offers any computational advantage over a full two–dimensional model for the mass, linear momentum and energy conservation equations, and for the degree of crystallization.

Fully steady–state, two–dimensional models that account for viscoelasticity, heat transfer, crystallization kinetics, and free–surface conditions have been developed by Joo et al. [60] and Sun et al. [61] who employed a non–isothermal, viscoelastic Giesekus constitutive model, the non–isothermal crystallization kinetics of Nakamura et al. [62] where the crystallization rate which is a function of both the temperature and molecular orientation, is evaluated according to the equation proposed by Ziabicki [6], the thermally–induced crystallization is approximated by a Gaussian function in Ziabicki's formulation, and the molecular orientation is determined from a stress–optical law that introduces stress–induced crystallization. It must be noted that Joo et al. [60] assumed that the solvent viscosity can be expressed as the product of two terms; the first term corresponds to amorphous polymers and depends exponentially on the temperature through an Arrhenius equation (cf. Eq. (6)), while the second one is an exponential function of the relativity crystallinity defined as the ratio of the volume fraction crystallinity to the maximum obtainable crystallinity. Moreover, their crystallization kinetics is based on the assumption of isokinetic conditions, for which the ratio of the linear crystal growth rate to the volume of the crystal grain does not depend on time; they also neglected secondary crystallization and overlapping of growing grains, and used an extension of the isothermal Avrami–Kolmogorov equation to describe non–isothermal crystallization, where the crystallinity is governed by an advection equation.

6.3 Orientation and Crystallization of Compound Fibers

In this section, we employ a generalization of the model developed by Forest et al. [52] for round fibers to analyze the effects of orientation and crystallization of the melting spinning of bicomponent or compound fibers. Such a generalization requires the numerical solution of the following equations (cf. Eqs. (54), (59) and (61))

$$\frac{\partial A_1}{\partial t}+\frac{\partial}{\partial x}(BA_1)=0, \qquad \frac{\partial A_2}{\partial t}+\frac{\partial}{\partial x}(BA_2)=0, \tag{111}$$

$$(\rho_1 A_1+\rho_2 A_2)\frac{DB}{Dt}=(\rho_1 A_1+\rho_2 A_2)\frac{\bar{R}}{\bar{F}}+\frac{\partial}{\partial x}\left((3\mu_{e,1}A_1+3\mu_{e,2}A_2)\frac{\partial B}{\partial x}\right)$$

$$+A_1\frac{1}{\bar{C}}\left(\frac{1}{R_{20}^2}\frac{\partial R_{20}}{\partial x}+\frac{\sigma_1}{\sigma_2}\frac{1}{R_{10}^2}\frac{\partial R_{10}}{\partial x}\right)+A_2\frac{1}{\bar{C}R_{20}^2}\frac{\partial R_{20}}{\partial x}, \tag{112}$$

$$(\rho_1C_1A_1+\rho_2C_2A_2)\bar{P}\frac{DE}{Dt}=\frac{\partial}{\partial x}\left((A_1K_1+A_2K_2)\frac{\partial E}{\partial x}\right)-\frac{\bar{B}}{K_2}R_{20}(E-T_\infty), \tag{113}$$

$$\frac{\partial s_i}{\partial t}+B\frac{\partial s_i}{\partial x}=(1-s_i)(2s_i+1)\frac{\partial B}{\partial x}-\frac{\sigma_i}{\lambda_i}U_i(s_i), \qquad i=1,2, \tag{114}$$

$$\frac{\partial \theta_i}{\partial t}+B\frac{\partial \theta_i}{\partial x}=k_{Ai}(s)(\theta_{\infty,i}-\theta_i), \qquad i=1,2, \tag{115}$$

where $k_{Ai}(s)=k_{Ai}(0)\exp(a_{2i}s_i^2)$ is the linearized growth rate, $k_{Ai}(0)$ is the amorphous growth rate [6], a_{2i} is a constant, $\theta_{\infty,i}$ is the ultimate degree of crystallinity,

$$U_i(s_i)=s_i\left(1-\frac{N_i}{3}(1-s_i)(2s_i+1)\right), \qquad i=1,2, \tag{116}$$

$$\mu_{e,1}=\mu_{10}(E)\exp\left(\beta_1\left(\frac{\theta_1}{\theta_{\infty,1}}\right)^{n_1}\right)+\frac{2}{3}\alpha_1\lambda_1s_1^2, \tag{117}$$

$$\mu_{e,2}=\mu_{20}(E)\exp\left(\beta_2\left(\frac{\theta_2}{\theta_{\infty,2}}\right)^{n_2}\right)+\frac{2}{3}\alpha_2\lambda_2s_2^2, \tag{118}$$

β_i is the crystallization viscosity rate and n_i is the crystallization viscosity index. It must be noted that μ_{i0} is given by Eq. (40) as $\mu_{i0}(E)=\psi_i\exp(G_i(1-E))$, and that μ_{i0}, β_i and n_i are material–dependent, e.g., $\beta_i=4.605$ and $n_i=12$ for nylon–66, $\beta_i=4$ and $n_i=2$ for PET.

The orientation (Eq. (114)) is affected by the axial velocity component, B, and affects the degree of crystallization through Eq. (115); both the orientation and crystallization affect the axial velocity component (Eq. (112)) through the effective dynamic viscosities (Eqs. (117) and (118)), and, of course, the compound fiber's geometry (Eq. (111)) and temperature (Eq. (113)) are affected by the axial velocity component. This implies that the orientation and crystallization of the compound fiber are nonlinearly coupled with the fiber's geometry, axial velocity component and temperature, and that Eqs. (111)–(115) must be solved numerically in an iterative fashion. However, if $\sigma_i=0$ in Eq. (114), then the resulting equation under steady state conditions can be written as $\frac{ds}{dy}=(1-s)(1+2s)$ where $y=\ln B$ and the subscript i has been dropped, and has fixed points at $s=1$ and $s=-\frac{1}{2}$ which are stable and unstable, respectively, and, since $s\geq 0$, it can be easily shown that the solution to this equation evolves towards $s=1$. Moreover, this equation can be integrated analytically subject to, say, $s(0)=0$, to yield the molecular orientation as a function of the leading–order axial velocity component as

$$B(x)=\frac{(1+2s)^2}{1-s}, \tag{119}$$

which is quadratic expression from which one can easily obtain $s(x) = s(B)$. Furthermore, under the same assumptions as discussed above, Eq. (115) can be integrated analytically subject to $\theta(0) = 0$ to yield

$$\frac{\theta_\infty}{\theta - \theta_\infty} = \int_0^x \frac{k_A(s)}{B(x)} dx, \tag{120}$$

where $k_A(s) = k_A(0)\exp(a_2 s^2)$ can be written as $k_A(B)$ upon using $s(B)$ from Eq. (119). Both Eq. (119) and Eq. (120) indicate that, under the above assumptions, the molecular orientation and crystallization depend on $B(x)$ which, in turn, depends in a nonlinear fashion on the solutions to Eqs. (111)–(113).

If Eqs. (114) and (115) were replaced by

$$A_i\left(\frac{\partial s_i}{\partial t} + B\frac{\partial s_i}{\partial x}\right) = (1-s_i)(2s_i+1)\frac{\partial B}{\partial x} - \frac{\sigma_i}{\lambda_i}U_i(s_i), \qquad i = 1,2, \tag{121}$$

$$A_i\left(\frac{\partial \theta_i}{\partial t} + B\frac{\partial \theta_i}{\partial x}\right) = k_{Ai}(s)(\theta_{\infty,i} - \theta_i), \qquad i = 1,2, \tag{122}$$

$\sigma_i = 0$, one could easily obtain that, for steady flows,

$$\frac{B(x)-1}{Q_i} + 1 = \frac{(1+2s_i)^2}{1-s_i}, \tag{123}$$

which is quadratic expression from which one can easily obtain $s(x) = s(B)$. This can be used in Eq. (122) to obtain the steady–state degree of crystallinity as

$$\frac{\theta_{\infty,i}}{\theta_i - \theta_{\infty,i}} = \frac{1}{Q_i}\int_0^x k_{Ai}(s)dx, \tag{124}$$

where $k_{Ai}(s) = k_{Ai}(0)\exp(a_{2i}s_i^2)$ can be written as $k_{Ai}(B)$ upon using $s_i(B)$ from Eq. (123).

6.4 Results for the Orientation and Crystallization of Compound Fibers

The steady state equations corresponding to Eqs. (111)–(115) were solved numerically for $N_i = 4$, $\sigma_i = 0.5$, $\alpha_i = 5$, $a_{2i} = 10$, $\lambda_i = 1$, $\theta_i(0) = 0$, $\theta_{\infty,i} = 0.8$, $s_i(0) = 0.5$, $n_i = 12$, $\beta_i = 4$, $\bar{R} = 1$, $p_e = 0$, $\frac{\bar{R}}{\bar{F}} = 1$, $\bar{C} = 10^{39}$, $\frac{\rho_1}{\rho_2} = 1$, $\frac{\psi_1}{\psi_2} = 1$, $\frac{\sigma_1}{\sigma_2} = 1$, $Q_1 = Q_2 = 0.5$, and $T_\infty = 0$, for the values of the parameters shown in Table 6.

Figure 16 illustrates the effects of the activation energy of the dynamic viscosity law of the cladding or outer jet on the compound fiber's geometry, axial velocity component and temperature, and on the crystallization degree, $X_c \equiv \theta$, orientation, s, and dynamic viscosity of the core and cladding. This figure shows that the higher the activation energy, the faster the fiber's cooling and molecular orientation which tends faster to the asymptotic value of one; however, although the degree of crystallization increases as G_2 is increased, the distance required to achieve its asymptotic value increases as G_2 increases. Figure 16 also shows that the axial velocity profile is a strong function of the temperature, molecular orientation and degree of crystallization, and its convexity is a strong function of G_2; small

Table 6: Values of the parameters used for steady–state, nonisothermal, compound fibers with orientation and crystallization.

Case	G_1	G_2	$\bar{P}_1$	$\bar{P}_2$	$\bar{B}$	$\frac{K_2}{K_1}$	$\frac{\Psi_2}{\Psi_1}$
1	10	10	100	100	50	1	1
2	50	10	100	100	50	1	1
3	100	10	100	100	50	1	1
4	50	50	100	100	50	1	1
5	500	100	100	100	50	1	1
6	50	10	10	100	50	1	1
7	50	10	1	100	50	1	1
8	50	10	100	10	50	1	1
9	50	10	100	1	50	1	1
10	50	10	100	100	10	1	1
11	50	10	100	100	5	1	1
12	50	10	100	100	50	10	1
13	50	10	100	100	50	0.1	1
14	50	10	100	100	50	1	10
15	50	10	100	100	50	1	0.1

values of G_2 result in convex axial velocity profiles analogous to those found under isothermal conditions without molecular orientation and crystallization. As G_2 is increased, $B(x)$ shows an upward concavity near the die's exit and a downward concavity at the take–up point. Analogous results to those shown in Figure 16 have been observed for Cases 1, 2 and 3 of Table 6.

Figure 16 also shows that the distance required to achieve full molecular orientation increases as G_2 is decreased and, for the smaller value of this parameter, the orientation reaches its asymptotic value at about $x = 0.8$ and then decreases slightly. By way of contrast the degree of crystallization is a monotonic function of the axial distance along the compound fiber, and the asymptotic value of the degree of crystallization is not reached for Case 3 of Table 6. Figure 16 also shows that full molecular orientation takes place before full crystallization.

The effects of the thermal Péclet number of the core on steady compound fibers with molecular orientation and crystallization are illustrated in Figure 17, and clearly indicate that the fiber's temperature decreases as the thermal Péclet number is decreased. This decrease has great effects on the compound fiber's geometry and axial distances for crystallization and orientation. In fact, the orientation increases as the thermal Péclet number is decreased, while the crystallization rate increases but takes a longer distance to achieve its asymptotic value as the thermal Péclet number is decreased.

Figure 17 indicates that Cases 4 and 6 produce similar axial velocity profiles and exhibit small differences in the molecular orientation and crystallization, but their small differences in temperature have large effects on the dynamic viscosities of both the core and the cladding. Analogous results to those of Figure 17 have been found for Cases 4, 8 and 9 of Table 6.

The effects of the Biot number on steady compound fibers are shown in Figure 18 which

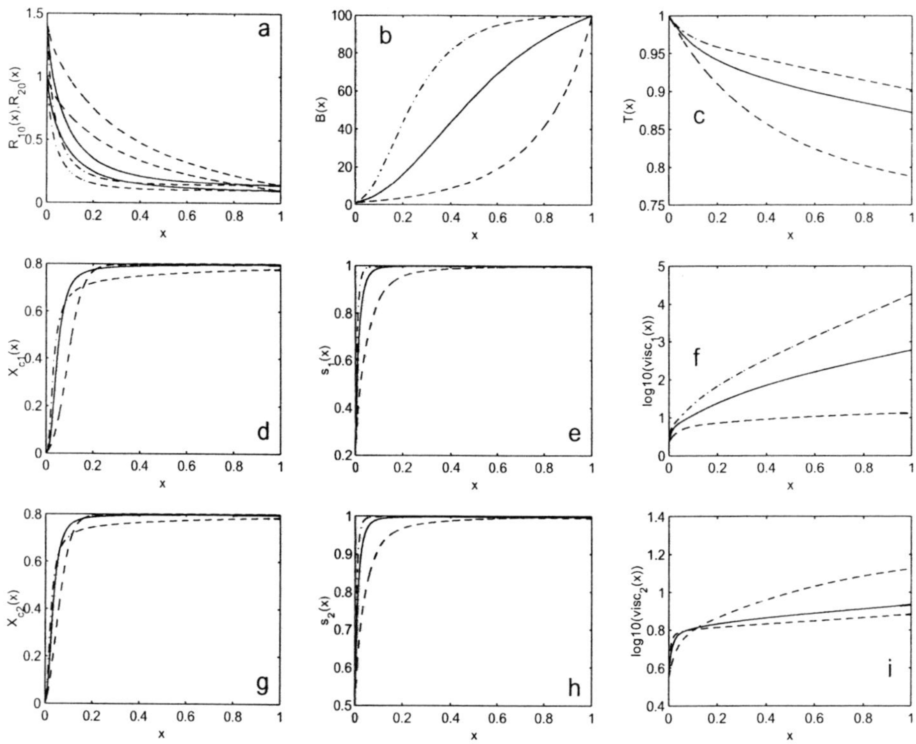

Figure 16: Steady compound's jet geometry (a), axial velocity component (b), temperature (c), degree of crystallization of the inner (d) and outer (g) jets, molecular orientation of the inner (e) and outer jets (h), and dynamic viscosity of the inner (f) and outer (i) jets for nonsisothermal, compound fibers. (Solid line: Case 4; dashed line: Case 3 ; dashed–dotted line: Case 5 of Table 6).

indicates that, for the thermal Péclet numbers used in the calculation of this figure, there are no significant differences between $\bar{B} = 10$ and 5 and these two values provide axial velocity profiles analogous to those found under isothermal conditions. Figure 18 also shows that the cooling of the compound fiber and the molecular orientation and crystallization increase as the Biot number, i.e., the heat loss, is increased.

The effects of the thermal conductivity ratio, i.e., the ratio of the thermal conductivity of the core to that of the cladding, are illustrated in Figure 19. This figure shows that the fiber solidification and molecular orientation increase as the thermal conductivity ratio is increased; the degree of crystallization increases initially but takes longer distances to reach its asymptotic value as the thermal conductivity ratio is increased. The results presented in Figure 19 are consistent with the formulation presented here, for an increase in the thermal conductivity of the cladding results in higher heat losses through the compound fiber's outer surface and, therefore, higher cooling rates. By way of contrast, the ratio of the pre–exponential terms of the dynamic viscosity laws, i.e., $\frac{\psi_2}{\psi_1}$, were not found to have large

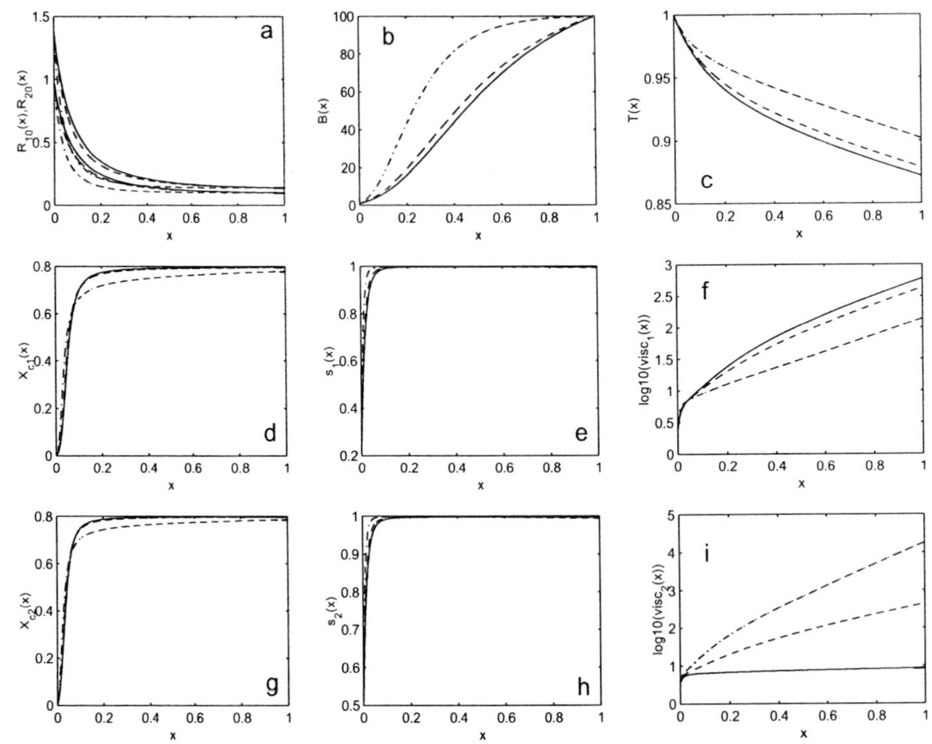

Figure 17: Steady compound's jet geometry (a), axial velocity component (b), temperature (c), degree of crystallization of the inner (d) and outer (g) jets, molecular orientation of the inner (e) and outer jets (h), and dynamic viscosity of the inner (f) and outer (i) jets for nonsisothermal, compound fibers. (Solid line: Case 4; dashed line: Case 6 ; dashed–dotted line: Case 7 of Table 6).

effects on the steady state behavior of non–isothermal compound fibers when the effects of orientation and crystallization were considered, when this parameter was varied from 0.1 to 1 as in Table 6. For this small variation, the exponential dependence of the dynamic viscosity on temperature, i.e., the activation energy, was found to have a larger effect on the dynamics of compound fibers as the results presented in Figure 16 indicate.

7 Conclusions and Discussion

In this chapter, perturbation methods based on the slenderness ratio have been employed to determine the leading–order fluid dynamics and heat transfer equations of axisymmetric, Newtonian, compound liquid jets at low Reynolds numbers. It has been shown that the leading–order equations are one–dimensional and correspond to the conservation of mass, global linear momentum conservation, energy, and kinematics of the interface between the outer annular jet and its surroundings. These (nondimensional) equations depend on the

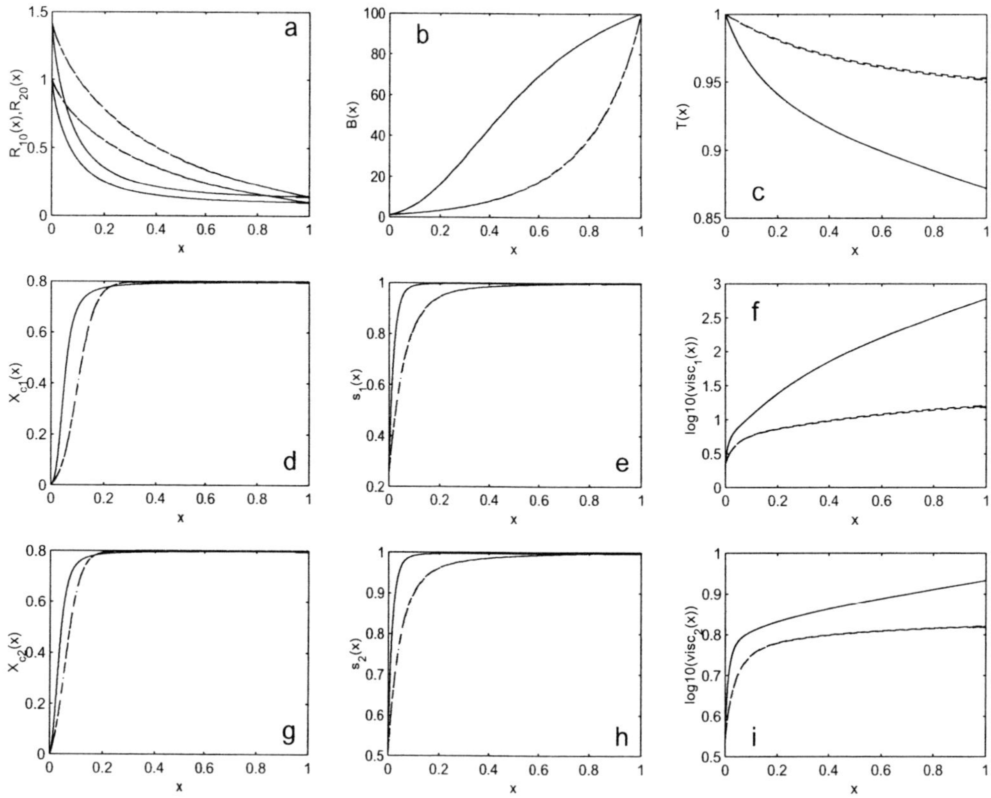

Figure 18: Steady compound's jet geometry (a), axial velocity component (b), temperature (c), degree of crystallization of the inner (d) and outer (g) jets, molecular orientation of the inner (e) and outer jets (h), and dynamic viscosity of the inner (f) and outer (i) jets for nonsisothermal, compound fibers. (Solid line: Case 4; dashed line: Case 10; dashed–dotted line: Case 11 of Table 6).

Reynolds, Froude, Prandtl and capillary numbers, pressure of the surroundings, and density, dynamic viscosity, thermal conductivity and thermal capacity, and surface tension ratios. They also depend on the non–dimensional pre–exponential factors and activation energies of the dynamic viscosity and the temperature of the surroundings.

Analytical solutions to the leading–order steady state equations have been obtained for zero Reynolds numbers, zero gravitational pull, or inertialess flows under isothermal conditions. For the steady, isothermal, viscous regime, a linear stability analysis has been performed and shown to be governed by the same eigenvalue equation as that for the spinning of isothermal, round jets, planar liquid sheets and annular liquid jets at zero Reynolds numbers.

Numerical results of the time–dependent leading–order equations for isothermal compound fibers indicate that as the axial velocity at the downstream boundary is increased, the radius of the inner round jet at the take–up point evolves from a fixed point to a limit cycle to

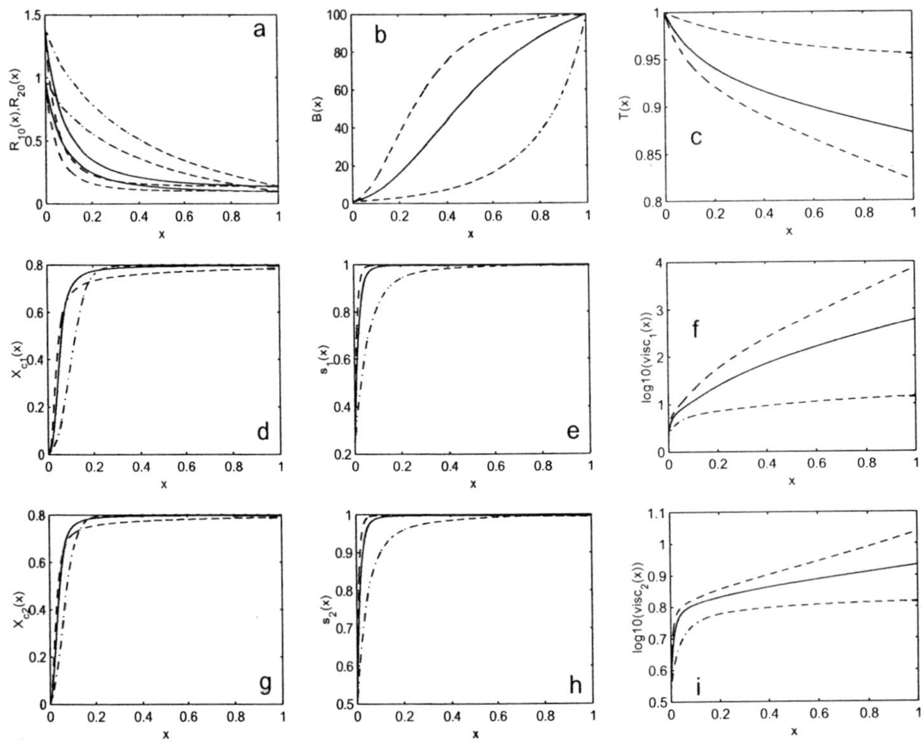

Figure 19: Steady compound's jet geometry (a), axial velocity component (b), temperature (c), degree of crystallization of the inner (d) and outer (g) jets, molecular orientation of the inner (e) and outer jets (h), and dynamic viscosity of the inner (f) and outer (i) jets for nonsisothermal, compound fibers. (Solid line: Case 4; dashed line: Case 12; dashed–dotted line: Case 13 of Table 6).

a broad spectrum. The phase diagrams of this radius may exhibit holes, i.e., regions which are not visited, when the axial velocity component at the take–up boundary is sufficiently large, thus indicating the presence of strange attractors and chaos.

It has also been shown that the radii and axial traction force at the take–up point of isothermal compound fibers exhibit very sharp spikes whose separation depends on the Reynolds, Froude and capillary numbers, density, viscosity and surface tension ratios, and location, amplitude and frequency of the imposed velocity perturbations. These spikes are somewhat modulated with a lower frequency which is of the order of that of the imposed axial velocity perturbation.

Analytical solutions for the temperature of steady compound fibers have been obtained when axial conduction is neglected. These solutions indicate that the temperature of the compound fiber decreases exponentially with the axial distance. Numerical solutions of the leading–order, one–dimensional equations of non–isothermal, compound fibers indicate that cooling of the fiber enhances their stability and the critical draw ratio for isothermal fibers can be exceeded without affecting the stability of freezing fibers. However, it has been found that sinusoidal perturbations of the axial velocity component at either the die's exit or the take–up point result in fiber's radii with oscillate sinusoidally in time with an amplitude that depends on that of the imposed perturbations.

The orientation and crystallization of polymers have also been accounted for by means of two additional one–dimensional equations for the conformation and crystallization without taking into consideration latent heat effects. It has also been assumed that the effects of orientation and crystallization can be accounted for by modifying the viscosity law through a multiplicative term that depends on the degree of crystallinity and an additive term that depends on the molecular orientation while preserving the leading–order equations derived in the asymptotic analysis. These modifications allow to study both amorphous and crystalline melt spinning processes of compound fibers, and reduce to the ones derived for non–isothermal fibers when the orientation and degree of crystallization is set to zero.

The assumptions and limitations used in the derivation of the leading–order equations have been discussed through the chapter. In the next paragraphs, we would like to emphasize some of the improvements that can be incorporated into these one–dimensional equations.

Equations (114) and (115) account for the orientation and crystallization of both the core and cladding of compound fibers, and surface tension, gravity, inertia, thermal conductivity, and cooling, but they do not account for air drag, viscous dissipation and latent heat effects. The absence of air drag in Eq. (112) is a consequence of the assumption made in Section 2 regarding the passive character of the gases surrounding the compound fiber. Air drag can be included in Eq. (112) by performing a boundary layer analysis of the gases surrounding the fiber and assuming that there are no confining walls. Under these conditions, one can use the von Kárman–Pohlhausen technique and derive an equation for the boundary layer thickness, δ, which determines the following (dimensional) friction force on the compound fiber's outer jet [64]

$$f_f = 2\pi M R_{20} \frac{\mu_g B}{\delta}, \tag{125}$$

where μ_g denotes the dynamic viscosity of the gases surrounding the fiber and M is a constant that depends on the polynomial used to approximate the axial velocity component of the gases surrounding the fiber. It must be noted that the above equation was derived under the condition that the axial velocity component of the gases surrounding the fiber is nil far away from the fiber's outer interface, and that the von Kárman–Pohlhausen technique provides a differential equation for $\delta(x)$. Equation (125) has to be modified when the gases surrounding the fiber are confined and natural and forced convection are important. It has also to be modified when gases are forced to increase the fiber's cooling.

For a single-phase, two–component system such as the one corresponding to Eqs. (111)–(116) where the two components are assumed to be the amorphous and crystalline phases of the polymer, one can easily include latent heat effects associated with crystallization by assuming that the mass diffusivities of these two phases are identical, and considering that the density of the mixture is constant [65]. Under these assumptions, latent heat effects may be accounted for by adding to the right–hand side of Eq. (113) the term $\bar{P}\frac{1}{C_0\Delta T}\left(\rho_1\Delta H_1 A_1\frac{D\theta_1}{Dt}+\rho_2\Delta H_2 A_2\frac{D\theta_2}{Dt}\right)$, where $\frac{D\theta_i}{Dt}=\frac{\partial\theta_i}{\partial t}+B\frac{\partial\theta_i}{\partial x}$, H is the heat of crystallization per unit mass, and the mass fraction of crystals is θ_i.

The energy equation, i.e., Eq. (61) of Section 3, was derived under the assumption that $Br=O(\varepsilon^4)$ although it was pointed out that this equation is also valid for $Br=O(\varepsilon^2)$. Taking into account Eq. (53) and Eqs. (51) and (52), Eq. (35) can be written as

$$\psi=3\varepsilon^2 Br\left(\frac{\partial B}{\partial x}\right)^2+O(\varepsilon^4), \tag{126}$$

which, for $Br=O(1)$, an asymptotic analysis analogous to that of Section 3 yields the following energy equation

$$(\rho_1 C_1 A_1+\rho_2 C_2 A_2)\bar{P}\frac{DE}{Dt}=\frac{\partial}{\partial x}\left((A_1K_1+A_2K_2)\frac{\partial E}{\partial x}\right)$$
$$+3Br(\mu_{e,1}A_1+\mu_{e,2}A_2)\left(\frac{\partial B}{\partial x}\right)^2-\frac{\bar{B}}{K_2}R_{20}(E-T_\infty), \tag{127}$$

which includes thermal inertia, axial conductivity, viscous dissipation, and heat losses due to convection and radiation. Moreover, by including the latent heat effects described previously, one can easily deduce the following energy equation

$$(\rho_1 C_1 A_1+\rho_2 C_2 A_2)\bar{P}\frac{DE}{Dt}=\frac{\partial}{\partial x}\left((A_1K_1+A_2K_2)\frac{\partial E}{\partial x}\right)$$
$$+3Br(\mu_{e,1}A_1+\mu_{e,2}A_2)\left(\frac{\partial B}{\partial x}\right)^2-\frac{\bar{B}}{K_2}R_{20}(E-T_\infty)$$
$$+\frac{\bar{P}}{C_0\Delta T}\left(\rho_1\Delta H_1A_1\frac{D\theta_1}{Dt}+\rho_2\Delta H_2A_2\frac{D\theta_2}{Dt}\right). \tag{128}$$

Acknowledgements

The research reported in this chapter was supported by Project PB97–1086 from the D.G.E.S. and by Project BFM2001-1902 from the Ministerio de Ciencia y Tecnología of Spain and Fondos FEDER. The author is also grateful to Mr. Frank Columbus, President and Editor–in–Chief at Nova Publishers, Inc., for his invitation to write this chapter. Such an invitation has served the author to dig for the research on compound jets and fibers that he has been performing since 1998. The author is also grateful to Dra. Carmen María López for her help in revising the proofs of this chapter.

References

[1] Pearson, J. R. A. *Mechanics of Polymer Processing*, Elsevier Applied Science Publishers, Ltd., NY, 1985.

[2] Geyling, F. T. *The Bell Syst. Tech. J.* 1976, 55, 1011–1056.

[3] White, J. L. *Polymer Eng.* 1981, 1, 297–362.

[4] Denn, M. M. *Ann. Rev. Fluid Mech.*, 1980, 12, 365–387.

[5] Denn, M. M. In *Computational Analysis of Polymer Processing* ; Eds. J. R. A. Pearson and S. M. Richardson; Applied Science Publishers, NY, 1983; pp. 179–387.

[6] Ziabicki, A. *Fundamentals of Fibre Formation* , John Wiley & Sons, NY, 1976.

[7] Yarin, A. L. *Free Liquid Jets and Films: Hydrodynamics and Rheology* , Longman, NY, 1993.

[8] Kerbage, C.; Eggleton, B. J. *Opt. Fiber Techn.* 2004, 10, 133–149.

[9] Zubia, J; and Arrue;, J. *Opt. Fiber Techn.* 2001, 7, 101–140.

[10] Reeve, H. M.; Mescher, A. M.; Emery, A. F. In *Proceeedings of IMECE: Symposium on Polymer and Composite Materials Processing* , NY, 2001.

[11] Reeve, H. M.; Mescher, A. M. *Optics Expr.* 2003, 11, 1770–1779.

[12] Hsu, J.–P.; Lin, S.–H. *Polymer* 2003, 44, 5461–5467.

[13] Pearson, J. R. A.; Matovich, M. A. *Ind. Chem. Eng. Fund.* 1969, 8, 512–520.

[14] Schultz, W. W.; Davis, S. H. *J. Rheol.* 1982, 26, 331–345.

[15] Park, C.–W. *AIChE J.* 1990, 36, 197–206.

[16] Lee, W.–S.; Park, C.–W. *ASME J. Appl. Mech.* 1995, 62, 511–516.

[17] Ji, C.–C. ; Yang, J.–C. *Polym. Eng. Sci.* 1996, 36, 1399–1409.

[18] Ji, C.–C. ; Yang, J.–C.; Lee, W.–S. *Polym. Eng. Sci.* 1996, 36, 2685–2693.

[19] Naboulsi, S. K.; Bechtel, S. E. *Phys. Fluids* 1999, 11, 807–820.

[20] Ramos, J. I. *Appl. Math. Comput.* 1999, 100, 223–240.

[21] Ramos, J. I. *Polymer* 2002, 43, 2889-2896.

[22] Ramos, J. I. *Int. J. Eng. Sci.* 2001, 39, 1949-1961.

[23] Pearson, J. R. A. ; Matovich, M. A. *Ind. Chem. Eng. Fund.* 1969, 8, 605–609.

[24] Schultz, W. W.; Davis, S. H. *ASME J. Appl. Mech.* 1984, 51, 1–5.

[25] Yarin, A. L. ; Gospodinov, P.; Roussinov, V. I. *Phys. Fluids* 1999, 11, 3201–3208.

[26] Gospodinov, P.; Yarin, A. L. *Int. J. Multiphase Flow* 1997, 23, 967–976.

[27] Ramos, J. I. *Comput. & Theor. Polymer Sci.* 2001, 11, 429-443.

[28] Ramos, J. I. *Int. J. Eng. Sci.* 2001, 39, 1289-1314.

[29] Schultz, W. W. ; Zebib, A.; Davis, S. H.; Lee, Y. *J. Fluid Mech.* 1984, 149, 455–475.

[30] Kase, S. ; Matsuo, T. J. *J. Polymer Sci. A* 1965, A3, 2541–2554.

[31] Gupta, G.; Schultz, W. W. *Int. J. Non–Linear Mech.* 1998, 33, 151–163.

[32] Pearson, J. R. A. ; Shah, Y. T.; Mhaskar, R. D. *Ind. Chem. Eng. Fund.* 1976, 15, 31–37.

[33] Bird, R. B.; Armstrong, R. C.; Hassager, O. *Dynamics of Polymeric Liquids: Fluid Mechanics*, John Wiley & Sons, NY, 1987 , Vol. 1.

[34] Bird, R. B.; Curtiss, C. F. ; Armstrong, R. C.; Hassager, O. *Dynamics of Polymeric Liquids: Kinetic Theory*, John Wiley & Sons, NY, 1987, Vol. 2.

[35] Schultz, W. W. *J. Rheol.* 1987, 31, 733–750.

[36] Ramalingam, S.; Armstrong, R. C. *J.Rheol.* 1993, 37, 1141–1169.

[37] Landau, L. D.; Lifshitz, E. M. *Statistical Physics: Part 1*, Course of Theoretical Physics, Pergamon Press, NY, 1980, Vol. 5.

[38] Landau, L. D.; Lifshitz, E. M. *Statistical Physics: Part 2*, Course of Theoretical Physics, Volume 9, Pergamon Press, NY, 1980, Vol. 9.

[39] Avrami, M. *J. Chem. Phys.* 1939, 7, 1103–1112.

[40] Avrami, M. *J. Chem. Phys.* 1940, 8 (1940) 212–224.

[41] Avrami, M. *J. Chem. Phys.* 1941, 9, 177–184.

[42] Kolmogorov, A. N. *Bull. USSR Acad. Sci., Math. Series* 1937, 1, 355–359.

[43] Eder, G.; Janeschitz–Kriegl, H.; Liedaues, S. *Progr. Polymer Sci.* 1990, 15, 629–714.

[44] Eder, G. In *Macromolecular design of polymeric materials* ; Eds. K. Hatada, T. Kitayama and O. Vogl, Marcel Dekker, NY, 1997; pp. 761–782,

[45] Burger, M.; Capasso, V.; Eder, G. *ZAMM, Z. angew. Math. Mech.* 2002, 82, 51–63.

[46] Capasso, V. *Mathematical Modelling for Polymer Processing* , Springer–Verlag, NY, 2003.

[47] Flory, P. J. *J. Chemis. Phys.* 1947, 15, 397–408.

[48] Rao, I. J. ; Rajagopal, R. K. *ZAMP, Z. angew. Math. Phys.* 2002, 53, 365–406.

[49] Forest, M. G.; Wang, Q.; Bechtel, S. E. *J. Rheol.* 1997, 41, 821–850.

[50] Forest, M. G.; Wang, Q.; Bechtel, S. E. *Phys. D* 1997, 99, 527–554.

[51] Forest, M. G.; Ueda, T. *J. Non–Newton. Fluid Mech.* 1999, 84, 109–121.

[52] Forest, M. G.; Zhou, H.; Wang, Q. *SIAM J. Appl. Math.* 2000, 60, 1177–1204.

[53] Doufas, A. K.; McHugh, A. J.; Miller, C. *J. Non–Newton. Fluid Mech.* 2000, 92, 27–66.

[54] Doufas, A. K.; McHugh, A. J.; Miller, C. *J. Non–Newton. Fluid Mech.* 2000, 92, 81–103.

[55] McHugh, A. J.; Doufas, A. K. *Composit.: Part A* 2001, 32, 1059–1066.

[56] Doufas, A. K.; McHugh, A. J. *J. Rheol.* 2001, 45, 403–420.

[57] Henson, G. M.; Cao, D.; Bechtel, S. E. *J. Rheol.* 1998, 42, 329–360.

[58] Bell, P. W.; Eddie, D. D. *J. Appl. Polymer Sci.* 1987, 33, 1073–1088.

[59] Doufas, A. K.; McHugh, A. J. *J. Rheol.* 2001, 45, 855–879.

[60] Joo, Y. L.; Sun, J.; Smith, M. D.; Armstrong, R. C.; Brown, R. A.; Ross, R. A. *J. Non–Newton. Fluid Mech.* 2002, 102, 37–70.

[61] Sun, J.; Subbiah, S.; Marchal, J. A. *J. Non–Newton. Fluid Mech.* 2000, 93, 133–151.

[62] Nakamura, K.; Katayama, K.; Amano, T. *J. Appl. Polymer Sci.* 1973, 17, 1031–1041.

[63] Ziabicki, A.; Jarecki, A.; Wasiak, A. *Comput. & Theor. Polymer Sci.* 1998, 8, 143–157.

[64] Ramos, J. I.; Pitchumani, R. *ZAMP, Z. angew. Math. Phys.* 1989, 40, 721–739.

[65] Williams, F. A. *Combustion Theory*, Second Edition, Addison–Wesley Publishing Company, NY, 1985.

In: Polymer Research Developments
Editor: Robert K. Bregg, pp. 187-214 ISBN 1-59454-742-4

Chapter 8

ANALYTICAL AND EXPERIMENTAL STUDIES ON MICROBIAL DEPOLYMERIZATION OF XENOBIOTIC POLYMERS

Masaji Watanabe
Faculty of Environmental Science and Technology, Okayama University 1-1, Naka 3-chome, Tsushima, Okayama 700-8530, Japan
Fusako Kawai
Research Institute for Bioresources, Okayama University 2-20-1 Chuo, Kurashiki 710-0046, Japan

1 Introduction

In general, microbial depolymerization is a degradation process of either an exogenous type or an endogenous type. In an exogenous depolymerization process, monomer units are separated stepwise from the terminals of molecules. Such depolymerization processes include the biodegradation processes of polyethylene (PE), which is structurally a long-chain alkane (basically normal-type). The initial step of the oxidation of *n*-alkanes is hydroxylation to produce the corresponding primary (or secondary) alcohol, which is oxidized further to an aldehyde (or ketone) and then to an acid. Those carboxylated *n*-alkanes are structurally analogous to fatty acids, and should be subject to β-oxidation processes to produce depolymerized fatty acids by two carbon units (acetic acid). The β-oxidation is one factor of PE biodegradation, which arises in its theoretical aspect. On the other hand, there is another factor that arises in its experimental aspect. It is shown by results based on gel permeation chromatography (GPC) analysis of PEwax obtained before and after cultivation of a bacterial consortium KH-12 that small molecules were consumed faster than large ones [2].

It is seen in the foregoing discussion that there are two essential factors in the mechanism of PE biodegradation: the gradual weight loss of large molecules due to the β-oxidation and the direct consumption or absorption of small molecules by cells. A mathematical model based on those factors was proposed, and the PE biodegradation was studied numerially with experimental data introduced into analysis [3, 4, 5, 11, 12]. In particular, the biodegradability of PE was studied comparatively between the microbial consortium KH-12 and the fungus *Aspergillus* sp. AK-3, which belong to two different classes of degraders, bacteria and fungi [11]. The weight distribution of PE after five weeks of cultivation was compared bewteen an experimental result and a numerical result based on the degradation rate that was obtained using the weight distribution before and after three weeks of cultivation [12].

The biodegradation of polyethylene glycol (PEG) is another example of an exogenous depolymerization process. PEG is depolymerized by liberating C_2 compounds, either aerobically or anaerobically [7, 8, 18]. The mathematical model originally developed for the PE biodegradation has been applied successfully to the biodegradation of PEG. The weight distribution of PEG with respect to the molecular weight, which was obtained before and after cultivation of a microbial consortium E-1, was introduced into analysis of the model to determine the oxidation rate numerically, and the transition of the weight distribution was simulated [16].

Unlike depolymerization processes of exogenous type in which oxidation plays an essential part, hydrolysis is often involved significantly in an endogenous depolymerization process. One of characteristics of an endogenous depolymerization process is the rapid breakdown of large molecules due to internal cleavage to yield small molecules in the early stages of depolymerization, whereas molecules lose their weight gradually throughout an exogenous depolymerization process. The depolymerization of polyvinyl alcohol (PVA) is an example of an endogenous depolymerization process. PVA is a carbon-chain polymer with hydroxyl groups attached alternately to carbon units, which are known to be degraded by random oxidation of hydroxyl groups and hydrolysis of mono/diketones [7, 18, 19]. A mathematical model based on the mechanism of random depolymerization was also proposed to study the enzymatic degradation of PVA [10, 5, 13, 17].

In the following sections, the mathematical models for exogenous and endogenous depolymerization processes are described. Inverse problems of initial value problems to determine degradation rates are formulated, and numerical techniques to solve those problems are illustrated with examples. In particular, the Simpson's rule [9] is applied, whenever it is appropriate, to solve equations whose solutions are the degradation rates. The numerical results to simulate the transition of the weight distributions are also presented.

2 Analysis of Exogenous Depolymerization

2.1 Modeling of Exogenous Depolymerization

Polyoleffins are considered as linear saturated hydrocarbons, and regarded chemically inert in the environment. However, it has been suggested that PE can be degraded although the process may be very slow, and its degradation is promoted by irradiation or oxidation. Measurement of $^{14}CO_2$ generation was taken to show the degradation of PE [14]. The linear paraffin molecules, whose molecular weight approximately equals 500 or less, were found to be utilized by several microorganisms [20]. Oxidation of n-alkanes up to tetratetracontane ($C_{44}H_{90}$, mass of 618) in 20 days was reported [15].

Commercially available PEwax was used as a sole carbon and energy source for soil microorganisms [2], and the microbial consortium KH-12 obtained from soil samples degraded PEwax, which was confirmed by significant weight loss (30-50 %). It was shown by GPC analysis of PEwax that small molecules were consumed faster than large ones in biodegradation process of PE.

On the other hand, PE can be considered structurally as hydrocarbon, and it should be subject to the following metabolic pathways.

- Terminal oxidation:

$$\mathrm{RCH_3} \longrightarrow \mathrm{RCH_2OH} \longrightarrow \mathrm{RCHO} \longrightarrow \mathrm{RCOOH}$$

- Diterminal oxidation:

$$\begin{aligned}&\mathrm{H_3CRCH_3} \to\to\to \mathrm{CH_3RCOOH} \longrightarrow \mathrm{HOH_2CRCOOH}\\&\longrightarrow \mathrm{OHCRCOOH} \longrightarrow \mathrm{HOOCRCOOH}\end{aligned}$$

- Subterminal oxidation:

$$\begin{aligned}&\mathrm{RCH_2CH_2CH_3} \longrightarrow \mathrm{RCH_2CH(OH)CH_3} \longrightarrow \mathrm{RCH_2C(O)CH_3} \longrightarrow\\&\mathrm{RCH_2OC(O)CH_3} \longrightarrow \mathrm{RCH_2OH} + \mathrm{CH_3COOH}\end{aligned}$$

A PE molecule which undergoes one of these oxidation processes is carboxylated at its terminal, and becomes structurally analogous to fatty acid. Such a molecule should be subject to the β-oxidation, and a series of weight loss due to terminal separation of a couple of carbon units should follow.

The foregoing theoretical and experimental aspects of PE biodegradation lead to the presumptions that each molecule loses its weight by a fixed amount per unit time, that some molecule are directly consumed by microorganisms, and that the consumption rate depends on the sizes of molecules. The mathematical model

$$\frac{dx}{dt} = -\alpha(M)x + \beta(M+L)\frac{M}{M+L}y, \tag{1}$$

where

$$\alpha(M) = \rho(M) + \beta(M),$$

based on those assumptions was proposed, and the biodegradability of PE was studied by analyzing it [3, 4, 5, 11, 12]. Here t and M denote the cultivation time and the molecular weight respectively, and $x = w(t,M)$ denotes the total weight of M-molecules (the PE molecules with molecular weight M) present at time t. The parameter L represents the amount of the weight loss due to the terminal separation. The variable y is given by $y = w(t,M+L)$, which is the total weight of $(M+L)$-molecules present at time t. The function $\rho(M)$ represents the direct consumption rate, and the function $\beta(M)$ represents the weight conversion rate from the class of M-molecules to the class of $(M-L)$-molecules. It follows that the first term of the right-hand side of (1) is the total weight loss per unit time in the class of M-molecules due to the direct consumption and the β-oxidation. Similarly, the second term represents the weight conversion per unit time from the class of $(M+L)$-molecules to the class of M-molecules due to the β-oxidation.

The mathematical model (1), which was originally developed for the PE biodegradation, can be viewed as a general biodegradation model for an exogenous depolymerization process, which covers not only the PE biodegradation but also other polymers such as PEG. As is shown in Figure 1, a PEG molecule is first oxidized at its terminal, and then an ether bond is split. This process corresponds to the β-oxidation in the PE biodegradation, and it is called oxidation, because oxidation is involved throughout its depolymerization process [8, 18]. It is seen that $L = 44$ (CH_2CH_2O) in the exogenous depolymerization of PEG.

Suppose that a function $f(M)$ represents the initial weight distribution. Then the solution $x = w(t,M)$ of (1) is subject to the initial condition

$$w(0,M) = f(M). \tag{2}$$

Given the total consumption rate $\alpha(M)$ and the β-oxidation rate $\beta(M)$, the equation (1) and the initial condition (2) form an initial value problem to find the unknown function $w(t,M)$. Suppose that a function $g(M)$ represents the weight distribution at $t = T > 0$. Then given the initial condition (2) and the additional condition:

$$w(T,M) = g(M), \tag{3}$$

an inverse problem is formulated to determine the degradation rates $\alpha(M)$ and $\beta(M)$ for which the solution $w(t,M)$ of the initial value problem (1), (2) also satisfies the condition (3).

The following condition is a sufficient condition for a unique positive total degradation rate $\alpha(M)$ to exist, given the β-oxidation rate $\beta(M+L)$ and the weight distribution $w(t,M+L)$ [11, 12]:

$$0 < g(M) < f(M) + \frac{M\beta(M+L)}{M+L}\int_0^T w(s,M+L)\,ds. \tag{4}$$

To see that this condition guarantees the unique existence of a positive total degradation rate $\alpha(M)$, note that the solution of the initial value problem (1), (2) can be written in the form:

$$w(t,M) = e^{-\alpha(M)t} f(M) + \frac{M\beta(M+L)}{M+L}\int_0^t e^{-\alpha(M)(t-s)} w(s,M+L)\,ds. \tag{5}$$

The condition (3) leads to the equation

$$g(M) = e^{-\alpha(M)T} f(M) + \frac{M\beta(M+L)}{M+L}\int_0^T e^{-\alpha(M)(T-s)} w(s,M+L)\,ds.$$

Define the function $\Phi(\eta)$ by

$$\begin{aligned}\Phi(\eta) &= v(T,M,\eta)-g(M)\\ &= e^{-\eta T}f(M)+\frac{M\beta(M+L)}{M+L}\int_0^T e^{-\eta(T-s)}w(s,M+L)\,ds-g(M),\end{aligned} \tag{6}$$

where $v=v(t,M,\eta)$ is the solution of the initial value problem

$$\frac{dv}{dt}=-\eta v+\frac{M\beta(M+L)}{M+L}w(t,M+L), \tag{7}$$

$$v(0,M,\eta)=f(M). \tag{8}$$

Then $\eta=\alpha(M)$ is a solution of the equation

$$\Phi(\eta)=0. \tag{9}$$

It is seen that

$$\Phi(0)=f(M)+\frac{M\beta(M+L)}{M+L}\int_0^T w(s,M+L)\,ds-g(M). \tag{10}$$

Since

$$\begin{aligned}0\le v(T,M,\eta) &\le e^{-\eta T}f(M)+w_{\max}\frac{M\beta(M+L)}{M+L}\int_0^T e^{-\eta(T-s)}ds\\ &\le e^{-\eta T}f(M)+w_{\max}\frac{M\beta(M+L)}{M+L}\cdot\frac{1}{\eta}\left(1-e^{-\eta T}\right),\end{aligned}$$

where $w_{\max}$ denotes the maximum value of the function $w(t,M+L)$ over the interval $[0,T]$,

$$\lim_{\eta\to\infty}v(T,M,\eta)=0,$$

and

$$\lim_{\eta\to\infty}\Phi(\eta)=-g(M). \tag{11}$$

On the other hand, the derivative of $\Phi(\eta)$, $\Phi'(\eta)$, is given by

$$\Phi'(\eta)=-Te^{-\eta T}f(M)-\frac{M\beta(M+L)}{M+L}\int_0^T(T-s)\,e^{-\eta(T-s)}w(s,M+L)\,ds \tag{12}$$

The equation (9) has at least one positive solution under the conditions

$$\Phi(0)>0,\ \lim_{\eta\to\infty}\Phi(\eta)<0. \tag{13}$$

In view of the equations (10), (11), these inequalities lead to the condition (4). Moreover, under the condition (4), either

$$f(M)>0,$$

or

$$w(t, M+L) > 0,$$

for some $t \in [0, T]$. Then the expression (12) shows that $\Phi'(\eta) < 0$ for all positive η. It follows that $\Phi(\eta)$ defined on $[0, \infty)$ is a strictly decreasing continuous function that satisfies the condition (13), and that the condition (4) is a sufficient condition for the equation (9) to have a unique positive solution $\eta = \alpha(M)$. Once the total degradation rate $\alpha(M)$ is determined, the solution $w(t, M)$ of the initial value problem (1), (2) is given by the expression (5).

Molecules must penetrate through the membranes to enter into the cells in order to become subject to the direct consumption, and the rate of the membrane penetration decreases, as the molecular size increases. It follows that the rate of the direct consumption must also decrease as the molecular size increases. Moreover, there should be a limit of the penetration with respect to the molecular size, and there is a $M_\rho > 0$ such that $\rho(M) = 0$ for $M \geq M_\rho$. Since $\alpha(M) = \rho(M) + \beta(M)$, it follows that

$$\alpha(M) = \beta(M) \text{ for } M \geq M_\rho. \tag{14}$$

The weight distribution of PEG with respect to the molecular weight M introduced in the following sections is given in the range $3.65 \leq \log M \leq 4$. It can be assumed that the molecular weight in this range should be greater than M_ρ.

2.2 Biodegradation of Polyethylene Glycol

Polyethers are utilized for constituents in products including lubricants, antifreeze agents, inks, and cosmetics, and also for raw materials to synthesize detergents or polyurethanes. Those polymers are either water soluble or oily liquid, and eventually discharged into the environment through sewage [7, 8]. Since they can not be incinerated or recycled, their biodegradability is an important factor in environmental protection. They include polyethylene glycol (PEG), polypropylene glycol (PPG), polytetramethylene glycol (PTMG), polybutylene glycol (PBG), polyglycerin, polyglycidol, and polyisobutylene oxide, polyglyoxylate. They are polymers whose chemical stuructures are represented by the expression $HO(R\text{-}O)_nH$, *e.g.*, PEG: $R = CH_2CH_2$, PPG: $R = CH_3CHCH_2$, PTMG: $R = (CH_2)_4$ [6].

PEG is produced in the largest quantity among polyethers, and its major part is consumed in production of nonionic surfactants. It is depolymerized exogenously by liberating C_2 compounds, either aerobically or anaerobically [7, 8, 18]. Figure 1 shows the metabolism of PEG.

Materials and Methods

Chemicals

All reagents used were of reagent grade.

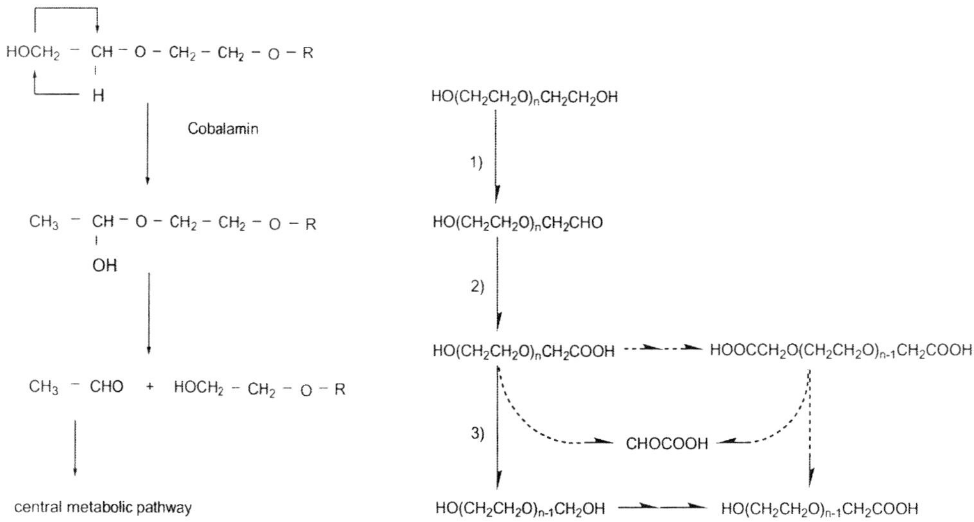

Figure 1: Anaerobic metabolism (left) and aerobic metabolism (right) of PEG. 1): PEG dehydrogenase, 2): PEG-aldehyde dehydrogenase, 3): PEG-carboxylate dehydrogenase.

Microorganisms and Cultivation

PEG 20,000-utilizing mixed culture was used as a PEG degrader [1]. The culture was centrifuged to remove cells and the resultant supernatant was subjected to HPLC analysis.

HPLC Analysis

The molecular weight of PEG before and after cultivation was measured by high performance liquid chromatography (HPLC), performed with Tosoh CCPM-II liquid chromatograph. The analytical conditions were as follows: detection, Tosho RI-8020; columns, Tosoh TSK-GEL2500PW (7.5 ϕ× 300 mm) connected with TSK-GEL3000PW (7.5 ϕ× 300 mm); eluent 0.3 M NaNO3; flow rate, 1 ml/min; column temperature, 40 °C. Molecular masses were measured using ethylene glycol, its oligomers and polyethylene glycols, and TSK standard polyethylene oxides (Tosoh).

Experimental Results

Figure 2 shows the HPLC patterns of PEG obtained before and after cultivation of a bacterial consortium E-1. Figure 3 shows the degradation ratio

$$y = \frac{f(x) - g(x)}{f(x)}$$

obtained from the HPLC patterns shown in Figure 2. Here $f(M)$ and $g(M)$ represent the initial weight distribution and the weight distribution after three day cultivation of the bac-

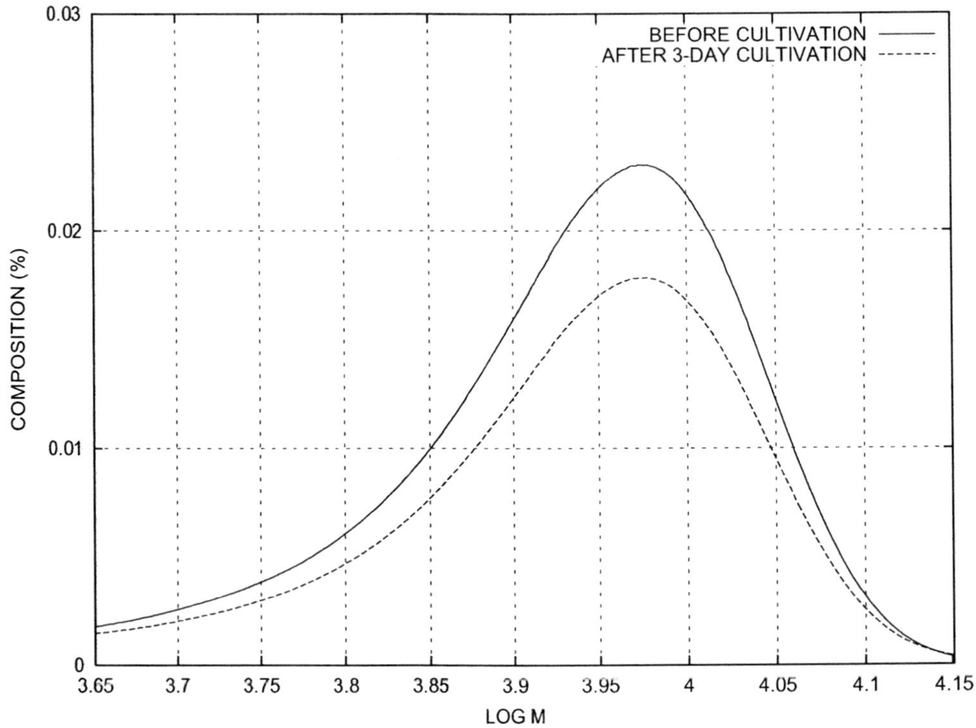

Figure 2: HPLC patterns of PEG obtained before and after cultivation of the bacterial consortium E-1 for three days.

terial consortium E-1, respectively, and M represents the molecular weight, and $x = \log M$.

No dependence of biodegradabilty on the molecular weight seems to be present in the result of HPLC analysis as far as the degradation ratio is concerned. On the contrary, it will be shown that a numerical result based on the HPLC patterns shown in Figure 2 apparently reflects a physical process of transport through membranes.

2.3 Numerical Study of Exogenous Depolymerization

2.3.1 Numerical Solutions of the Inverse Problem

The data shown in Figure 1 was introduced into analysis based on a numerical scheme developed in [12], and the inverse problem was solved in order to find the oxidation rate $\beta(M)$ for which the solution of the initial value problem (1) and (2) also satisfies the condition (3). The numerical result is introduced in this section.

First, the numerical technique to solve the inverse problem is illustrated. Suppose that the initial weight distribution $f(M)$ and the weight distribution at $t = T$, $g(M)$, are defined

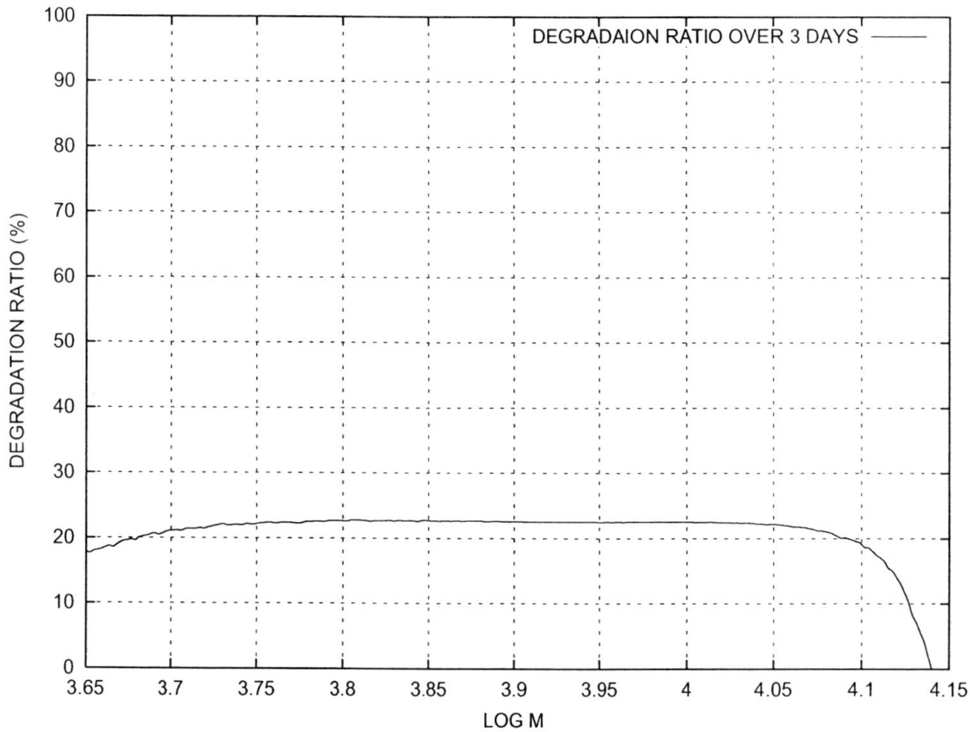

Figure 3: Biodegradation ratios of PEG based on HPLC patterns obtained before and after 3-day cultivation of the bacterial consortium E-1 (cf. Figure 2).

on the interval $[a,b]$. In view of the equation (14), it may be assumed that

$$\alpha(M) = \beta(M), \quad M \in [a,b]. \tag{15}$$

Suppose that the positive integer l satisfies the condition

$$\frac{b-a}{L} \leq l < \frac{b-a}{L} + 1.$$

The interval $[a,b]$ can be covered with the intervals of length L: $[a+iL, a+(i+1)L]$, $i = 0, \ldots, l-1$, extending the definition of $f(M)$ and $g(M)$ if necessary. Choose a positive integer m, and let

$$\delta M = \frac{L}{m}.$$

Set

$$M_{i,j} = a + iL + j\delta M, \quad i = 0, 1, \ldots, l-1, \quad j = 0, 1, \ldots, m,$$

and divide each interval into m intervals of length δM:

$$[M_{i,j-1}, M_{i,j}], \quad i = 0, 1, \ldots, l-1, \quad j = 1, \ldots, m.$$

Suppose that $\alpha_{i,j}$ and $\beta_{i,j}$ denote approximate values of $\alpha(M_{i,j})$ and $\beta(M_{i,j})$, respectively:

$$\alpha_{i,j} \approx \alpha(M_{i,j}), \quad \beta_{i,j} \approx \beta(M_{i,j}).$$

The equation (15) shows that

$$\alpha_{i,j} = \beta_{i,j}, \quad 0 \leq i \leq l, \quad 0 \leq j \leq m.$$

Choose a positive integer n, and set

$$\delta t = \frac{T}{n}.$$

Let

$$t_k = k\delta t, \quad k = 0, 1, 2, \ldots, n.$$

The time interval $[0, T]$ is divided into n intervals: $[t_{k-1}, t_k]$, $k = 1, 2, \ldots, n$. Suppose that $w_{i,j,k}$ denotes an approximate value of $w(t_k, M_{i,j})$:

$$w_{i,j,k} \approx w(t_k, M_{i,j}),$$

and that $w_{i+1,j,k}$, $j = m-1, m-2, \ldots, 0$, $k = 0, 1, \ldots, n$, and $\beta_{i+1,j}$, $j = m-1, m-2, \ldots, 0$ are given. Then $\eta = \alpha_{i,j}$ is a solution of the equation

$$\Phi_{i,j}(\eta) = 0, \tag{16}$$

where the function $\Phi_{i,j}(\eta)$ and its derivative $\Phi'_{i,j}(\eta)$ are obatained by applying a numerical integration formula to the integrals of the equations (9) and (12). In particular, the application of the Simpson's rule leads to the following expression:

$$\begin{aligned} \Phi_{i,j}(\eta) = e^{-\eta T} \Bigg[f_{i,j} + \frac{\beta_{i+1,j} M_{i,j}}{M_{i+1,j}} \times \\ \frac{\delta t}{3} \sum_{k=1}^{n/2} \left(e^{\eta t_{2k-2}} w_{i+1,j,2k-2} + 4 e^{\eta t_{2k-1}} w_{i+1,j,2k-1} + e^{\eta t_{2k}} w_{i+1,j,2k} \right) \Bigg] - g_{i,j}, \end{aligned} \tag{17}$$

$$\begin{aligned} \Phi'_{i,j}(\eta) = -e^{-\eta T} \Bigg[T f_{i,j} + \frac{\beta_{i+1,j} M_{i,j}}{M_{i+1,j}} \times \frac{\delta t}{3} \sum_{k=1}^{n/2} \{ (T - t_{2k-2}) e^{\eta t_{2k-2}} w_{i+1,j,2k-2} \\ +4 (T - t_{2k-1}) e^{\eta t_{2k-1}} w_{i+1,j,2k-1} + (T - t_{2k}) e^{\eta t_{2k}} w_{i+1,j,2k} \} \Bigg], \end{aligned} \tag{18}$$

where

$$f_{i,j} = f(M_{i,j}), \quad g_{i,j} = g(M_{i,j}).$$

An approximate solution of the equation (16) can be found by the Newton's method

$$\eta_{p+1} = \eta_p - \frac{\Phi_{i,j}(\eta_p)}{\Phi'_{i,j}(\eta_p)}, \quad p = 0, 1, 2, \ldots. \tag{19}$$

The numerical process is initiated at $i = l-1$ setting appropriate values of $w_{l,j,k}$, $j = 0, 1, \ldots, m-1$, $k = 0, 1, \ldots, n$, and proceeds backward: $i = l-1, l-2, \ldots, 0$. At each step, a numerical solution $\eta = \alpha_{i,j}$ of the equation (16) is found by the Newton's method (19) for $j = m-1, m-2, \ldots, 0$, examining the condition

$$0 < g_{i,j} < f_{i,j} + \frac{\beta_{i+1,j} M_{i,j}}{M_{i+1,j}} \times \frac{\delta t}{2} \sum_{k=1}^{n} \left(w_{i+1,j,k-1} + w_{i+1,j,k} \right) \tag{20}$$

for the solvability of the equation (16). To see that this condition guarantees the solvability of the equation (16), note that

$$\Phi_{i,j}(0) > 0, \quad \lim_{\eta \to \infty} \Phi_{i,j}(\eta) < 0,$$

under the condition (20). Moreover, the equation (18) shows that $\Phi'_{i,j}(\eta) < 0$ for all positive η, and $\Phi_{i,j}(\eta)$ defined on the interval $[0, \infty)$ is a strictly decreasing function. Consequently the condition (20) is a sufficient condition for the equation (16) to have a unique positive solution $\eta = \alpha_{i,j}$. The condition (20) corresponds to the condition (4), which guarantees the solvability of the equation (9). Once an approximate solution $\eta = \alpha_{i,j}$ of the equation (16) is determined, approximate values $w_{i,j,2k-1}$ and $w_{i,j,2k}$ of $w(t_{2k-1}, M_{i,j})$ and $w(t_{2k}, M_{i,j})$are given by

$$w_{i,j,2k-1} = e^{-\eta t_{2k-1}} f_{i,j} + \frac{\beta_{i+1,j} M_{i,j}}{M_{i+1,j}} e^{-\eta t_{2k-1}} \left\{ \frac{\delta t}{3} \sum_{p=1}^{k-1} \left(e^{\eta t_{2p-2}} w_{i+1,j,2p-2} \right. \right.$$
$$\left. \left. + 4 e^{\eta t_{2p-1}} w_{i+1,j,2p-1} + e^{\eta t_{2p}} w_{i+1,j,2p} \right) + \frac{\delta t}{2} \left(e^{\eta t_{2k-2}} w_{i+1,j,2k-2} + e^{\eta t_{2k-1}} w_{i+1,j,2k-1} \right) \right\}$$
$$w_{i,j,2k} = e^{-\eta t_{2k}} f_{i,j} + \frac{\beta_{i+1,j} M_{i,j}}{M_{i+1,j}} e^{-\eta t_{2k}} \times \frac{\delta t}{3} \sum_{p=1}^{k} \left(e^{\eta t_{2p-2}} w_{i+1,j,2p-2} \right.$$
$$\left. + 4 e^{\eta t_{2p-1}} w_{i+1,j,2p-1} + e^{\eta t_{2p}} w_{i+1,j,2p} \right)$$

with $\eta = \alpha_{i,j}$ for $k = 0, 1, \ldots, n$.

The inverse problem was solved numerically to find the exogenous degradation rate $\beta(M)$ based on the initial weight distribution and the final weight distribution shown in Figure 1 with the following values of the parameters.

$$\begin{aligned} L &= 44, & T &= 3, \\ n &= 1200, & \Delta t &= 0.0025. \\ m &= 10. \end{aligned}$$

The following values of the parameters

$$a = 10^{3.63} \approx 4265.795188, \quad b = 10^{4.15} \approx 14125.375446, \quad l = 225$$

were also set. The numerical result is given in Figure 4 which shows the graph of the exogenous degradation rate generated numerically.

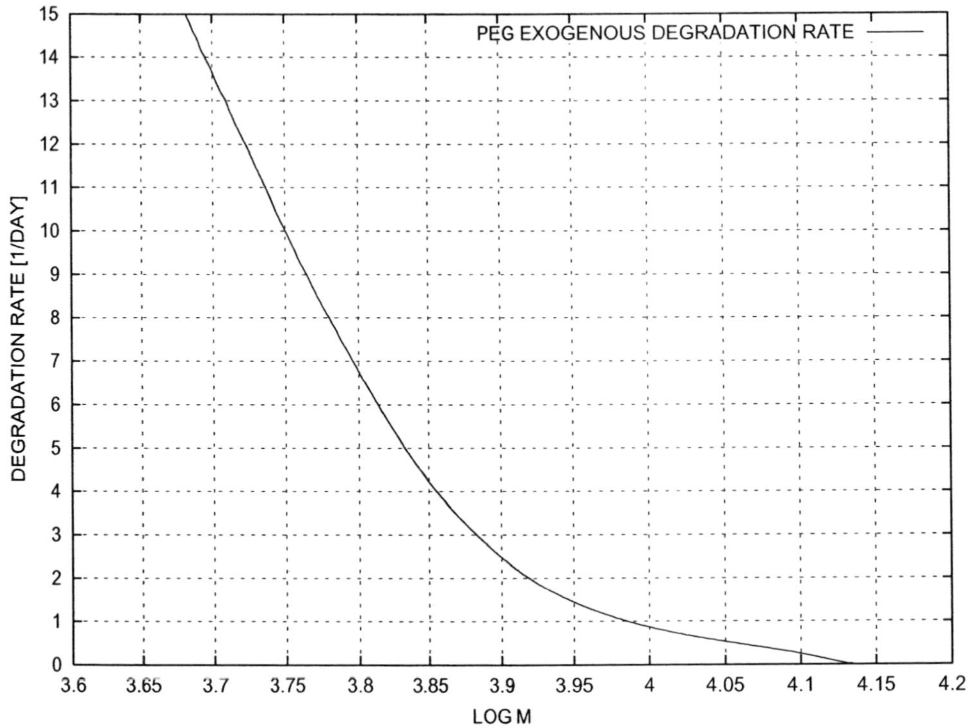

Figure 4: The exogenous degradation rate $\beta(M)$ based on the result of the HPLC analysis shown in Figure 2.

2.3.2 Numerical Simulation of Exogenous Depolymerization

In the process of evaluating the exogenous degradation rate, the approximate values $w_{i,j,k}$ of $w(t,M)$ at $t = t_k$ and $M = M_{i,j}$ are also determined. On the other hand, once the approximate values of the total consumption rates and the β-oxidation rates are found, the initial value problem (1) and (2) can be solved directly to see how appropriately the numerical results and the experimental results agree. The initial value problem was solved with numerical techniques developed previously [4, 11, 12].

First, the numerical techniques are illustrated. Choose a positive integer N and let

$$\Delta M = \frac{b-a}{N}.$$

Define the $N+1$ equally spaced points:

$$M_i = a + i\Delta M, \qquad i = 0, 1, 2, \cdots, N.$$

Suppose that an approximate solution of the differential equation (1) at $M = M_i$ is denoted by $w_i = w_i(t)$ $(i = 0, 1, 2, \cdots, N)$. That is, $w_i(t)$ is an approximate value of $w(t, M_i)$.

There are a non-negative integer K and a constant R such that

$$L = K\Delta M + R, \ \ 0 \leq R < \Delta M,$$

and

$$M_{i+K} \leq M_i + L < M_{i+K+1}.$$

Then an approximate value of $w(t, M_i + L)$ can be given by the expression:

$$\begin{aligned} w(t, M_i + L) &\approx \frac{M_{i+K+1} - (M_i + L)}{\Delta M} w(t, M_{i+K}) + \frac{(M_i + L - M_{i+K})}{\Delta M} w(t, M_{i+K+1}) \\ &= \left(1 - \frac{R}{\Delta M}\right) w(t, M_{i+K}) + \frac{R}{\Delta M} w(t, M_{i+K+1}) \end{aligned}$$

Similarly, an approximate value of $\beta(M_i + L)$ can be given by the expression:

$$\beta(M_i + L) \approx \left(1 - \frac{R}{\Delta M}\right) \beta(M_{i+K}) + \frac{R}{\Delta M} \beta(M_{i+K+1}).$$

Substituting these approximate values in the diffrential equation (1) and setting $M = M_i$, the following linear system of ordinary differential equations with constant coefficients is obtained:

$$\frac{dw_i}{dt} = -\alpha_i w_i + \beta_i w_{i+K} + \gamma_i w_{i+K+1}, \ \ i = 0, 1, 2, \cdots, N, \tag{21}$$

where

$$\alpha_i = \alpha(M_i),$$

$$\beta_i = \sigma_i \frac{M_i}{M_i + L}\left(1 - \frac{R}{\Delta M}\right),$$

$$\gamma_i = \sigma_i \frac{M_i}{M_i + L} \cdot \frac{R}{\Delta M},$$

$$\sigma_i = \left(1 - \frac{R}{\Delta M}\right) \beta(M_{i+K}) + \frac{R}{\Delta M} \beta(M_{i+K+1}).$$

Approximate values of the degradation rates $\alpha(M_i)$ and $\beta(M_i)$ can be given in terms of the approximate values $\alpha_{j,k}$ and $\beta_{j,k}$ obtained by solving the inverse problem:

$$\alpha(M_i) = \frac{M_{j,k} - M_i}{\delta M} \alpha_{j,k-1} + \frac{M_i - M_{j,k-1}}{\delta M} \alpha_{j,k},$$

$$\beta(M_l) = \frac{M_{j,k} - M_i}{\delta M} \beta_{j,k-1} + \frac{M_i - M_{j,k-1}}{\delta M} \beta_{j,k},$$

for

$$M_{j,k-1} \leq M_i < M_{j,k}.$$

For all sufficiently large M, the degradation rate equals 0. In particular, it may be assumed that the last two terms on the right-hand side of the equation (21) are absent when

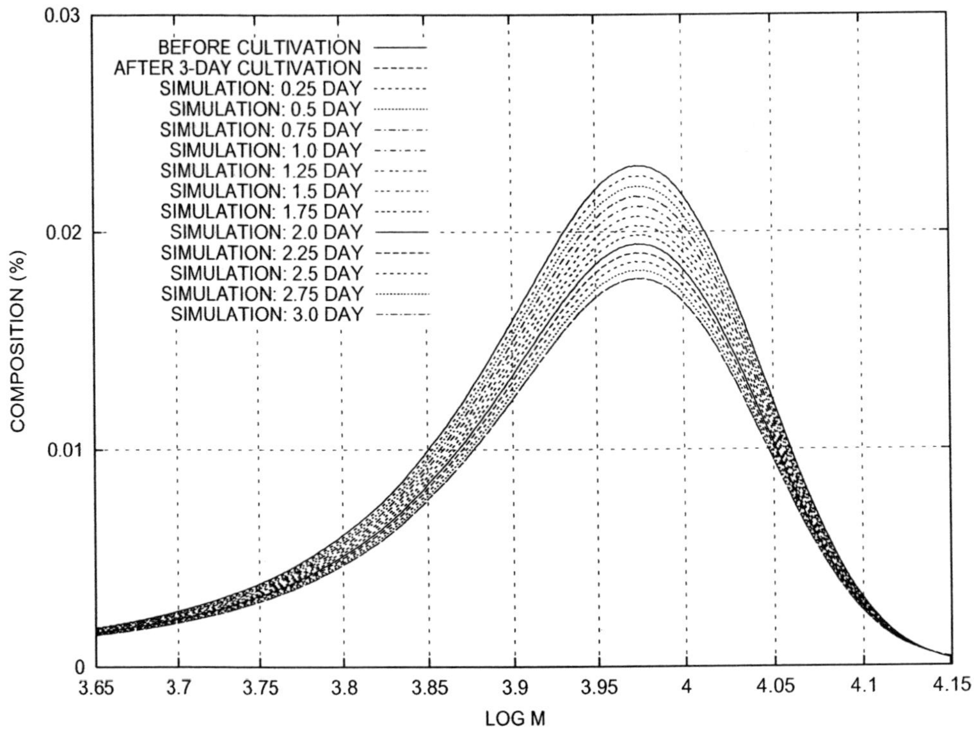

Figure 5: Numerical result based on the exogenous degradation rate $\beta(M)$ shown in Figure 4 to simulate the transition of the weight distribution of PEG in 3day cultivation of the microbial consortium E-1.

$i+K$ exceeds N. Thus those variables whose subscripts exceed N can be set equal to 0, and the system (21) becomes a closed system to be solved for unknown functions $w_i = w_i(t)$, $i = 0, 1, 2, \ldots, N$. In view of the initial condition (2), these solutions are subject to the initial condition

$$w_i(0) = f_i = f(M_i). \tag{22}$$

Given the initial weight distribution shown in Figure 1 and the corresponding degradation rates shown in Figure 3, the initial value problem (21), (22) was solved numerically implementing the forth-order Adams-Bashforth-Moulton predictor-corrector in PECE mode (P: predictor, E: evaluation, C: corrector) in conjunction with the Runge-Kutta method to generate approximate solutions in the first few steps [9] by setting $N = 10000$, and $\Delta t = 0.00125$, where Δt is the steplength. Figure 4 shows the numerical result in which the transition of the weight distribution during cultivation of the microbial consortium E-1 is simulated.

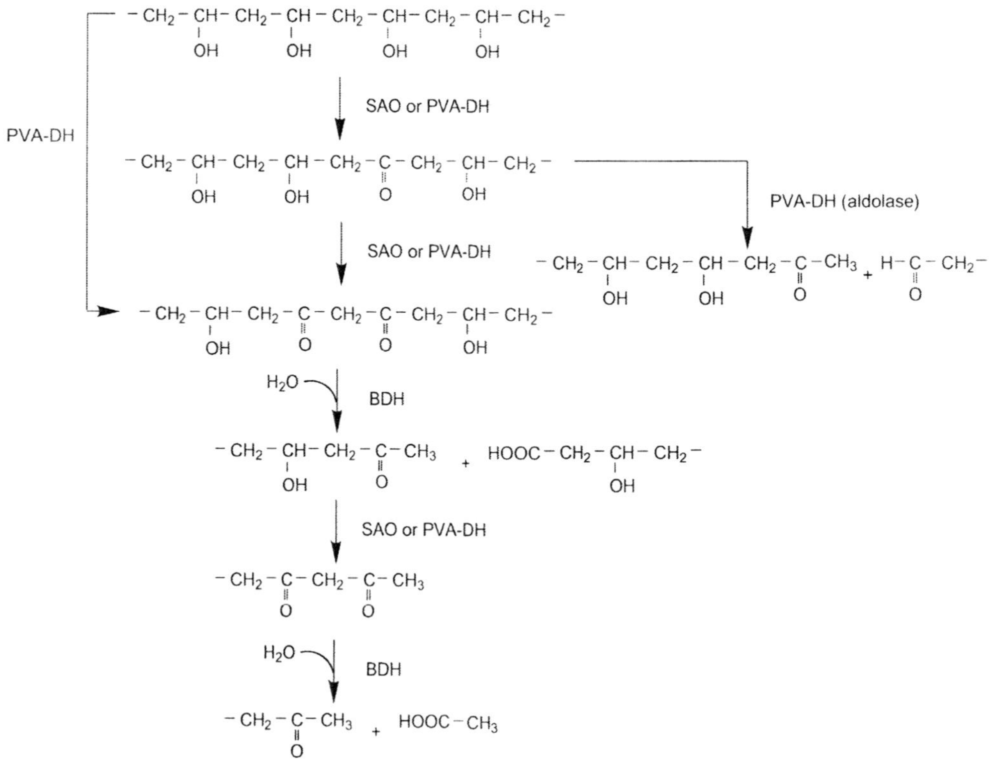

Figure 6: Metabolic pathways of PVA [18, 19]. SAO: secondary alcohol oxidase, PVA-DH: PVA dehydrogenase, BDH: β-diketone hydrolase.

3 Analysis of Endogenous Depolymerization

3.1 Modelling of Endogenous Depolymerization

PVA is degraded in a succession of two processes: oxidation of a couple of pendant hydroxyl groups either by oxidase or dehydrogenase follow by hydrolysis. The sequence of reactions results in a seperation of carbon-carbon chain at a carbonyl group and the adjacent methyne group [7, 18]. Matsumura *et al.* proposed a new metabolism of PVA by oxidation of a hydroxyl group and aldolase reaction of a monoketone structure, which results in a separation of carbon-carbon chain between a methyne group next to a carbonyl group and an adjacent hydroxymethyne group [19]. Those metabolic pathways are shown in Figure 6. Irrespective of different metabolic pathways, PVA is in general depolymerized by oxidation and the resultant separation of carbon-carbon chain between two carbonyl groups/ a carbonyl group and an adjacent hydroxymethyne group, which produces smaller molecules of random sizes. Thus, the weight is shifted from classes of large molecules to classes of small molecules in the process of enzymatic degradation. GPC patterns of PVA obtained before and after 2-day enzymatic reaction are given in Figure 7, which shows a substan-

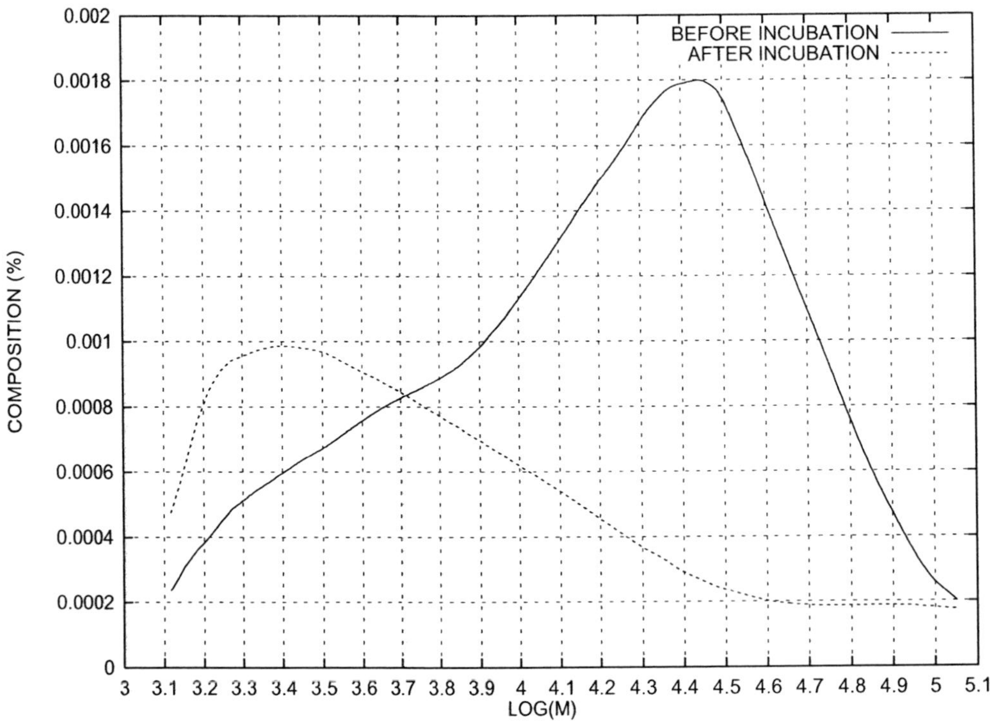

Figure 7: Weight distributions of PVA-23000 obtained before and after 2-day enzymatic degradation. Based on data in [19] (with permission by Prof. S. Matsumura).

tial weight shift from large molecules (molecular weight between 10^4 and 10^5) to small molecules (molecular weight between 10^3 and 10^4).

In order to model an enzymatic degradation process of a polymeric compound such as PVA mathematically, let $w(t,M)$ be its weight distribution with respect to the molecular weight M at time t. Suppose that the class of all molecules whose molecular sizes lie between A and B is denoted by $C(A,B)$. Then the total weight present at time t in $C(A,B)$ is the integral of $w(t,M)$ with respect to M over the interval $[A,B]$:

$$y = \int_A^B w(t,M)\, dM.$$

For $K \le M$, let $p(t,K,M)$ denote the rate of transition from $w(t,M)$ to $w(t,K)$ due to enzymatic degradation. Then the transition of the weight from $C(A,B)$ to $C(D,E)$ per unit time is given by the integral of $p(t,K,M)$ with respect to (K,M) over the region R:

$$\int\int_R p(t,K,M)\, dM\, dK,$$

where

$$R = \{(K,M) \,|\, K \le M, D \le K \le E, A \le M \le B\}.$$

On the other hand, the total weight decrease in $C(A,B)$ per unit time is given by

$$\int_A^B \int_0^M p(t,K,M)\,dK\,dM,$$

whereas the total weight increase per unit time is given by

$$\int_A^B \int_M^\infty p(t,M,K)\,dK\,dM.$$

The rate of change of y:

$$\frac{dy}{dt} = \frac{d}{dt}\int_A^B w(t,M)\,dM = \int_A^B \frac{\partial w}{\partial t}(t,M)\,dM$$

should be equal to their difference:

$$\int_A^B \frac{\partial w}{\partial t}(t,M)\,dM = -\int_A^B \int_0^M p(t,K,M)\,dK\,dM + \int_A^B \int_M^\infty p(t,M,K)\,dK\,dM.$$

Dividing the both sides of the equation by $B-A$, letting $B \longrightarrow A$, and replacing A with M, the equation

$$\frac{\partial w}{\partial t} = -\int_0^M p(t,K,M)\,dK + \int_M^\infty p(t,M,K)\,dK. \tag{23}$$

is obtained.

Next, suppose that the molecules are degraded randomly. In particular, suppose that the amount lost in $w(t,M)$ per unit time is proportional to itself:

$$\lambda(M)\,w(t,M),$$

where $\lambda(M)$ is a function that represents the degradation rate. Note that the number of the degraded molecules is proportional to

$$\frac{1}{M}\lambda(M)\,w(t,M).$$

Those molecules are cleaved to produce smaller molecules, and the number of those smaller molecules should be proportional to the number of the original molecules. Suppose that those cleaved molecules are distributed uniformly over the weight interval $[0,M]$. Then the number of molecules of any molecular weight $K \in [0,M]$ should be a constant multiple of the number of the original molecules:

$$\frac{c}{M}\lambda(M)\,w(t,M),$$

where the constant c is independent of K. It follows that

$$p(t,K,M) = c\frac{K}{M}\lambda(M)\,w(t,M).$$

The total weight loss in $C(A,B)$:

$$\int_A^B \lambda(M)\, w(t,M)\, dM \tag{24}$$

should equal the total weight gain in $C(0,M)$ for $A \le M \le B$:

$$\int_A^B \int_0^M p(t,K,M)\, dK\, dM = \int_A^B \int_0^M c\frac{K}{M}\lambda(M)\, w(t,M)\, dK\, dM, \tag{25}$$

and the equation

$$\int_A^B \left(1 - c\frac{M}{2}\right) \lambda(M)\, w(t,M)\, dM = 0$$

is obrained from the equations (24) and (25). Since this equation must hold for any interval $[A,B]$, the equation

$$c = \frac{2}{M}$$

must hold, and it follows that

$$p(t,K,M) = \frac{2K}{M^2}\lambda(M)\, w(t,M).$$

Substituting this expression of $p(t,K,M)$ in Eq. (23), the equation

$$\frac{\partial w}{\partial t} = -\lambda(M)\, w + \int_M^\infty \frac{2M}{K^2}\lambda(K)\, w(t,K)\, dK. \tag{26}$$

is obtained [10]. This model, in its equivalent form, was originally developed for the enzymatic degradation of PVA, but it can be regarded as a general model for an endogenous depolymerization process that follows the following rule: smaller molecules produced by internal separations are uniformly distributed over smaller weight classes. Differentiating the both sides of (26) with respect M, the equation:

$$\begin{aligned}
\frac{\partial^2 w}{\partial M \partial t} &= -\frac{\partial}{\partial M}[\lambda(M)\, w] + \int_M^\infty \frac{2}{K^2}\lambda(K)\, w(t,K)\, dK - \frac{2}{M}\lambda(M)\, w \\
&= -\frac{\partial}{\partial M}[\lambda(M)\, w] + \frac{1}{M}\left[\frac{\partial w}{\partial t} + \lambda(M)\, w\right] - \frac{2}{M}\lambda(M)\, w \\
&= -\frac{\partial}{\partial M}[\lambda(M)\, w] + \frac{1}{M}\left[\frac{\partial w}{\partial t} - \lambda(M)\, w\right]
\end{aligned}$$

is obtained. The foregoing results are summarized in Proposition 1.

Proposition 1 *The weight distribution function of a polymeric compound:*

$$w = w(t,M)$$

is a solution of the linear second-order hyperbolic partial differential equation:

$$\frac{\partial}{\partial M}\left[\frac{\partial w}{\partial t}+\lambda(M)\,w\right]=\frac{1}{M}\left[\frac{\partial w}{\partial t}-\lambda(M)\,w\right] \tag{27}$$

where $\lambda(M)$ *is the degradation rate of* $w\,(t,M)$.

When the initial weight distribution is given in terms of a prescribed function $f(M)$, the initial condition:

$$w(0,M)=f(M), \tag{28}$$

is imposed on the solution of the equation (27). Given the degradation rate in terms of a prescribed function $\lambda(M)$, the equation (27) and the initial condition (28) form an initial value problem to determine its solution $w(t,M)$. In addition to the initial condition (28), given the weight distribution at $t=T>0$ in terms of a prescribed function $g(M)$:

$$w(T,M)=g(M), \tag{29}$$

the equation (27), the initial condition (28), and the condition (29) form an inverse problem to determine the degradation rate $\lambda(M)$ for which the solution of the initial value problem (27), (28) also satisfies the condition (29). In Section 3.2, it is shown that the inverse problem can be reduced to an ordinary differential equation.

3.2 Reduction to an Ordinary Differential Equation

In Section 3.1, the inverse problem was formulated to determine the unknown degradation rate $\lambda(M)$ for which the solution of the initial value problem (27), (28) also satisfies final condition (29). This problem can be reduced to an ordinary differential equation. Let

$$\mu=b-M.$$

Then the equation (27), the initial condition (28), and the condition (29) become the equations:

$$\frac{\partial}{\partial \mu}\left(\frac{\partial v}{\partial t}+\kappa(\mu)v\right)=-\frac{1}{b-\mu}\left(\frac{\partial v}{\partial t}-\kappa(\mu)v\right), \tag{30}$$

$$v(0,\mu)=\phi(\mu), \tag{31}$$

$$v(T,\mu)=\psi(\mu), \tag{32}$$

where $v(t,\mu)=w(t,b-\mu)$, $\kappa(\mu)=\lambda(b-\mu)$, $\phi(\mu)=f(b-\mu)$, and $\psi(\mu)=g(b-\mu)$. It is assumed that the values of the functions $f(M)$ and $g(M)$ are always positive. It follows that the values of $\phi(\mu)$ and $\psi(\mu)$ are also positive.

The equation (30) can be viewed as the ordinary differential equation

$$\frac{\partial u}{\partial \mu}=r(\mu,u,t), \tag{33}$$

where

$$u = \frac{\partial v}{\partial t} + \kappa(\mu)v, \tag{34}$$

$$r(\mu,u,t) = -\frac{1}{b-\mu}\left(\frac{\partial v}{\partial t} - \kappa(\mu)v\right). \tag{35}$$

Note that the variable u corresponds to the integral on the right-hand side of the equation (26), that it represents the weight flux into the class of all the molecules of size $b-M$ from the classes of all larger molecules, and that the value of such a function should always be nonnegative. In the equation (33), the temporal variable t is treated as a parameter, not as an independent variable. Then, for a fixed but arbitrary $t \in [0,T]$, it is an ordinary differential equation whose solution is a function of the variable μ, and it should be associated with an initial condition in terms of a prescribed function $h(t)$:

$$u(\mu_0,t) = h(t). \tag{36}$$

Suppose that $v(\mu,t)$ is a solution of the equation (30), which satisfies the conditions (31) and (32) with a prescribed function $\kappa(\mu)$. Define $u = u(\mu,t)$ and $r(\mu,u,t)$ by Eq. (34) and Eq. (35) with $v = v(\mu,t)$, respectively. Then, for a fixed but arbitrary $t \in [0,T]$, $u = u(\mu,t)$ is a solution of the equation (33). For the problem to determine the degradation rate, given a non-negative continuous function $u = u(\mu,t)$, the function $r(\mu,u,t)$ can be determined as follows. In view of the expression (34) of u, one seeks a positive value of a parameter η, for which the unique solution $v(t,\mu,u,\eta)$ of the initial value problem (37), (38) also satisfies the condition (39):

$$\frac{\partial v}{\partial t} + \eta v = u(\mu,t), \tag{37}$$

$$v(0,\mu,u,\eta) = \phi(\mu), \tag{38}$$

$$v(T,\mu,u,\eta) = \psi(\mu). \tag{39}$$

It will be shown that such value is necessarily unique whenever it exists. Then $\kappa_u(\mu)$ will be set equal to that unique value. Then the function $r(\mu,u,t)$ is defined by

$$r(\mu,u,t) = -\frac{1}{b-\mu}\{u(\mu,t) - 2\kappa_u(\mu)v(t,\mu,u,\kappa_u(\mu))\}, \tag{40}$$

in view of the expression (35).

For a fixed but arbitrary positive η, the solution $v = v(t,\mu,u,\eta)$ of the initial value problem (37), (38) is given by

$$v(t,\mu,u,\eta) = e^{-\eta t}\left(\phi(\mu) + \int_0^t e^{\eta s}u(\mu,s)\,ds\right), \tag{41}$$

and $\eta = \kappa_u(\mu)$ is a solution of the equation:

$$\Phi_u(\mu,\eta) = 0, \tag{42}$$

where

$$\Phi_u(\mu,\eta) = v(T,\mu,u,\eta) - \psi(\mu) = e^{-\eta T}\left(\phi(\mu) + \int_0^T e^{\eta s} u(\mu,s)\,ds\right) - \psi(\mu). \quad (43)$$

It is easily seen that

$$\Phi_u(\mu,0) = \phi(\mu) + \int_0^T u(\mu,s)\,ds - \psi(\mu),$$

and that

$$\lim_{\eta\to\infty} \Phi_u(\mu,\eta) = -\psi(\mu) < 0.$$

It follows that the equation (42) has a positive solution, if

$$\Phi_u(\mu,0) = \phi(\mu) + \int_0^T u(\mu,s)\,ds - \psi(\mu) > 0. \quad (44)$$

On the other hand, since the partial derivative of $\Phi_u(\mu,\eta)$ with respect to η:

$$\frac{\partial \Phi_u}{\partial \eta}(\mu,\eta) = \frac{\partial v}{\partial \eta}(T,\mu,u,\eta) = -e^{-\eta T}\left(T\phi(\mu) + \int_0^T (T-s)\,e^{\eta s} u(\mu,s)\,ds\right), \quad (45)$$

is always negative, a positive solution of Eq. (42) is unique. It follows that condition (44) is a necessary and sufficient condition for the unique positive solution of the equation (42) to exist. Note that the ordinary differential equation (33) reflects the nature of the inverse problem. The evaluation of $r(\mu,u,t)$ involves the solution of the equation (42), and its dependence on u is nonlinear and implicit, whereas the equation (27) is linear.

3.3 Numerical Study of the Inverse Problem

The formulation of the inverse problem in the ordinary differential equation makes it tractable to its numerical study. Choose a positive integer K, and let

$$t_i = i\Delta t \quad (i = 0,1,2,\ldots,K),$$

where

$$\Delta t = \frac{T}{K}.$$

Given a nonnegative function $u(\mu,t)$, denote an approximate value of $u(\mu,t_i)$ by $u_i = u_i(\mu)$. Then the initial value problem (33), (36) becomes the following initial value problem for the $K+1$ unknown variables $u_0, u_1, \ldots, u_K$:

$$\frac{du_i}{d\mu} = r_i(\mu, u_0, u_1, \ldots u_K), \quad (46)$$

$$u_i(\mu_0) = h_i \equiv h(t_i). \quad (47)$$

On the other hand, an approximate value $v_i = v_i(\mu, u_0, u_1, \ldots, u_M, \eta)$ of $v(\mu, t, u, \eta)$, which is the solution of the initial value problem (37), (38), at $t = t_i$ can be obtained by application of a numerical integration to the integral on the right-hand side of (41), and an approximate value of the function $\Phi_u(\mu, \eta)$ can be given by setting $\Phi_u(\mu, \eta) \approx v_K - \psi(\mu)$. In particular, the Simpson's rule leads to

$$v_{2i-1} = e^{-\eta t_{2i-1}} \left\{ \phi(\mu) + \frac{\Delta t}{3} \sum_{j=1}^{i-1} \left(e^{\eta t_{2j-2}} u_{2j-2} + 4 e^{\eta t_{2j-1}} u_{2j-1} + e^{\eta t_{2j}} u_{2j} \right) \right.$$
$$\left. + \frac{\Delta t}{2} \left(e^{\eta t_{2k-1}} u_{2k-1} + e^{\eta t_{2k}} u_{2k} \right) \right\}$$

$$v_{2i} = e^{-\eta t_{2i}} \left\{ \phi(\mu) + \frac{\Delta t}{3} \sum_{j=1}^{i} \left(e^{\eta t_{2j-2}} u_{2j-2} + 4 e^{\eta t_{2j-1}} u_{2j-1} + e^{\eta t_{2j}} u_{2j} \right) \right\}$$

Similarly, an approximate value of the partial derivative of $\Phi_u(\mu, \eta)$ with respect to η can also be given by application of a numerical integration to the integral in its definition (45). In particular, Simpson's rule leads to

$$\frac{\partial \Phi_u}{\partial \eta}(\mu, \eta) = -e^{-\eta T} \left(T \phi(\mu) + \frac{\Delta t}{3} \sum_{j=1}^{K/2} \{ (T - t_{2j-2}) e^{\eta t_{2j-2}} u_{2j-2} \right.$$
$$\left. + 4 (T - t_{2j-1}) e^{\eta t_{2j-1}} u_{2j-1} + (T - t_{2j}) e^{\eta t_{2j}} u_{2j} \} \right)$$

for $i = 1, 2, \ldots, K/2$. Once those values are given, an approximate value $\tilde{\kappa}$ of the solution of the equation (42), $\kappa_u(\mu)$, can be obtained by the Newton's method, and then an approximate value $r_i(\mu, u_0, u_1, \cdots, u_K)$ of the function $r(t, u, t)$, defined by the expression (40), at $t = t_i$ can be given by

$$r_i(\mu, u_0, u_1, \cdots, u_K) = -\frac{1}{b - \mu} \{ u_i - 2 \tilde{\kappa} v_i(u_0, u_1, \ldots, u_K, \mu, \tilde{\kappa}) \}.$$

In order to solve the initial value problem (46), (47) numerically, the Adams-Bashforce-Moulton predictor and corrector in PECE mode was applied in conjunction with Runge-Kutta method to generate approximate solutions for the first three steps [9]. Here, P, E, and C stand for an application of the predictor, an evaluation of a function, and an application of the corrector, respectively. Here, the values of the parameters T and K were set equal to 2 and 1000, respectively. The steplength $\Delta\mu$ in the computation to solve the initial value problem (46), (47) numerically was set equal to

$$\frac{b - a}{1000},$$

where

$$a \approx 1309.990059, \quad b = 112647.355314.$$

It took at most six iterations for the absolute difference between two consecutive approximate solutions of the equation (42) to become less than 10^{-12}. The parameter μ_0 was set

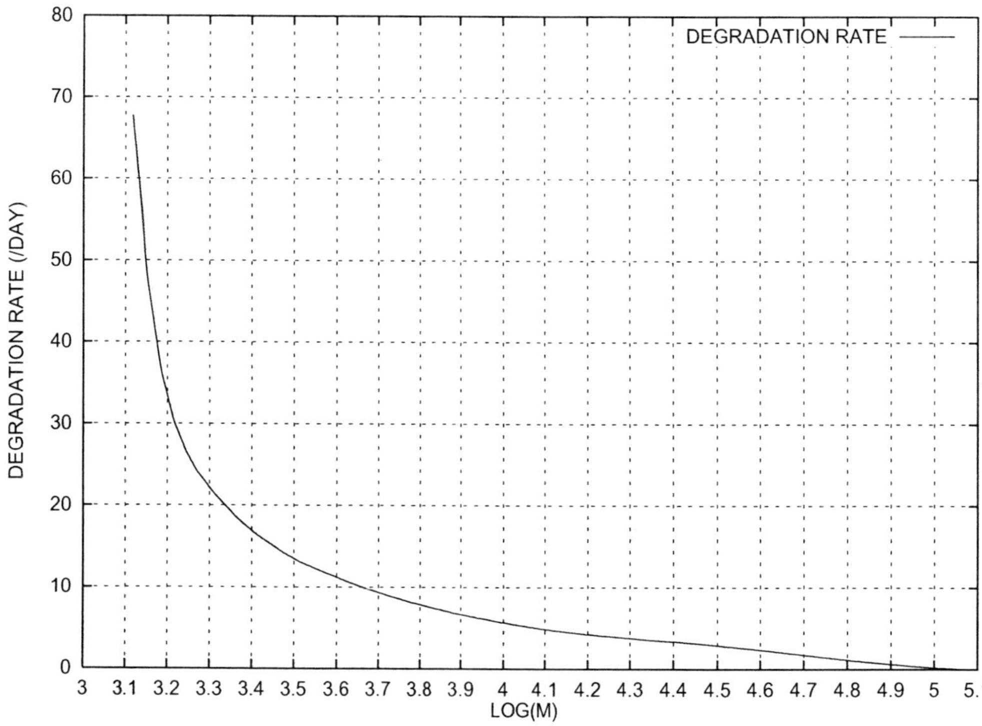

Figure 8: Enzymatic degradation rate of PVA-23000 based on the GPC profiles obtained before and after enzymatic degradation shown in Figure 7.

equal to 0, and the function $h(t)$ was also set equal to 0 $\forall t \in [0, T]$. The numerical result is given in Figure 8 which shows the graph of a degradation rate, $\lambda(M) = \kappa_u(b-M)$, generated numerically by applying the GPC patterns obtained before and after 2-day enzymatic degradation of PVA shown in Figure 7.

Note that the weight distribution $w(t,M)$ equals the solution $v(t,\mu,u,\eta)$ of the initial value problem (37), (38) for $\eta = \kappa_u(\mu)$ and $\mu = b-M$. Figure 9 shows numerical solutions of the initial value problem (27), (28) solved with the degradation rate shown in Figure 8 for $t = 0.2,\ 0.4,\ 0.6,\ \ldots,\ 2$ [day] based on the initial value problem (37), (38). Properties of a predictor-corrector method such as stability and convergence depend on the steplength Δt. Note that the GPC pattern obtained after 2-day enzymatic reaction and the numerical result for $t = 2.0$ [day] can hardly be distingushed. Judging from this numerical result, the choice of Δt for this particular case should be appropriate.

Once the degradation rate $\lambda(M)$ is determined, the weight distribution $w(t,M)$ can be obtained by solving the initial value problem (26), (28) directly by applying a numerical integration formula to the integral on the right side of the equation (26). In particular, the application of the trapezoidal rule to each interval $[M_{i-1}, M_i]$, $i = 1, 2, \cdots, N$, for the integral

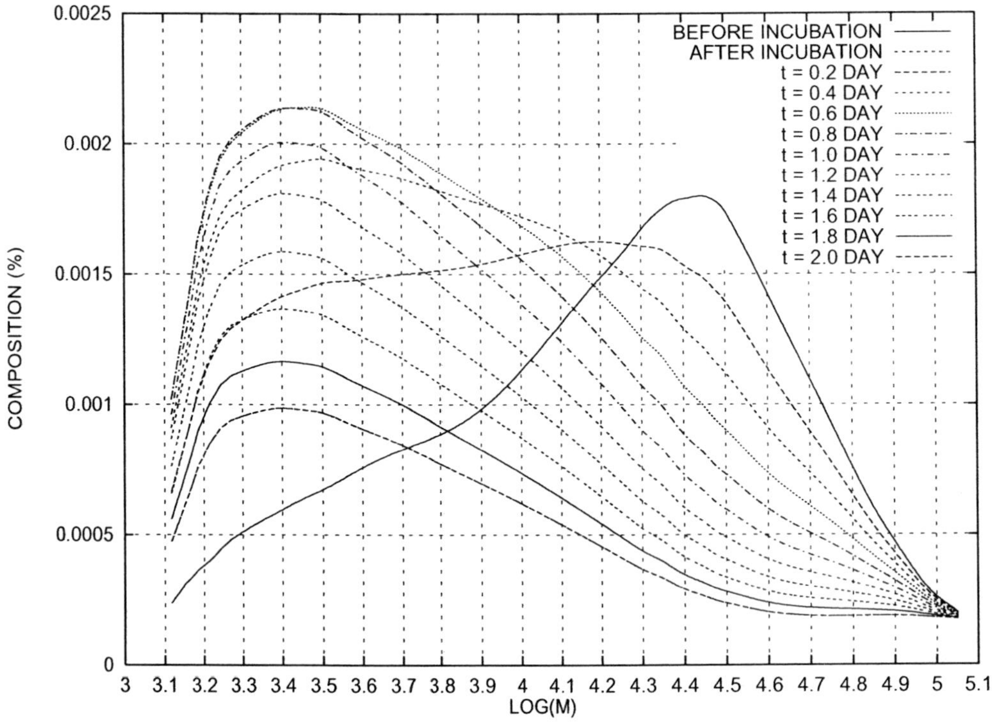

Figure 9: Numerical simulation of enzymatic degradation of PVA-23000 based on the initial value problem (37), (38) with the degradation rate shown in Figure 8.

on the right side of the equation (26) leads to the following system [10, 5]:

$$\frac{dw_i}{dt} = -\alpha_i w_i + M_i \Delta M F_i, \quad i = 0, 1, 2, \cdots, N, \tag{48}$$

where

$$\begin{aligned} \Delta M &= \frac{b-a}{N}, \\ M_i &= a + i\Delta M, \\ \lambda_i &= \lambda(M_i), \\ \alpha_i &= \lambda_i \left(1 - \frac{\Delta M}{M_i}\right), \\ F_i &= 2\sum_{j=i+1}^{N-1} \frac{1}{M_j^2}\lambda_j w_j + \frac{1}{M_N^2}\lambda_N w_N, \end{aligned} \tag{49}$$

where the unknown variable $w_i = w_i(t)$ denotes an approximate value of $w(t, M_i)$. According to the initial condition (28), the system (48) is associated with the initial condition:

$$w_i(0) = f_i = f(M_i), \quad i = 0, 1, 2, \cdots, N. \tag{50}$$

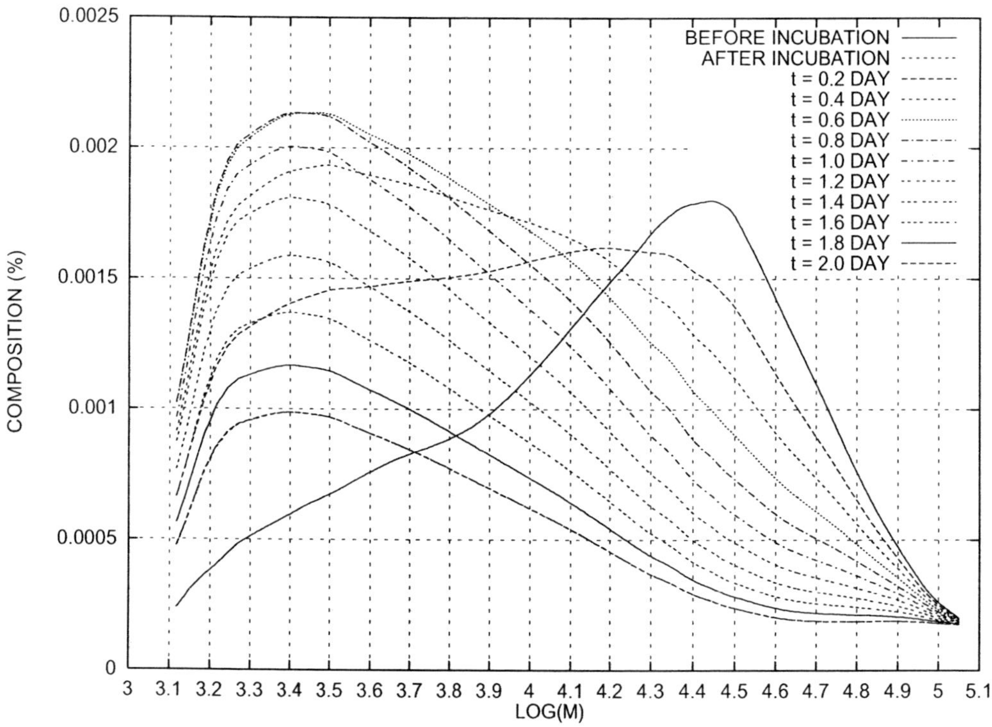

Figure 10: Numerical simulation of the transition of the weight distribution based on the initial value problem (26), (28) with the degradation rate shown in Figure 8.

The parameter N was set equal to 1000, and as was done to solve the initial value problem (46), (47), the Adams-Bashforce-Moulton predictor and corrector in PECE mode was applied in conjunction with Runge-Kutta method with the steplength 0 .002 to solve the initial value problem (48), (50) numerically. Figure 10 shows the numerical result. Although the numerical solutions in Figures 9 and 10 are hardly distingushed, yet there are slight differences between them. Although both the results seem to be acceptable, the difference between them may trace back to the truncation error in the numerical integration.

4 Discussion

The β-oxidation rate $\beta(M)$ of the exogenous depolymerization model is the ratio of the total weight of M-molecules that are degraded per unit time. It also represents the ratio of the number of M-molecules subject to exogenous depolymerization processes. It might be assumed that the β-oxidation rate is independent of the molecular size M, for the exogenous depolymerization processes take place only at the terminals of molecules. In practice, however, that is the case only for molecules of moderate sizes. As the metabolic enzymes are located in cell membranes, the exogenous depolymerization processes take place inside cells, and at least one end of a molecule must penetrate through its outer membrane for a molecule to become subject to an exogenous depolymerization process in the periplasm.

In the presence of sufficient number of microorganism, it may be assumed that the number ratio of molecules in contact with microorganisms in a fixed period of time is independent of the molecular weight. When a molecule makes contact with a microorganism, its part of a fixed length should be taken into a cell in the fixed period. If that part contains one of the terminals, the enzyme takes effect and the exogenous biodegradation process is initiated. The possibility for the part to contain a terminal becomes less when the molecule becomes large. This should be the reason for the β-oxidation rate to be a decreasing function of the molecular weight. However no dependence of the biodegradability on the molecular weight can be concluded from the degradation ratio shown in Figure 3. On the contrary, the result shown in Figure 4 clearly indicates the dependence of the rate of membrane transport on the molecular weight, and our mathematical analysis has revealed the role of membrane transport in exogenous depolymerization processes, which is unpredictable from experimental results alone.

The exogenous depolymerization model was originally developed for PE biodegradation, and it has been applied successfully to PEG biodegradation. The numerical results suggest how molecules are incorporated into cells in exogenous depolymerization processes. The validity of the numerical result concerning the oxidation rate has been confirmed by the numerical simulation (Figure 5). This is a typical microbial depolymerization process of exogenous type, in which monomer units are split from the teminals of molecules.

The only factor assumed in construction of the endogenous depolymerization model was internal separations of molecules. There are other factors of the weight change in tne process through the metabolic pathways of PVA as was described, but those factors should be negligible when compared to the weight shift due to the internal separations of molecules due to enzymatic degradation. As a mathematical model for the enzymatic degradation of PVA, a linear second-order hyperbolic partial differential has been derived from the original model. Given the degradation rate and an initial condition, it forms an initial value problem to determine the weight distribution. On the other hand, given the initial condition and an condtion that presicribes the weight distribution after degradation, an inverse problem to determine the degradation rate for which the solution of the initial value problem also satisfies the additional condition.

It is shown that the inverse problem can be reduced to a nonlinear ordinary differential equation whose unknown variable represents the weight flux into the class of all molecules of a fixed size from the classes of all larger molecules. Note that one can hardly predict the existence of its solution at the outset. On the contrary, that the inverse problem was numerically solved successfully applying the GPC data obtained before and after enzymatic degradation of PVA. The result of numerical simulation indicate that the reduction technique of the inverse problem to the ordinary differential equation is practically acceptable.

5 Acknowledgement

This study was supported in part by a JSPS (MEXT) Grant-in-Aid for Scientific Research (C) (16540106). Authors' gratitude is expressed to Ms. Y. Shimizu (Research Institute for Bioresources, Okayama University) for her technical support.

References

[1] F. Kawai and H. Yamanaka. Biodegradation of polyethylene glycol by symbiotic mixed cultue (obligate mutualism), *Arch. Microbiol.*, **146** (1986) 125-129.

[2] F. Kawai, M. Shibata, S. Yokoyama, S. Maeda, K. Tada, and S. Hayashi. Biodegradability of Scott-Gelead photodegradable polyethylene and polyethylene wax by microorganisms. *Macromol. Symp.*, **144** (1999) 73-84.

[3] Fusako Kawai, Masaji Watanabe, Masaru Shibata, Shigeo Yokoyama, Yasuhiro Sudate, Experimental analysis and numerical simulation for biodegradability of polyethylene, *Polymer Degradation and Stability* **76** (2002) 129-135.

[4] Masaji Watanabe, Fusako Kawai, Masaru Shibata, Shigeo Yokoyama, Yasuhiro Sudate, Computational method for analysis of polyethylene biodegradation, *Journal of Computational and Applied Mathematics*, **161** (2003) 133-144.

[5] Masaji Watanabe and Fusako Kawai, Analysis of polymeric biodegradability based on experimental results and numerical simulation, *Environmental Research and Control*, **25** (2003) 25-32 (in Japanese).

[6] Fusako Kawai, *Biodegradability and chemical structure of polyethers* , *Kobunshi Ronbunshu*, **50** (10) (1993) 775-780 (in Japanese).

[7] Fusako Kawai, Breakdown of plastics and polymers by microorganisms, *Advances in Biochemical Engineering/Biotechnology*, **52** (1995) 151-194.

[8] F. Kawai, Microbial degradation of polyethers, *Appl Microbiol Biotechnol* (2002) 58:30-38.

[9] Lambert, J. D., *Computational Methods in Ordinary Differential Equations* , John Wiley Sons, Chichester, 1973.

[10] Masaji Watanabe, Fusako Kawai, Numerical Simulation for Enzymatic Degradation of Poly(vinyl Alcohol), *Polymer Degradation and Stability*, **81** (2003) 393-399.

[11] Fusako Kawai, Masaji Watanabe, Masaru Shibata, Shigeo Yokoyama, Yasuhiro Sudate, Shizue Hayashi, Comparative study on biodegradability of polyethylene wax by bacteria and fungi, *Polymer Degradation and Stability*, **86** (2004), 105-114.

[12] Masaji Watanabe, Fusako Kawai, Masaru Shibata, Shigeo Yokoyama, Yasuhiro Sudate, Shizue Hayashi, Analytical and computational techniques for exogenous depolymerization of xenobiotic polymers, *Mathematical Biosciences*, **192** (2004) 19-37.

[13] Masaji Watanabe, Fusako Kawai, Mathematical modelling and computational analysis of enzymatic degradation of xenobiotic polymers, submitted for publication.

[14] A.-C. Albertsson, S.O. Andersson and S. Karlsson, The mechanism of biodegradation of polyethylene, *Polym. Degr. Stabil.* **18** (1987), 73-87.

[15] J. R. Haines and M. Alexander, Microbial degradation of high-molecular-weight alkanes, *Appl. Microbiol.* **28** (1974), 1084-1085.

[16] Masaji Watanabe, Fusako Kawai, Analysis of biodegradability for polyethylene glycol via numerical simulation, *Environmental Research and Control*, **26** (2004), 17-22 (in Japanese).

[17] Masaji Watanabe, Fusako Kawai, Numerical and experimental study on mechanism of microbial depolymerization, International Symposium on Numerical Simulation of Environmental Problems, November 22-23, 2004 at Faculty of Environmental Science and Technology, Okayama University), *Journal of Environmental Science and Technology*, (Special Edition) (2004), 17-29.

[18] F. Kawai, Xenobiotic polymers, in: T. Imanaka, ed., *Great Development of Microorganisms*, (NTS. Inc., Tokyo, 2002) 865-870 (in Japanese).

[19] S. Matsumura, N. Tomizawa, A. Toki, K. Nishikawa and K. Toshima, Novel Poly(vinyl alcohol)-degrading enzyme and the degradation mechanism, *Macromolecules* **32** (1999), 7753-7761.

[20] J. E. Potts, R. A. Clendinning, W. B. Ackart and W. D. Niegisch, The biodegradability of synthetic polymers, *Polymer Preprints* **13** (1972), 629-634.

INDEX

A

B

E

F

J

K

L

M

N

O

P

Q

R

S

T

U

V

W

Y